Watching the River Run

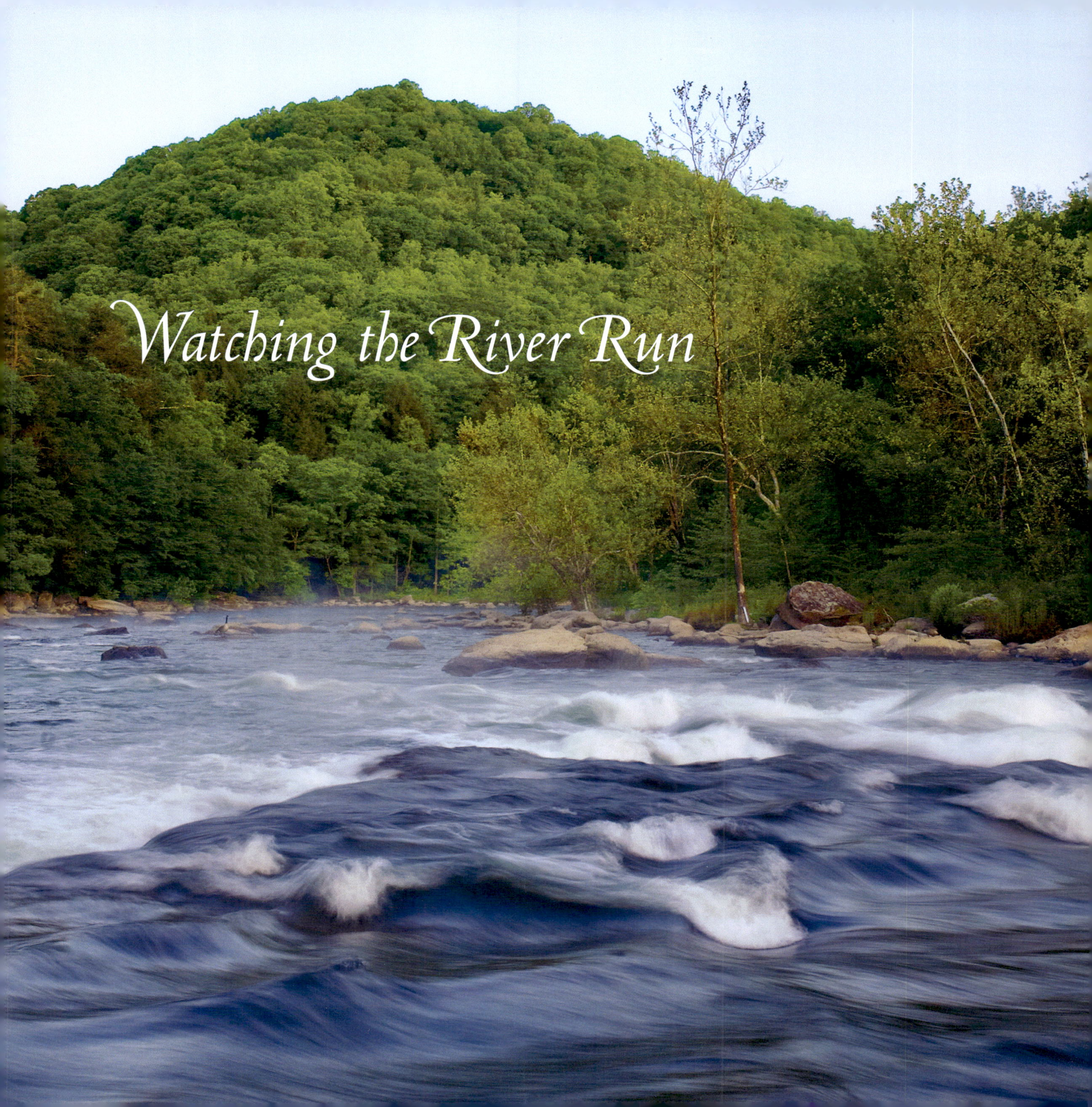

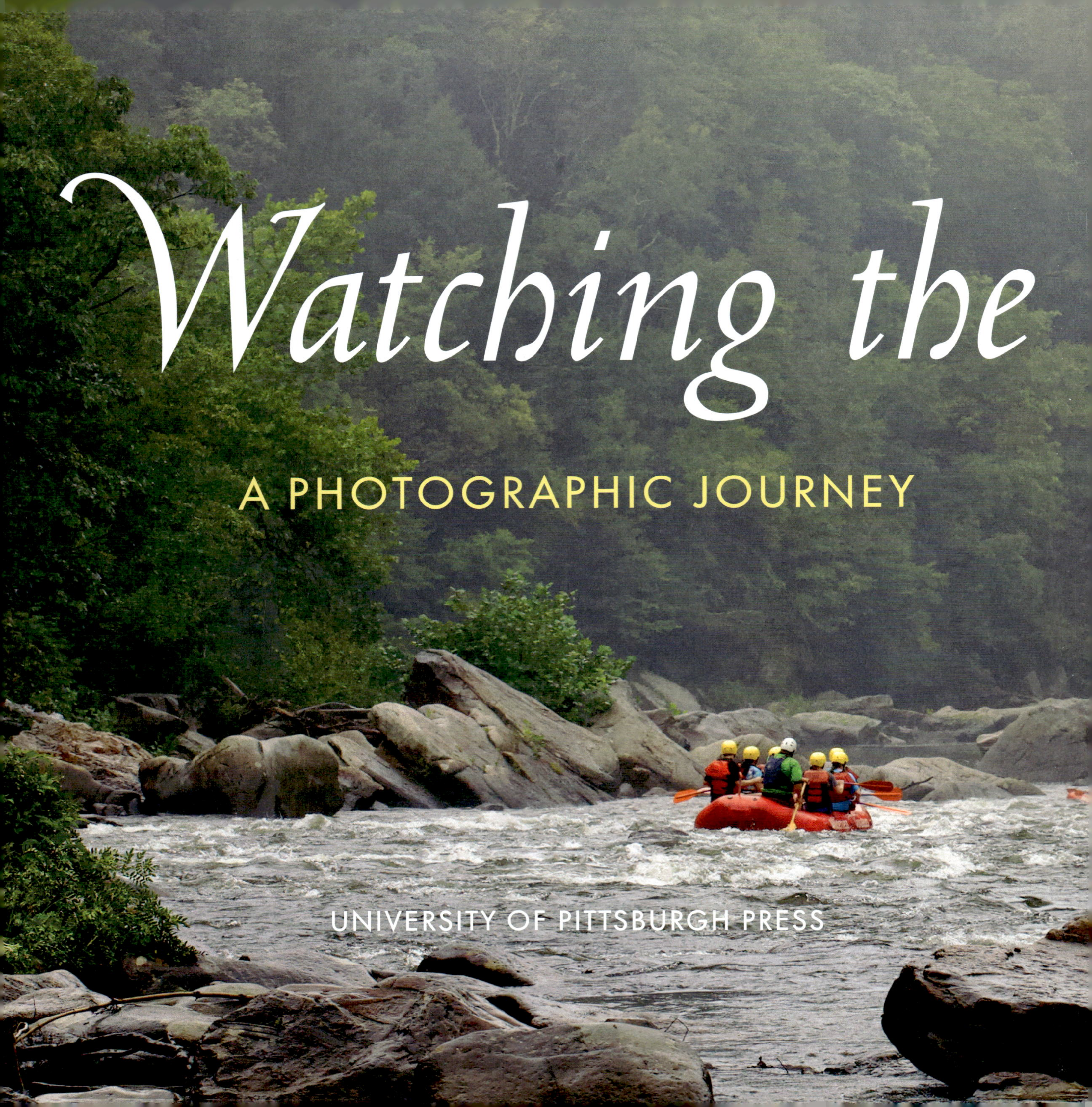

Watching the

A PHOTOGRAPHIC JOURNEY

UNIVERSITY OF PITTSBURGH PRESS

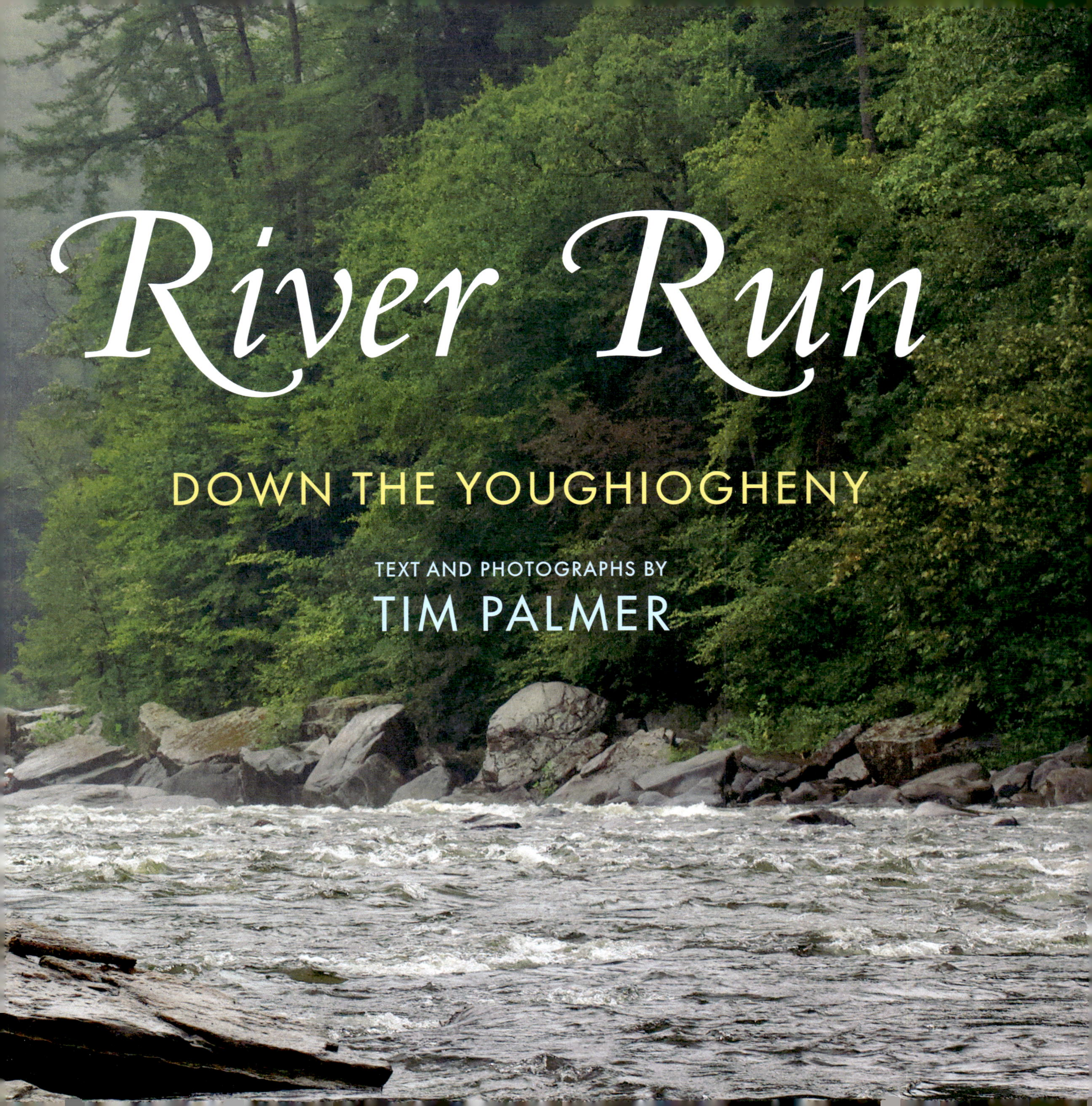
River Run
DOWN THE YOUGHIOGHENY
TEXT AND PHOTOGRAPHS BY
TIM PALMER

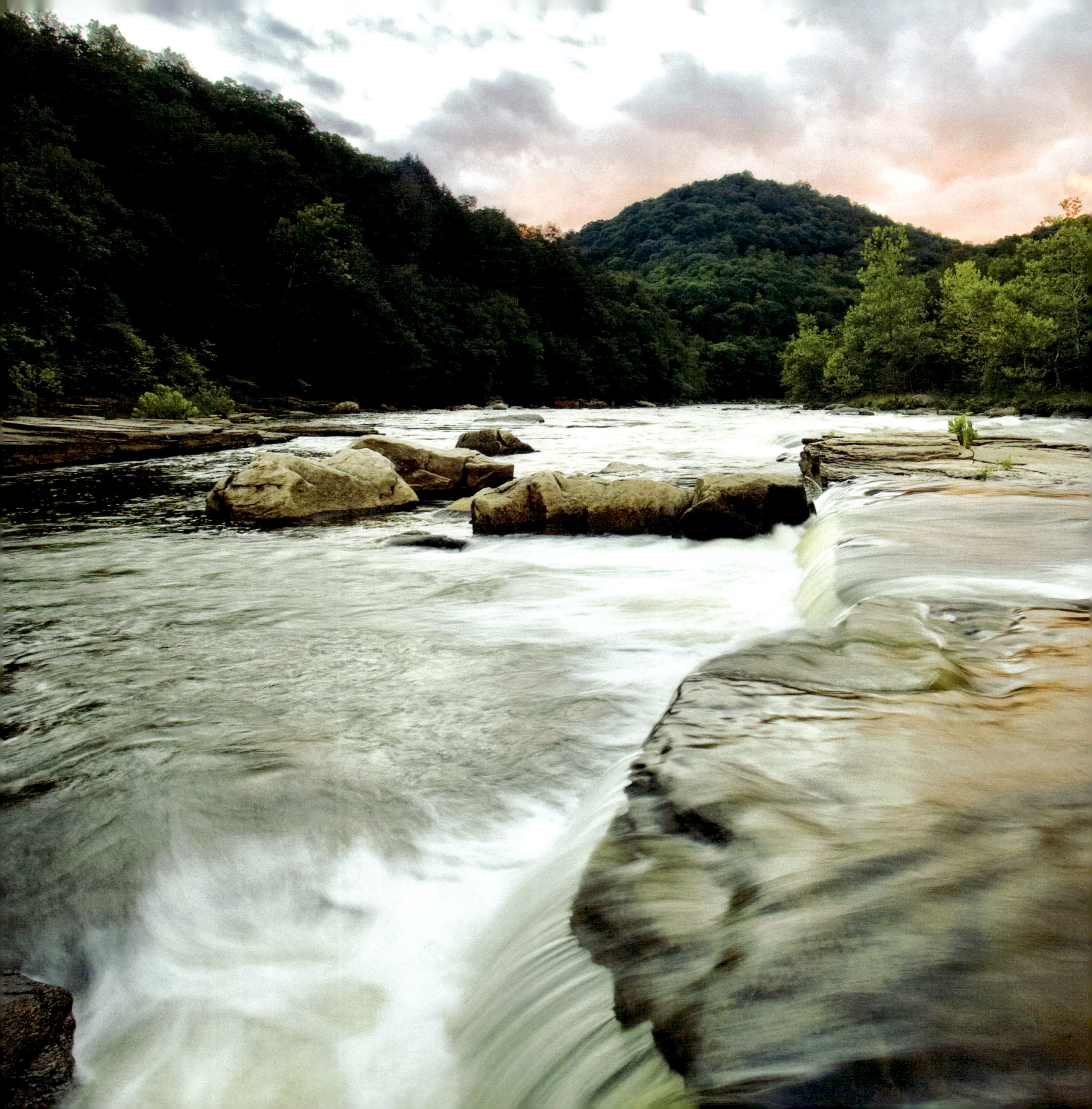

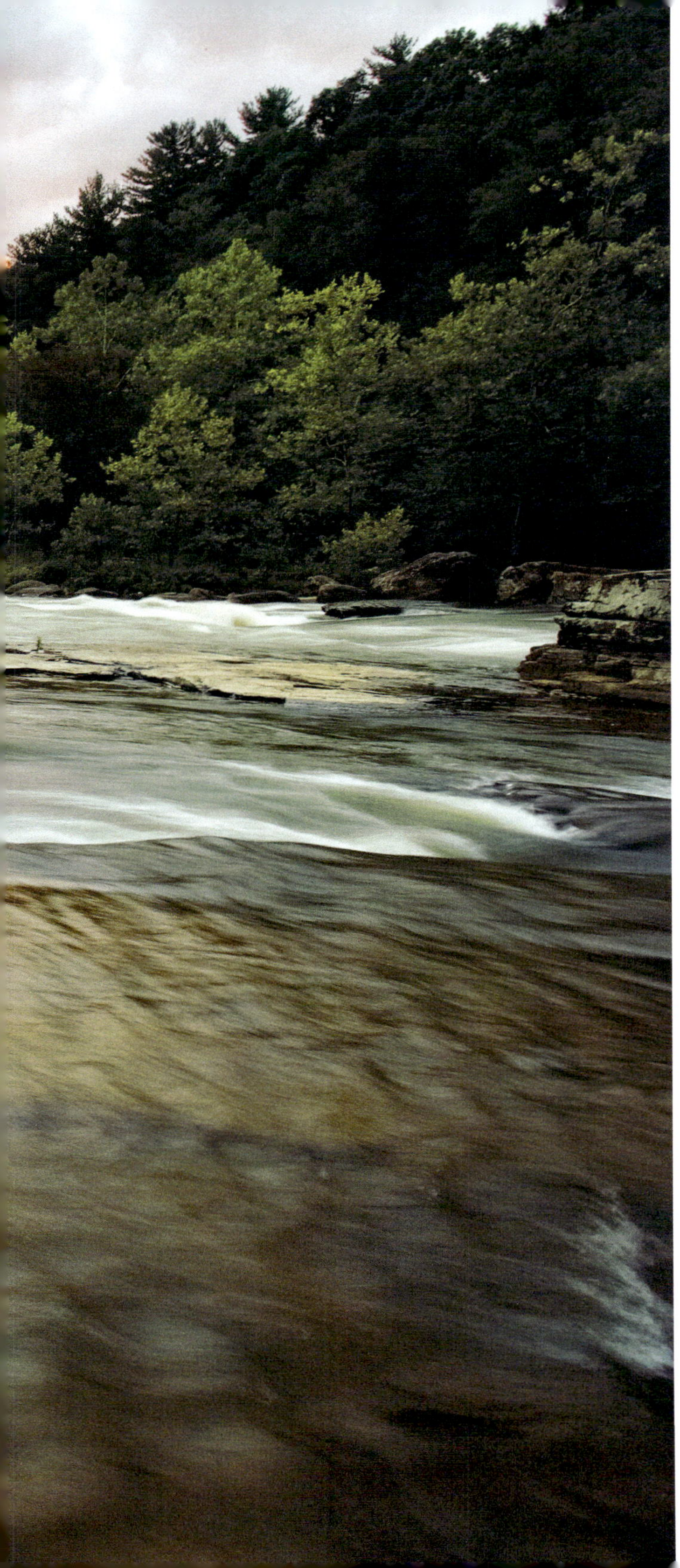

Published by the University of Pittsburgh Press, Pittsburgh, Pa., 15260

Manufactured in the United States of America
Printed on acid-free paper
10 9 8 7 6 5 4 3 2 1

Cataloging-in-Publication data is available from the Library of Congress

ISBN 13: 978-0-8229-4841-4
ISBN 10: 0-8229-4841-9

Cover photograph: The Youghiogheny River rushes through Entrance Rapid downstream from Ohiopyle.

Cover design: Alex Wolfe

Additional frontmatter image credits:
pp. ii–iii: With a rainstorm misting the sky on a summer afternoon, rafters enter Cucumber Rapid in Ohiopyle State Park.
pp. iv–v: Sunset glows over the river downstream from Ohiopyle.
p. vi–vii: The Youghiogheny churns over Ohiopyle Falls and foams into the deep pool below.
pp. viii–ix: The Youghiogheny pours over ledges at Entrance Rapid.

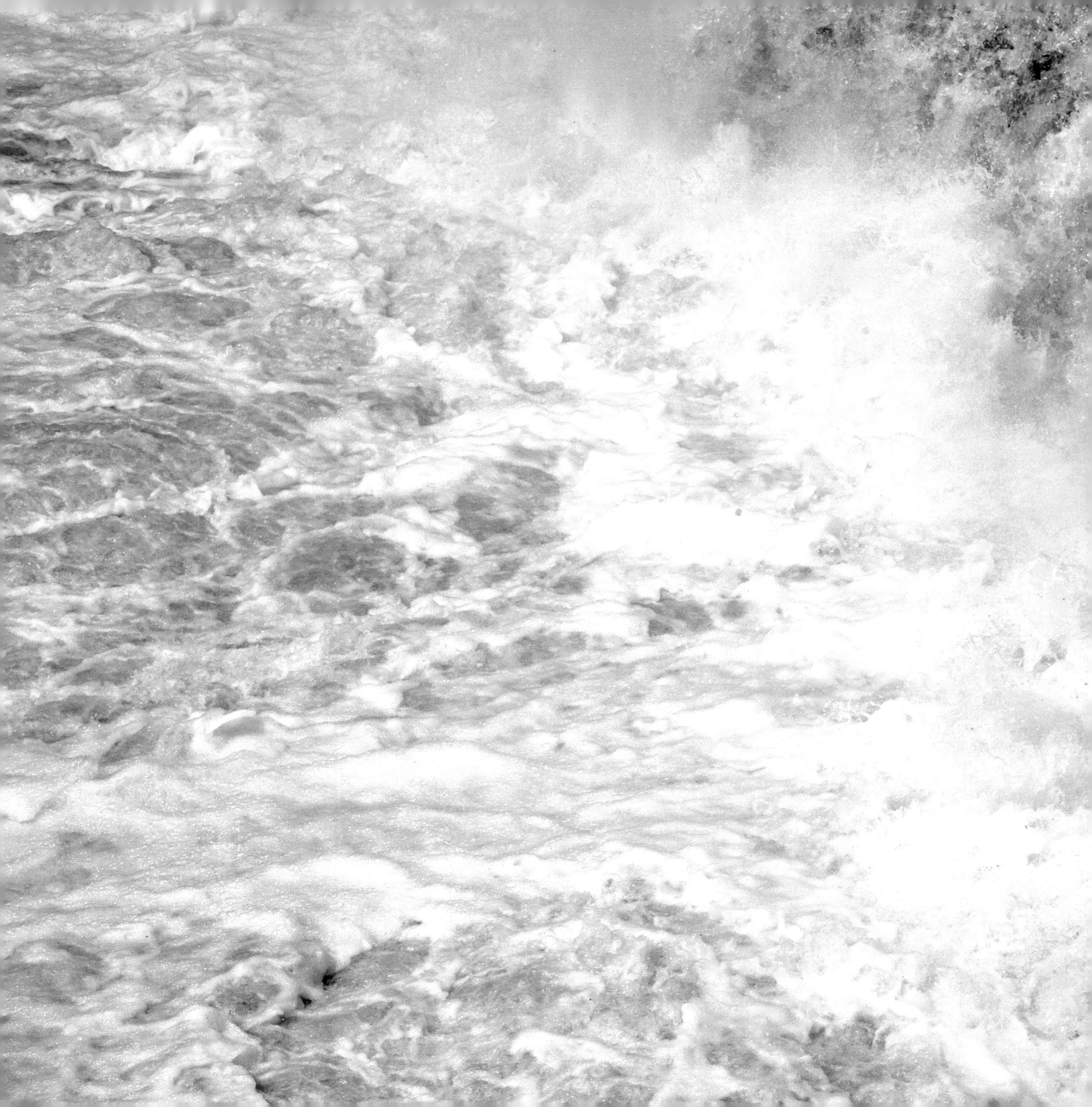

Contents

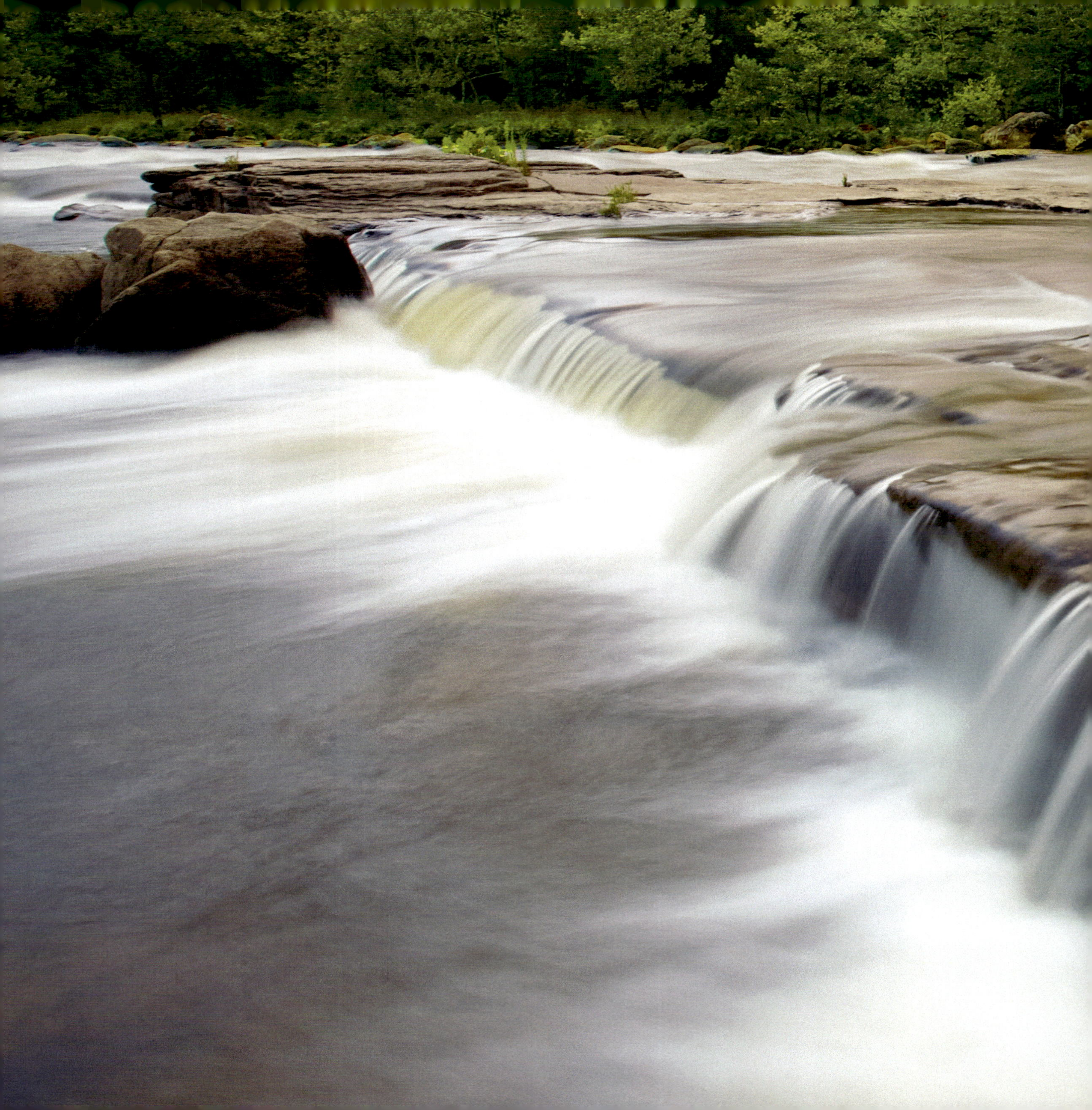

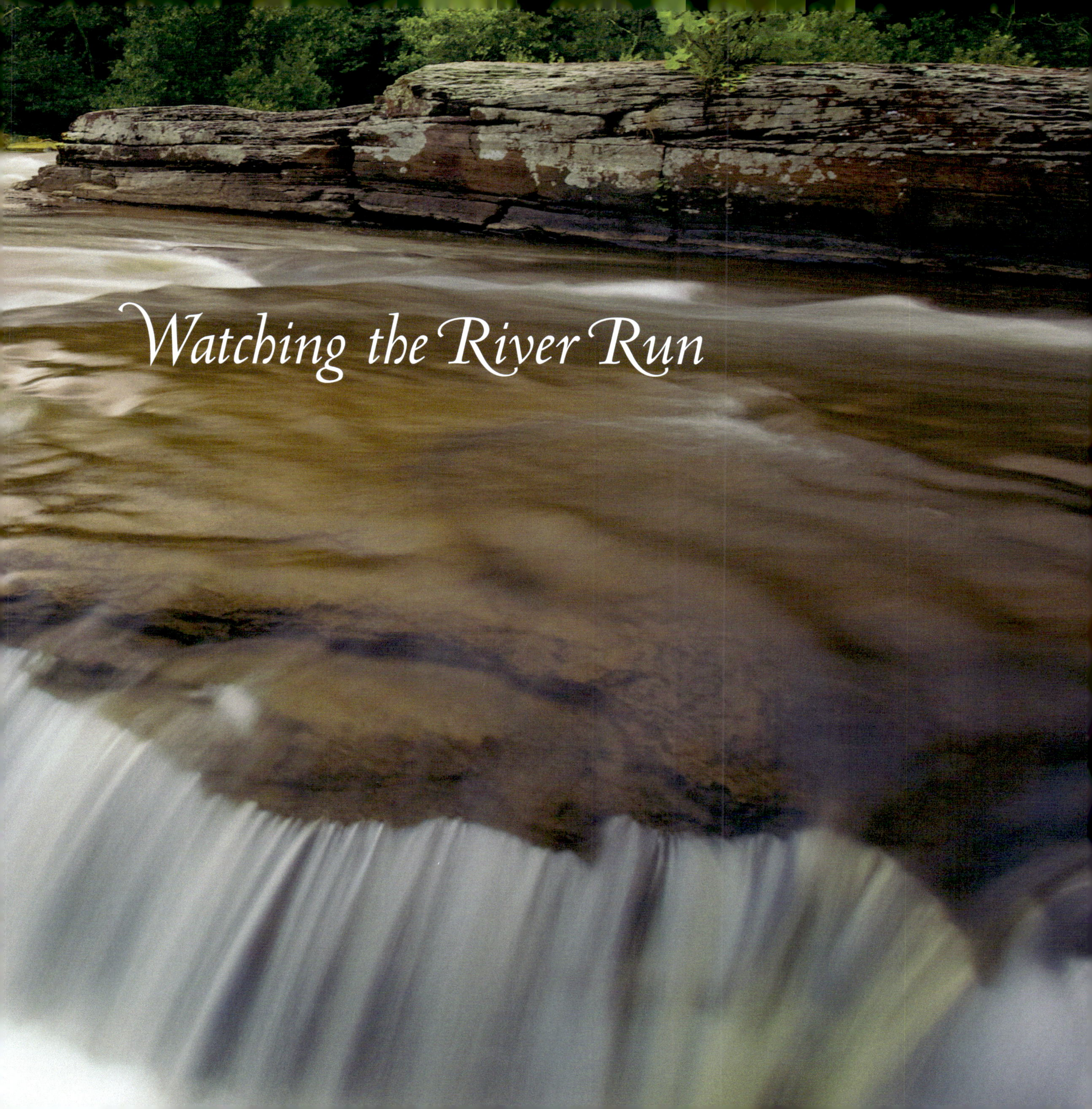
Watching the River Run

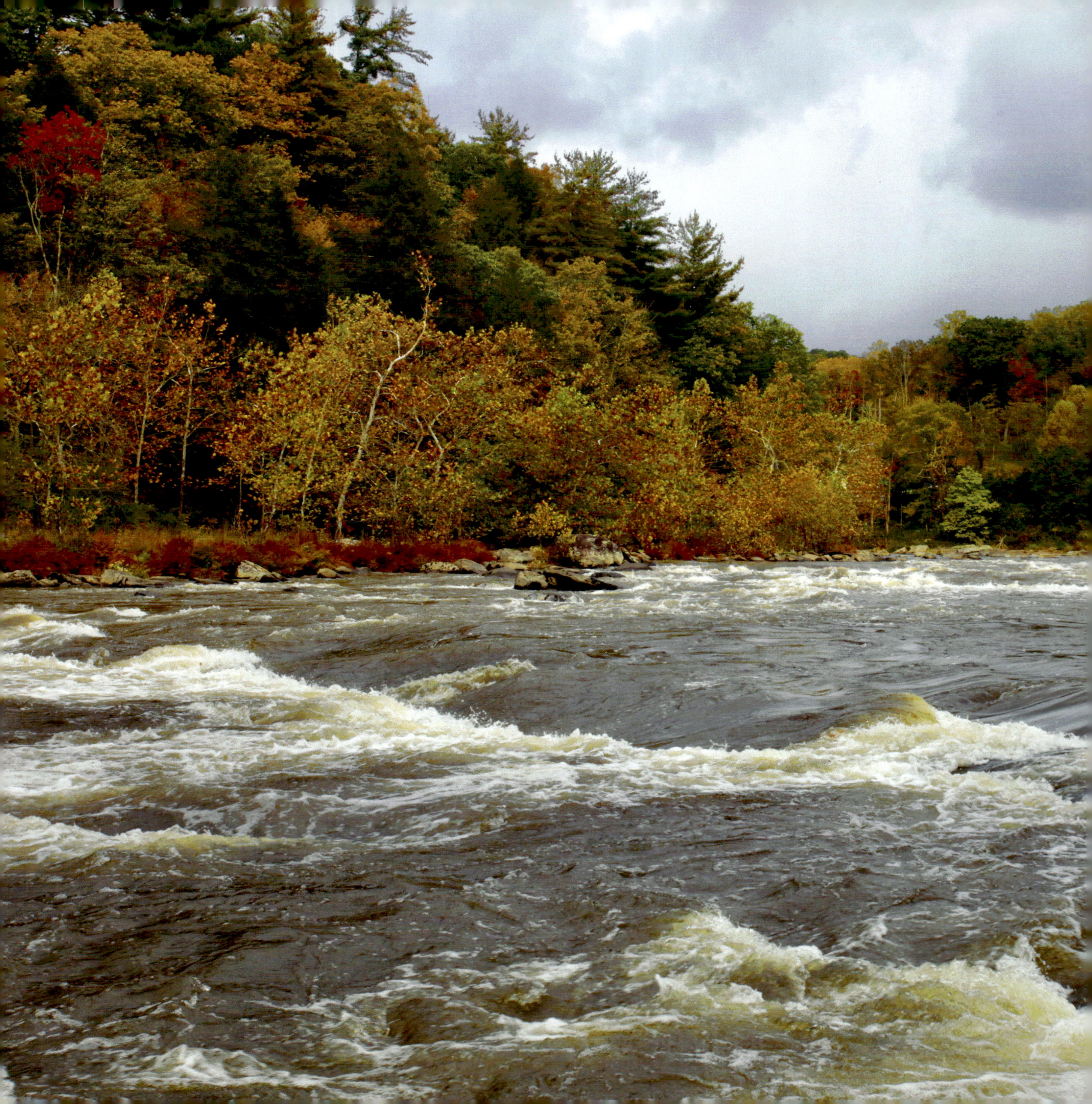

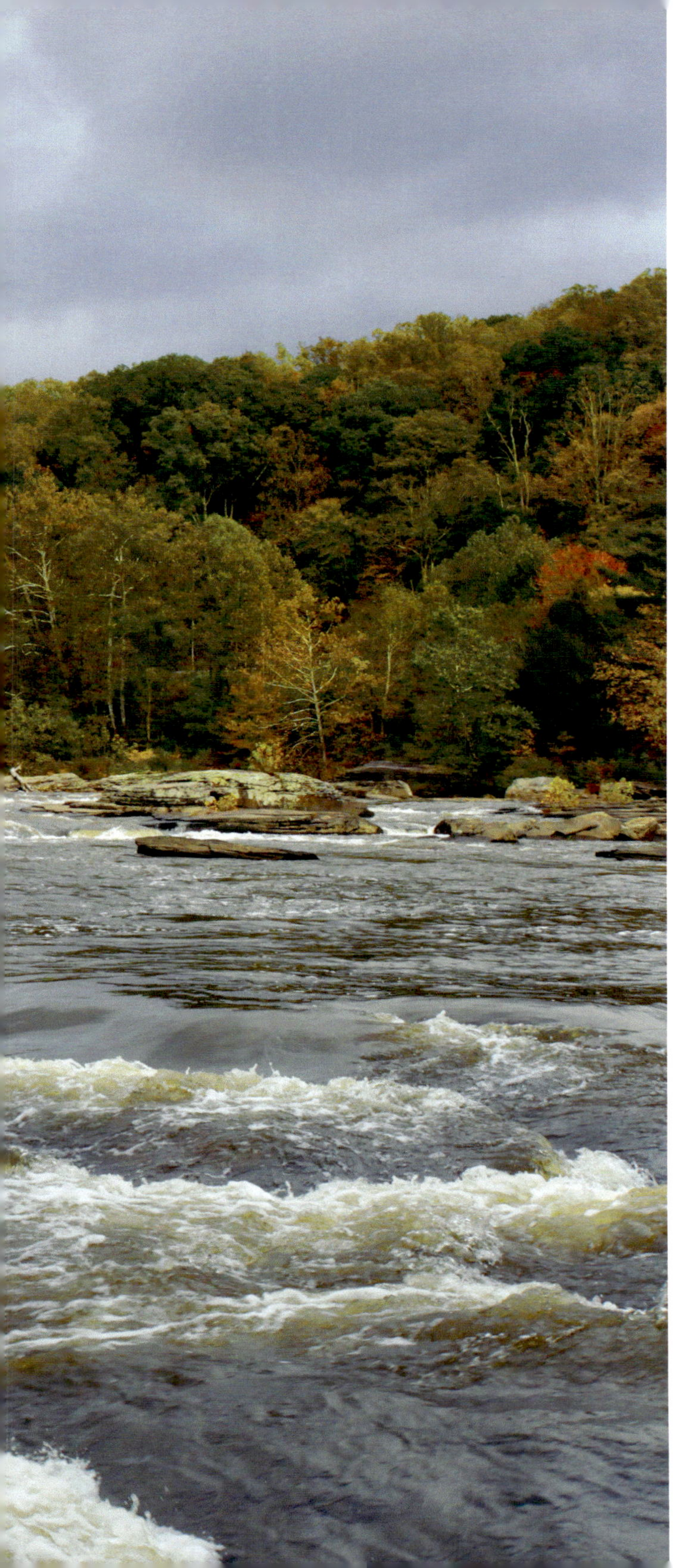

Chapter One

To the River

Any one of us who ventures to the edge of the Youghiogheny River in Ohiopyle State Park, or for that matter anywhere along this tumbling stream in Pennsylvania and Maryland, might sense the special nature of an extraordinary waterway along with that of the emerald mountains it intersects.

Staring into the flow, and paging back to some of my earliest and most formative childhood memories, I'm awestruck at the beauty, the power, and the captivating appeal of this Appalachian artery. After learning more about its exceptional place on earth, I'm also impressed at how important this ribbon of free-flowing water is to all who live along it and to all who visit from near and far.

This destination pulls many of us back again and again. Beyond the limited boundaries of two state parks, the entire river—spring-fed source bubbling down from lofty heights to languid finale in the Monongahela River's flatwater south of Pittsburgh—the Youghiogheny (yaw-ki-GAY-nee) offers a kaleidoscope of scenery delivering both high adventure and deep solace.

Consider the surprising superlatives here in what was once a forgotten backwater of our eastern mountains. Ohiopyle is Pennsylvania's largest state park in land area and among the top

With high runoff from recent storms, the Youghiogheny drops through Entrance Rapid while autumn colors of the Appalachian forest peak in October.

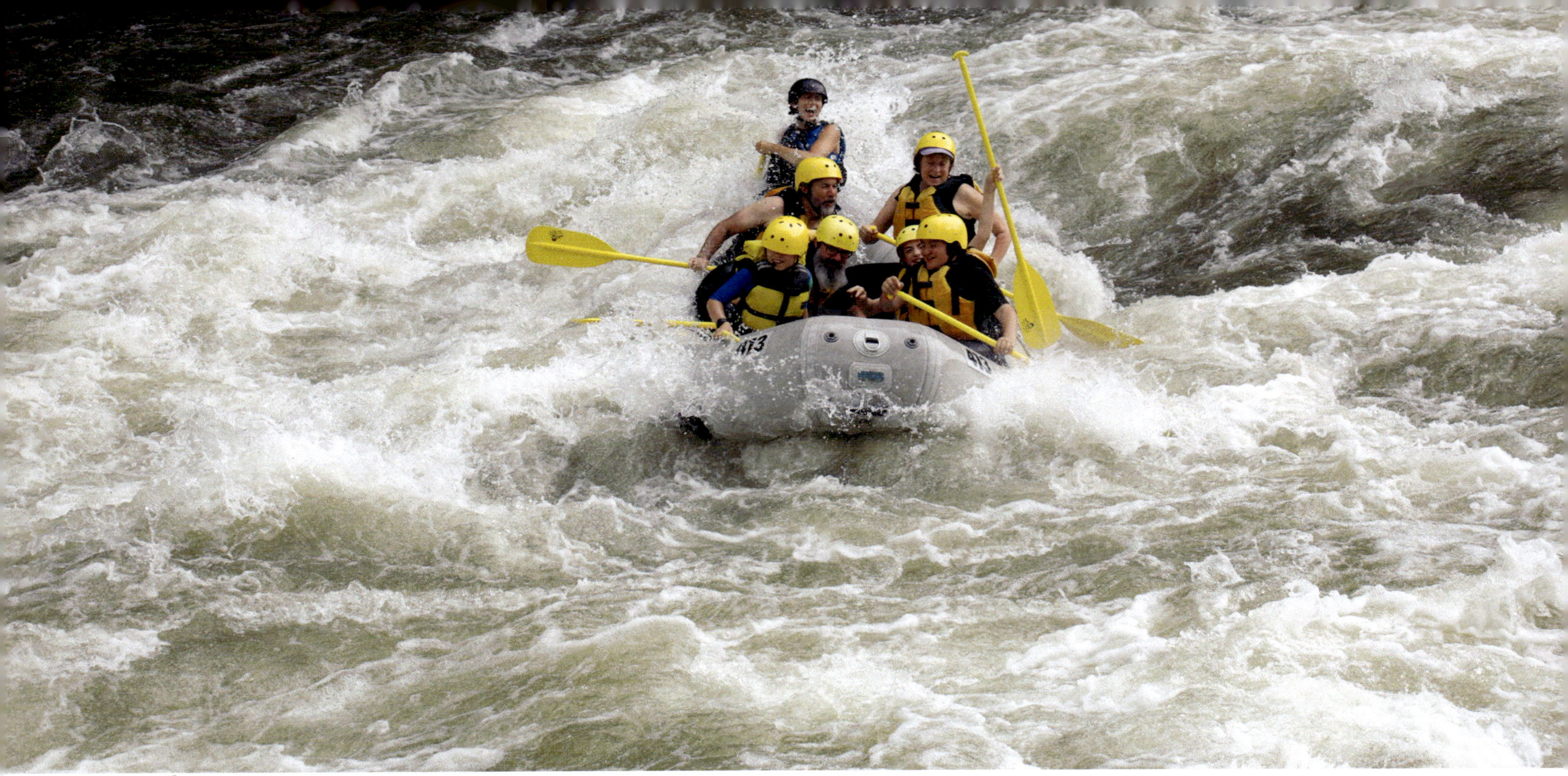

Rafters plunge through Cucumber Rapid. The Youghiogheny is among the most paddled whitewater rivers in America.

five in numbers of visitors, more than one million per year. Whitewater here ranks among the best and the most paddled in America. The river includes not only the extraordinarily popular rapids below the town of Ohiopyle but also gentler flows in the park above, intensely cascading mileage farther upstream in Maryland, and calming riffles in the lower half of the river as it emerges from mountain gorges to drift among westward foothills.

From the 1960s to the early 2000s, brilliant whitewater was the Youghiogheny's major claim to recreational fame, but now Ohiopyle and much of the river's length are also known for a world-class bicycle trail, the Great Allegheny Passage. This traverses the Ohiopyle State Park, follows the river down to its confluence with the larger flow of the Monongahela, and continues in a short urban stint to the glass and steel towers of Pittsburgh. Going the other way, this bicycling route runs from Ohiopyle upstream and eastward to the town of Confluence, climbs the Youghiogheny's principal tributary, the Casselman River, then crosses the Eastern Continental Divide for a total of 150 miles to the North Branch Potomac River at Cumberland, Maryland. And there's more. The seamless trail continues downstream along the historic Chesapeake and Ohio (C&O) Canal and main stem Potomac for an uninterrupted peddle-powered odyssey of 333 miles, from Pittsburgh to Washington, DC. We can bike across the entire Appalachian range, from hilly western foothills to the largest estuary of the Atlantic Ocean.

However, there's no need for boats or bikes if one simply wants to absorb, appreciate, and be nourished by the essence of this river. State parks at both Ohiopyle and upstream at Maryland's Swallow Falls lure visitors to stroll along watery edges, swim in chilling pools, or relax at the interface of what is wild with what is tame. The Youghiogheny invites us to come, stay, and eagerly return, as I've been doing for a lifetime.

Even more vital, the river and its centerpiece park at Ohiopyle provide essential habitat for an intricate web of life where plant and animal species mix and migrate in timeless patterns. Towering hardwoods and hemlocks blend in seamless shades of green bridging the biological treasure house of the Southern Appalachians with the scenic Northern Hardwoods array of trees with their contrasting remnants of glacial epochs from cooler climates, a varied community that, though imperiled, endures in sequestered gorges.

The recreational appeal of the river, the whitewater, the bike trail, the footpaths, the campgrounds, and the postcard-perfect scenery have long drawn people to cherish this world apart. However, among the festive opportunities for daily, weekend, or vacation escape, the more essential workings of nature can easily be taken for granted. For that reason, the main purpose of this book is to highlight the natural world of the Youghiogheny and all that it offers to everyone.

The river and its surroundings surely reflect beauty and wildness, though not without their share of modern problems and plagues. Coal mining for the past two hundred years with the resulting water pollution that sterilized streams if not whole rivers still continues, though at far reduced levels compared to the basin's checkered past. Old problems have become aggravated by fracking for "natural" gas along some tributaries, all while the age of fossil-fuel-burning enters its diminished but lingering stages. Likewise, land development creeps ever outward from the cities, towns, and brightly lit highway interchanges with effects felt even at remote retreats. More ubiquitous but less noticed by many people, exotic plants and pests have invaded here by means as surreptitious as ornamental shrubbery imported for yards and gardens but unknowingly infected by pathogens promising a death sentence to native life. Courageously confronting these problems, dedicated caretakers along the Youghiogheny strive to safeguard the best that remains and to restore the workings of the natural world where possible. Their success at dozens of streams and mountainsides has been promising, enriching, and inspiring.

In my earlier book, *Youghiogheny: Appalachian River* (first published in 1984, then updated in 2023), I focused on the big picture of this intriguing watershed while delving deep into history and cultural intricacies, from journeys of George Washington to the

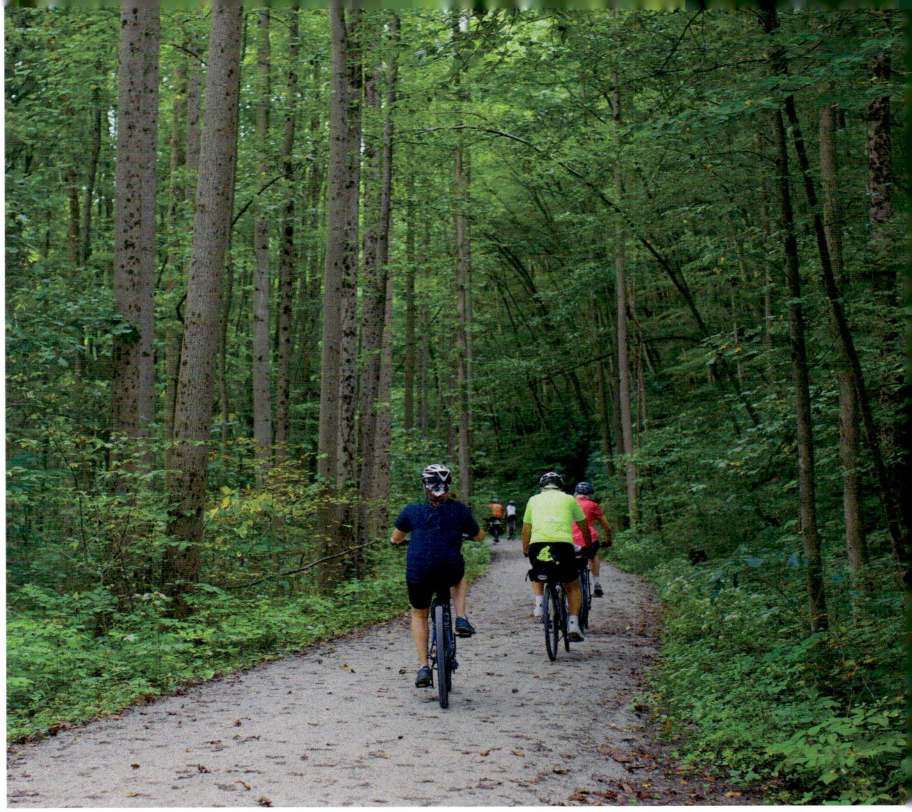

Bicyclists breeze through Ohiopyle State Park on the Great Allegheny Passage bike trail. Combined seamlessly with the C&O Canal towpath along the Potomac River, this is America's longest continuous bike trail that does not use segments of roads.

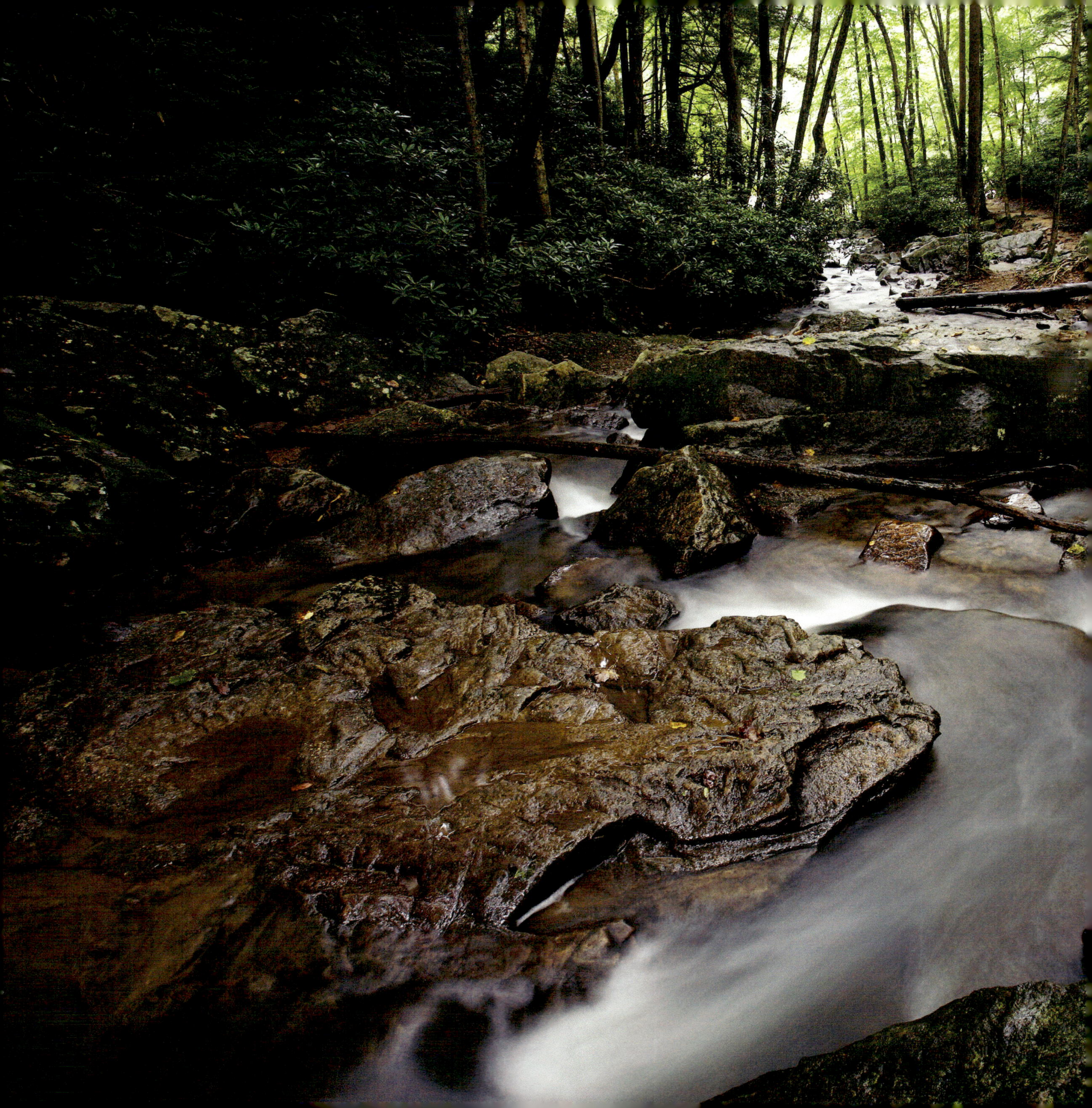

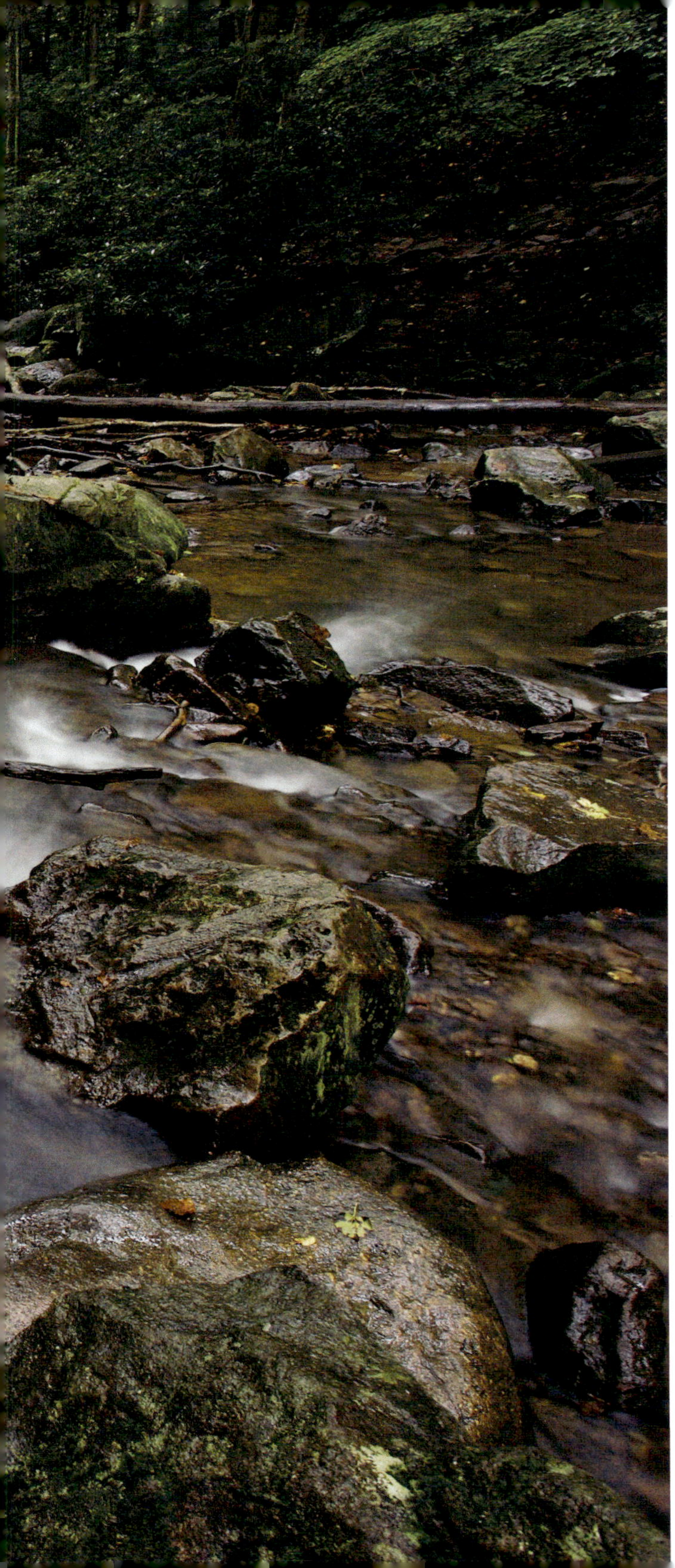

complexities of today's world, unimagined just one lifetime or even a few decades ago. The Youghiogheny's mix of what has long been wild with what has become a popular playground now motivates me to write this follow-up book featuring photographs of the river's surviving natural wonders along with scenes that have been restored by visionary stewards of these borderlands between the urban maw of the East Coast and that of Pittsburgh with its industrial outliers fingering out toward the Midwest.

Knowing that the Youghiogheny River and Ohiopyle State Park are remarkable might be self-evident, reinforced in the words of chapters that follow. But seeing is believing, and a picture, as they say, can be worth a thousand words. So the photos take center stage in this book that accompanies my earlier story of the river and its people.

Exploring, once again, the stream's shores, forests, and rugged mountain slopes, I sought out scenes that show what is truly irreplaceable and eternally natural about this part of the earth in its whitewater and in its quiet pools, its intimate edges and boldly towering ridges, its rocks, trees, illuminating sunsets and darkening storms. To me, every bit is enticing. However, photos offer only an introduction to what's really here. So in this book I invite you to go, to look, and to find for yourself all that this river has to offer, from rock-hard fossils of giant tree ferns thriving three hundred million years ago to bubbling rapids that change hourly with the weather and the water, renewed every day as we stand along these shores, watching the river run.

Cucumber Run approaches its mouth at the Youghiogheny River. Treatment of mine drainage has reclaimed this stream from the orange stains of sterilizing acid and toxic iron pollution that prevailed in the 1960s.

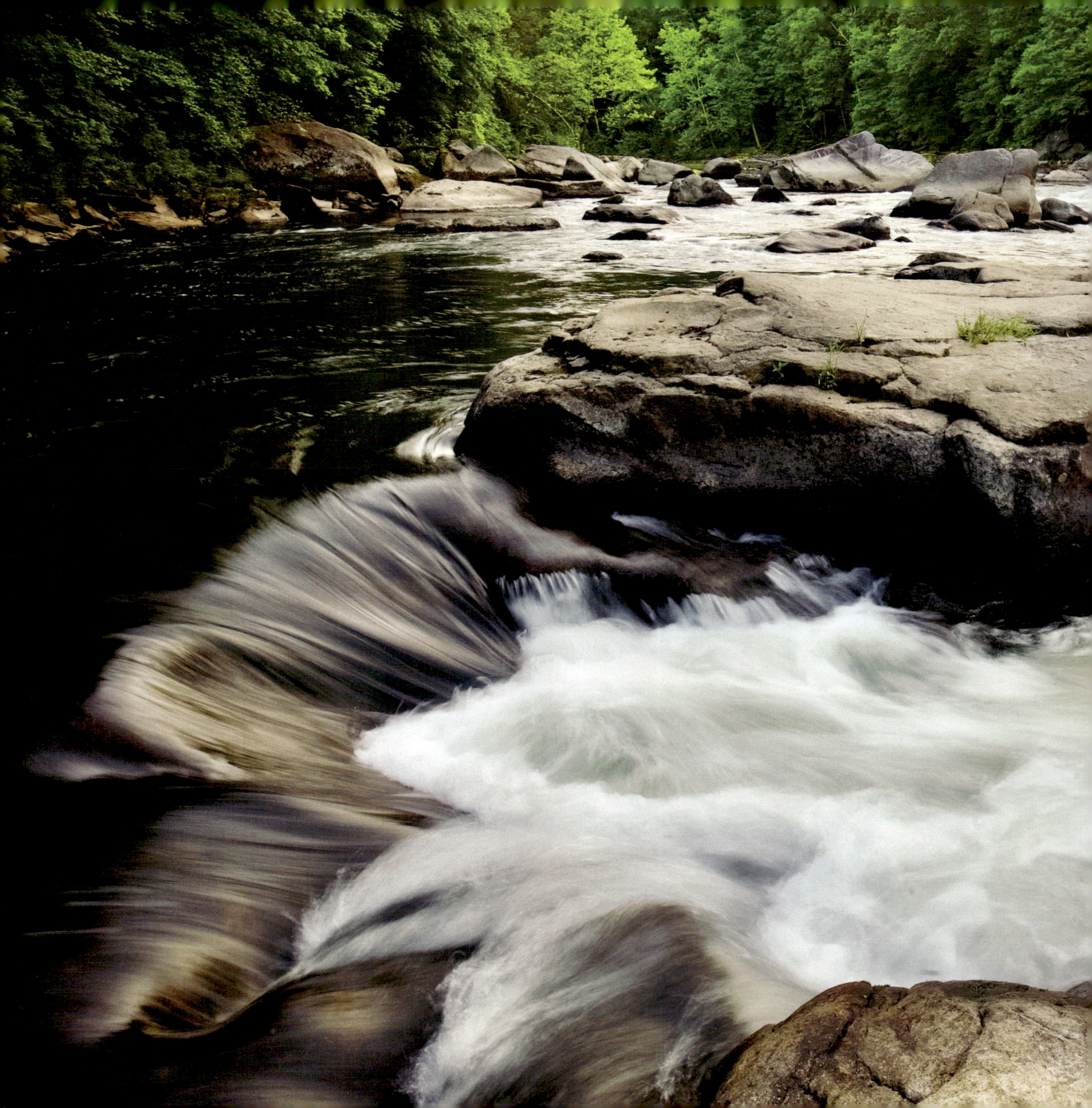

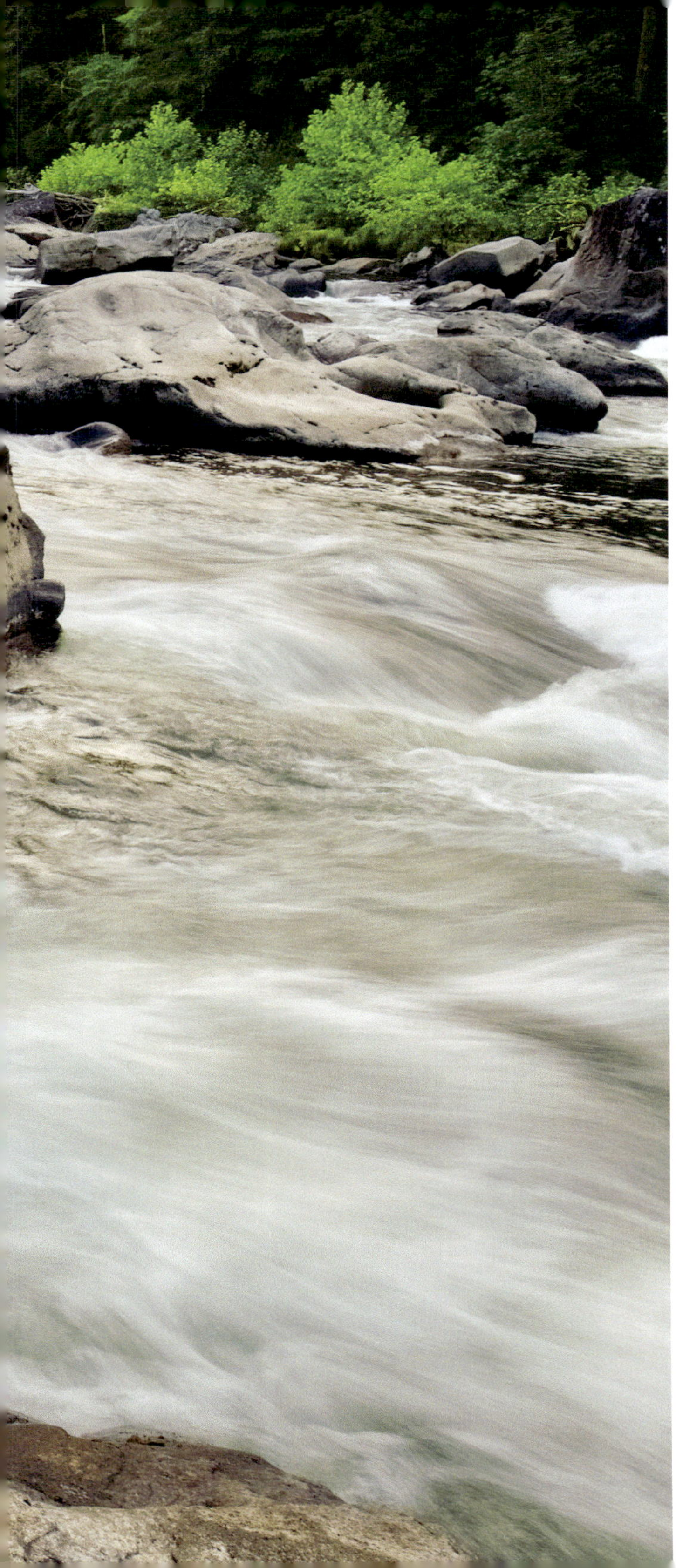

Chapter Two

Watching the River Run

Some people come to the Youghiogheny River to go rafting, kayaking, or canoeing. Some come to bike and some to hike. But everyone comes to look. *Seeing* the river comes first, no matter what else we do.

Whether we paddle, peddle, or walk even a short distance, just making eye contact with nature might be the main draw here—hitting the spot, so to speak. Simply looking can be enough. It's something that nearly all of us can do, at any time in our lives, any season of the year, any moment of the day. So what does it mean to truly watch a river run? It sounds easy, but how can we do it best?

Earlier I hadn't thought much along this line, though as a writer and especially as a photographer, I had to figure it out. "Look more" got me only so far toward describing a place that's important to many people. I wanted to see clearly, to find not only what's obvious but also what's hidden, to be surprised by unexpected realities, to know what is most vital, essential, and alive about this river that we visit, whether coming from a short drive away or from the other side of the continent or globe.

We see a river, of course, simply by coming here and opening our eyes. Yet, beyond that, what do we look *for*? How can we grasp, as a

Railroad Rapid speeds over sandstone bedrock downstream from Ohiopyle.

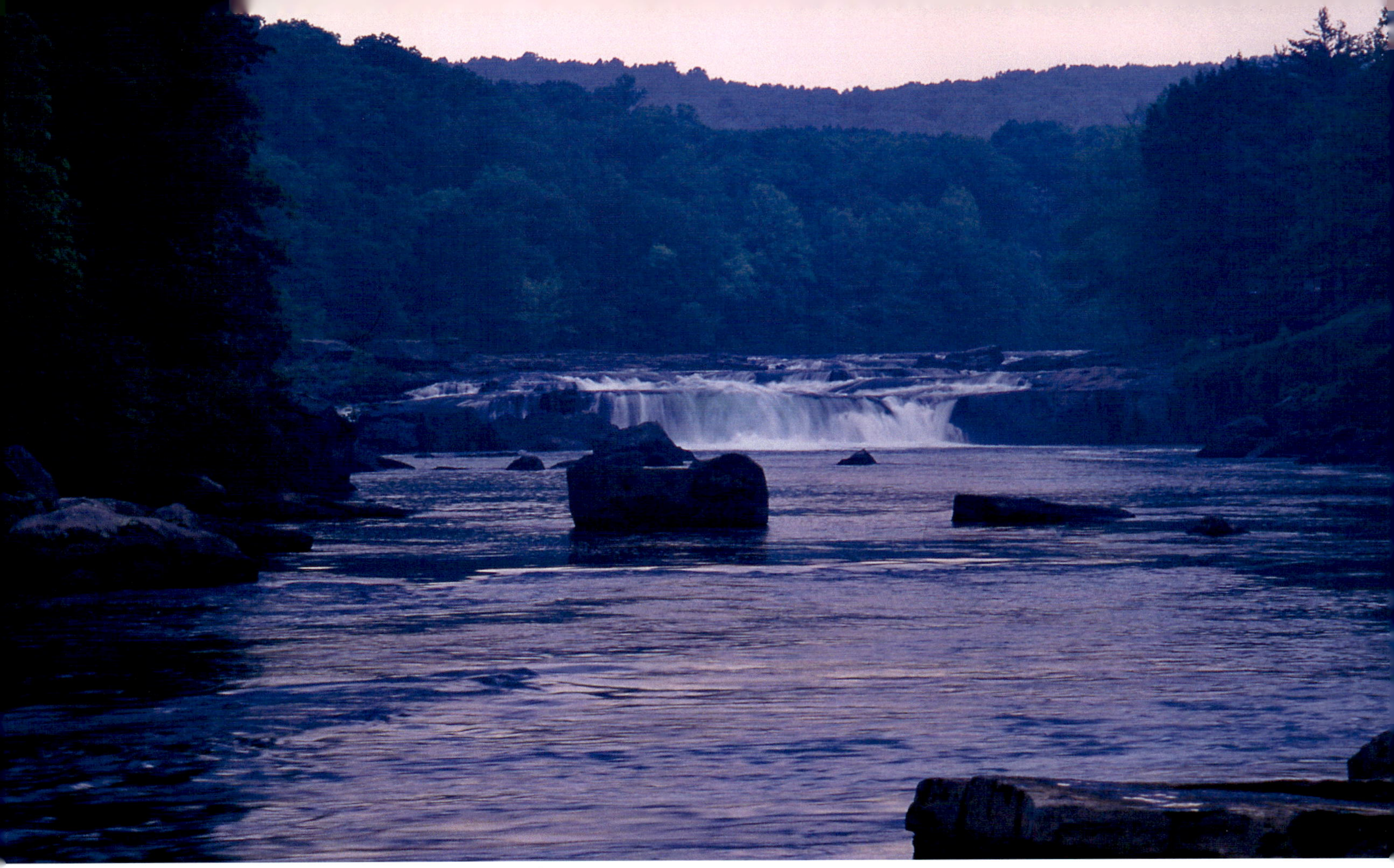

The Youghiogheny thunders over Ohiopyle Falls, seen here from downstream at the mouth of Meadow Run.

picture or an idea, more than just a piece of the place, but all of it? When I stand along the shore with both curiosity in mind and camera in hand, I want to reach beyond the postcard-perfect view. A snapshot of Ohiopyle Falls might top any recall of memories for a day invested here. Yet, if we've come a long way to spend a whole afternoon, or weekend, or (for the extremely fortunate) a whole week or month along this river, we want to see more, see further, see deeper. In essence, and in the end, this means *knowing* the place. So, how can we know it better?

Taking photos might lead us to see more thoroughly, because doing that pushes us to appreciate essential views and to recognize actual shapes and colors. But to see more than we usually do, we mainly need an open mind along with a desire to stretch the boundaries of how we look at the place.

Certainly we can see more of the river by going to more spots along it. That's one of the easiest ways to expand our horizons, literally and figuratively. This might involve just a short walk, or maybe a full day or week

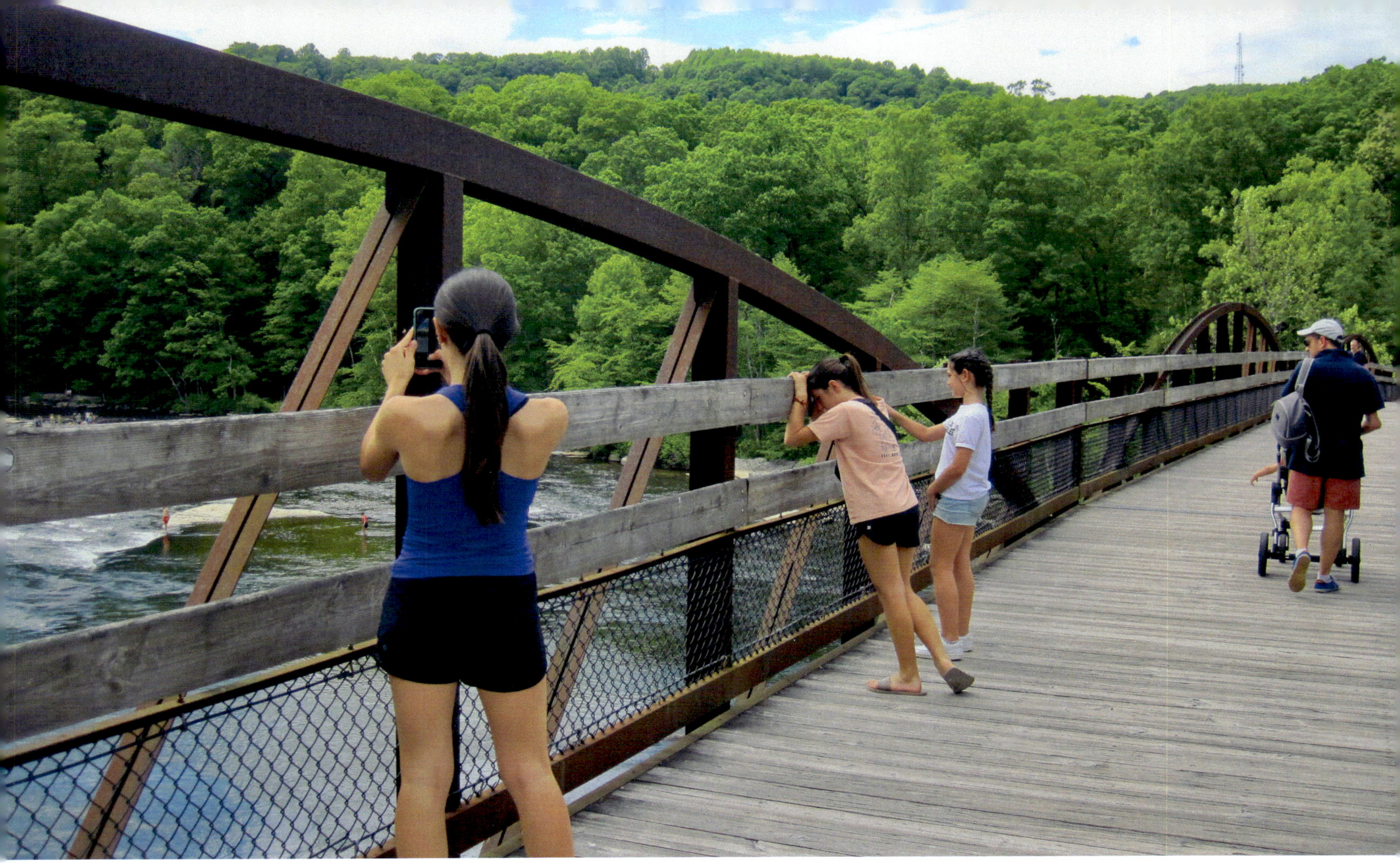

of travel upstream or down. Seeing where the water comes from and where it goes is interesting and perhaps necessary for knowing the full Youghiogheny. Doing this led me upward to the remote spring-fed source on Backbone Mountain, and it led me down to the terminal flatwater of the dammed-up industrial artery at McKeesport, 16 miles from Pittsburgh. I went the distance because I wanted to know and write about the entire river, though you don't need to do that. Even the first short Youghiogheny stroll of my life, downstream from Ohiopyle Falls, opened enticing views and new worlds of natural wonders that left me wanting more.

Explorations upstream or down can also reach outward along the river's tributaries, each an uphill extension of the river to its hundreds of sources. If I had to pick a favorite, it might be Meadow Run, which tumbles through layers of sandstone before merging into the Youghiogheny a short walk from Ohiopyle. Cucumber Run, just another hop downstream, treats us to one of Pennsylvania's most exquisite waterfalls.

The Youghiogheny bridge for pedestrians and bicyclists at Ohiopyle has become a favorite photo location.

The Casselman River is the largest branch, perceptively called "Little Youghiogheny" by George Washington during his explorations here—though a smaller stream, at Oakland, Maryland, has that name today. Coal mine drainage for years stained the 55-mile-long Casselman to a dayglow orange. Mudflows from strip-mined sources still murk the Youghiogheny downstream after drenching storms. But targeted efforts plus public investments have transformed much of the Casselman by fixing abandoned coal mines, restoring cold currents, and nurturing shaded shores where native brook trout might once again thrive, provided the improvements in habitat continue.

Paths upstream along smaller tributaries take us farther into forests, farther onto mountains, together accounting for everyplace the water originates. Meanwhile, to follow the river downstream shows us where the water ultimately joins the Monongahela and then the Ohio River, which merges with the Mississippi 980 miles beyond—all one great hydraulic megalith. If we travel to the mouth of that largest river in America as it enters the Gulf of Mexico, some of the water we see would be from the Youghiogheny. But there are less epic ways to expand one's view of the river.

Short walks around Ferncliff Peninsula at Ohiopyle State Park are easy enough for most people and can be done in a couple of hours, any day. Side strolls lead to river views, each opening new perspectives. Furthermore, what we see can be expanded anywhere along the water by focusing downward to details. There's always more than we can realize at a glance. Check out the waves and eddies. Look closely at

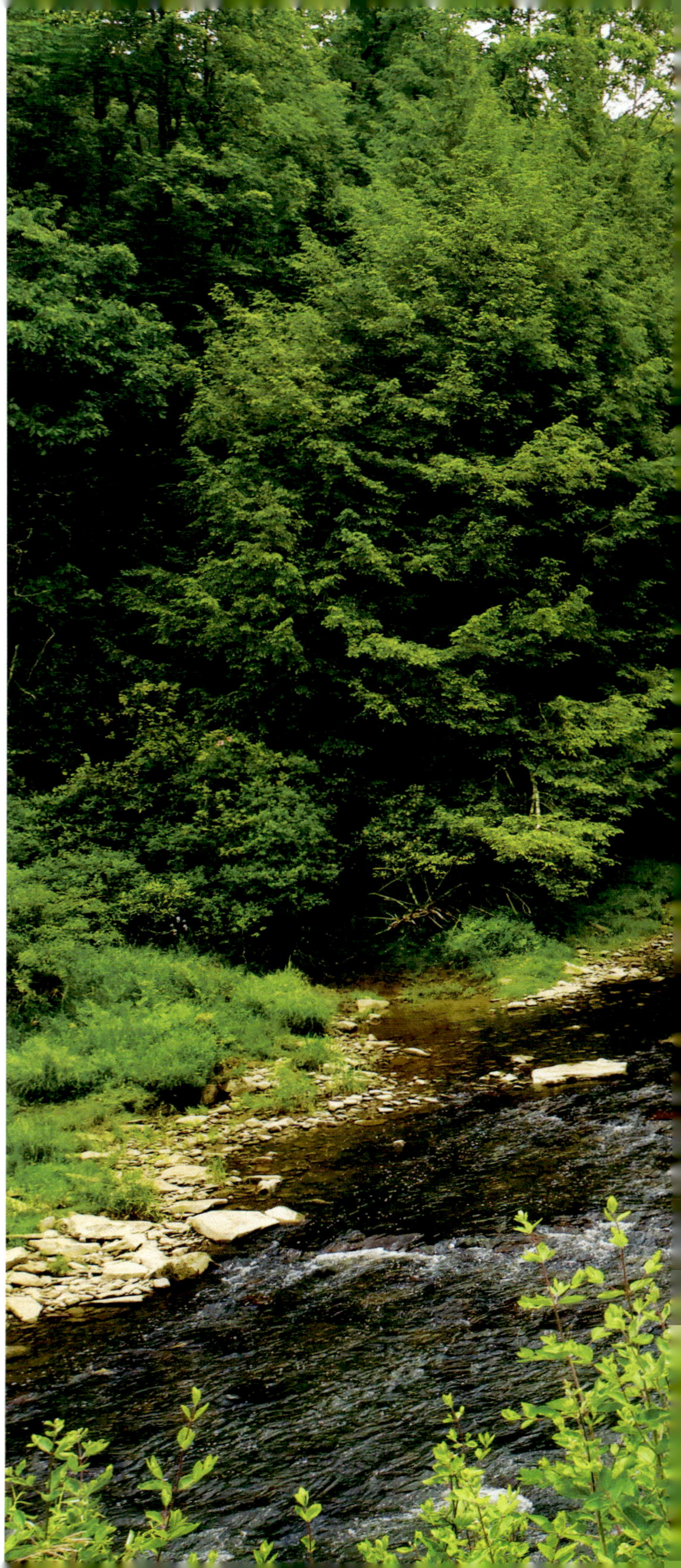

After gathering from sources on Backbone Mountain, the Youghiogheny grows with headwater tributaries downstream from Oakland, Maryland.

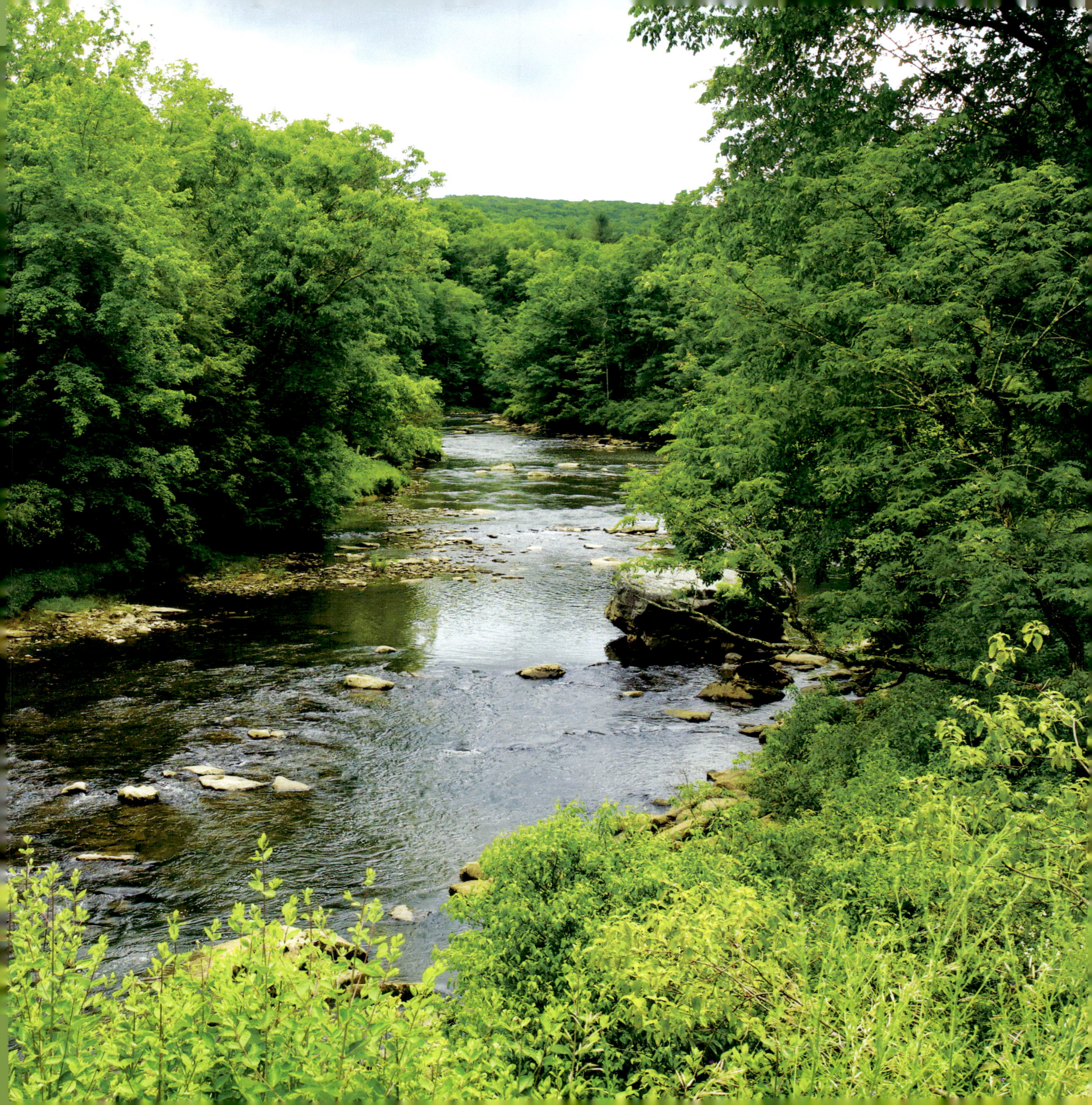

riverbank flowers, leaves, boulders, and stones: each one shines with its own beauty when you focus on what's small. Birds included. Speak of which, within a stone's throw of the Ferncliff parking lot I once spotted the incomparable brilliance of a scarlet tanager, the sunshine yellow of a goldfinch, the stunning orange and black of a Baltimore oriole, and the sky-blue shine of a bluebird, *all in nearly one single view*! Never since then have I seen such color all at once in nature.

For other details I've held my waterproof camera under the surface to capture the river's riot of bubbles as they infuse nourishing oxygen into the water with each bursting micro-dome of air. In the process I unexpectedly discovered crowds of tiny native midges that crawl underwater as larvae, to hatch out later as winged adults hovering like gnat-sized helicopters above the flow. These along with other insects serve as fundamental building blocks in the river's food chain linked upward to fish, otters, eagles, and anyone who eats a trout or bass caught here. If you fry that fish up, you might think about appreciating the midges—almost invisible—that made it what it was.

Knowing all this is one thing, but to look closer and see what's there is to believe in nature in a deeper, more enriching way. Kneeling at the water's edge or climbing high above it on rocks each lends a different perspective on what the river is. Also on what the river does.

Those tiny midges with see-through wings hover within inches of the flow while osprey gliding on four-foot black-and-white wingspans soar high above. To feel what that's like, I walk across the hiker-biker High

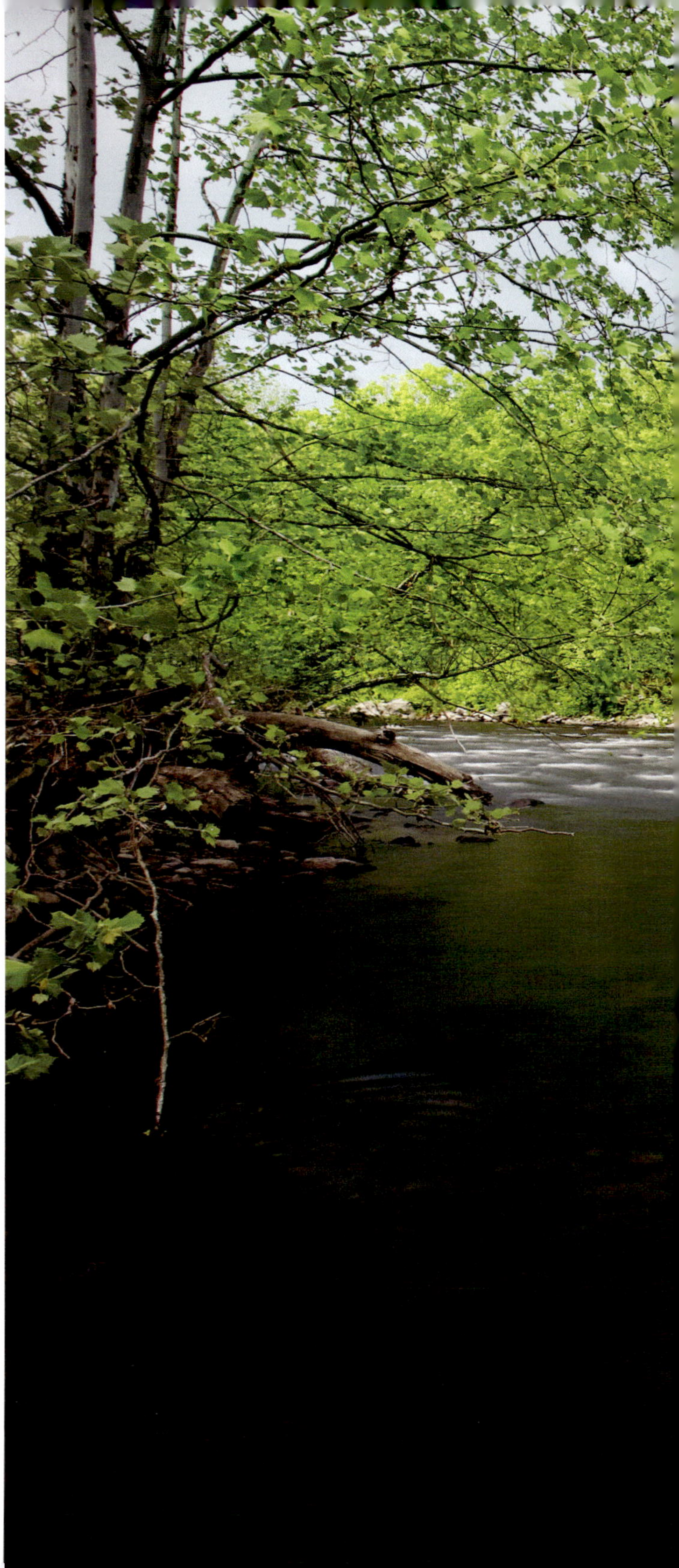

Largest among Youghiogheny tributaries, the Casselman River can be seen near the Route 523 bridge upstream from the town of Confluence.

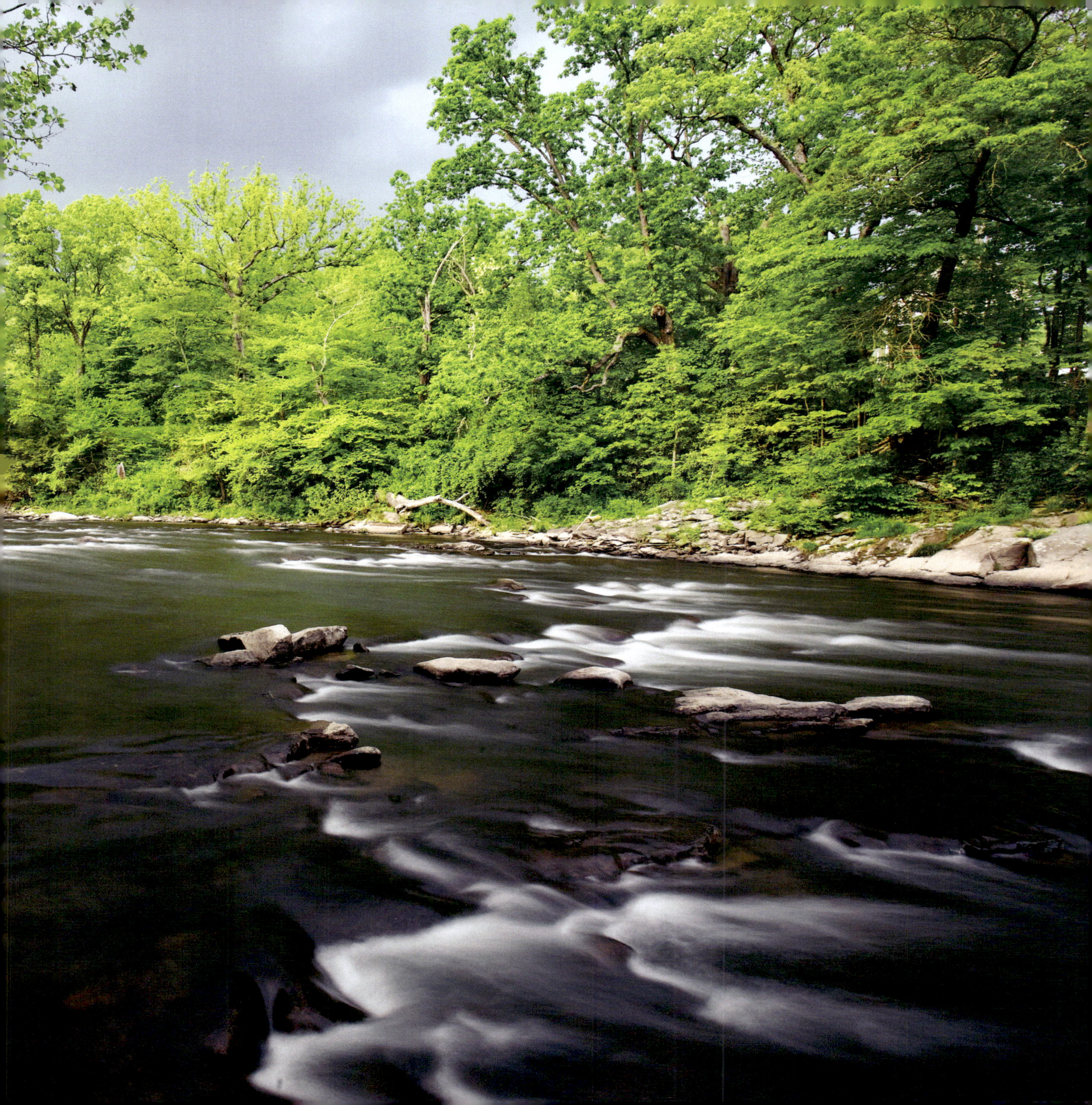

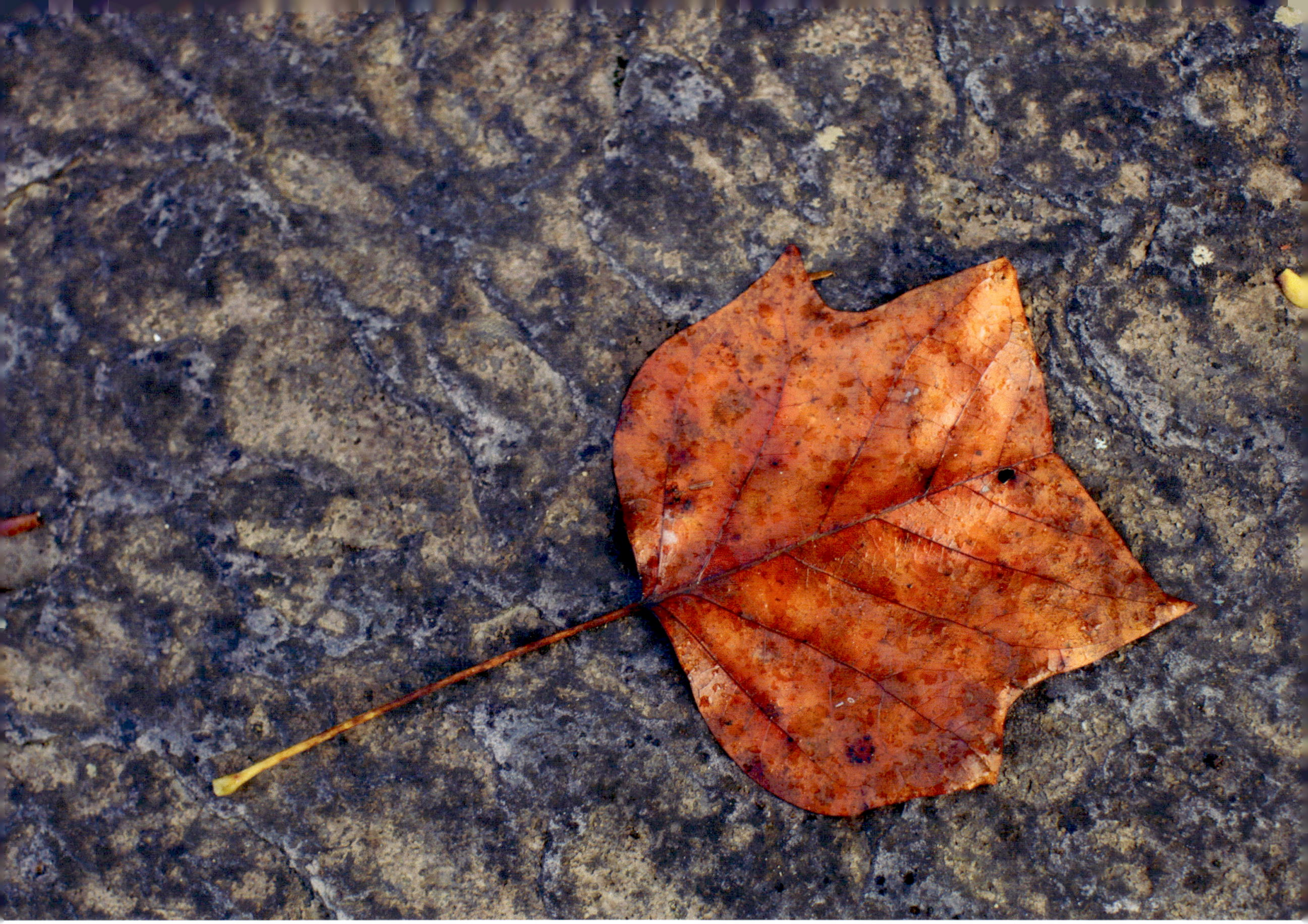

Tuliptree leaves have fallen at the river's edge and brighten rocky waterfront trails.

Bridge west of Ohiopyle where I can stare straight down a hundred feet to rocks, rapids, and transparent pools. I try to imagine seeing, as those graceful hawks do with vision eight times better than ours, fish swimming underwater. At either end of that bridge, where I had to flee from trains during my youth but now stare at the river with total abandon, we can also look with rarely seen bird's-eye perspectives down into the canopy of the Appalachian forest, complete with its hustle of squirrels chasing each other in what looks like fun; also to songbirds in springtime treetops singing with hopes that a mate will answer and come forth.

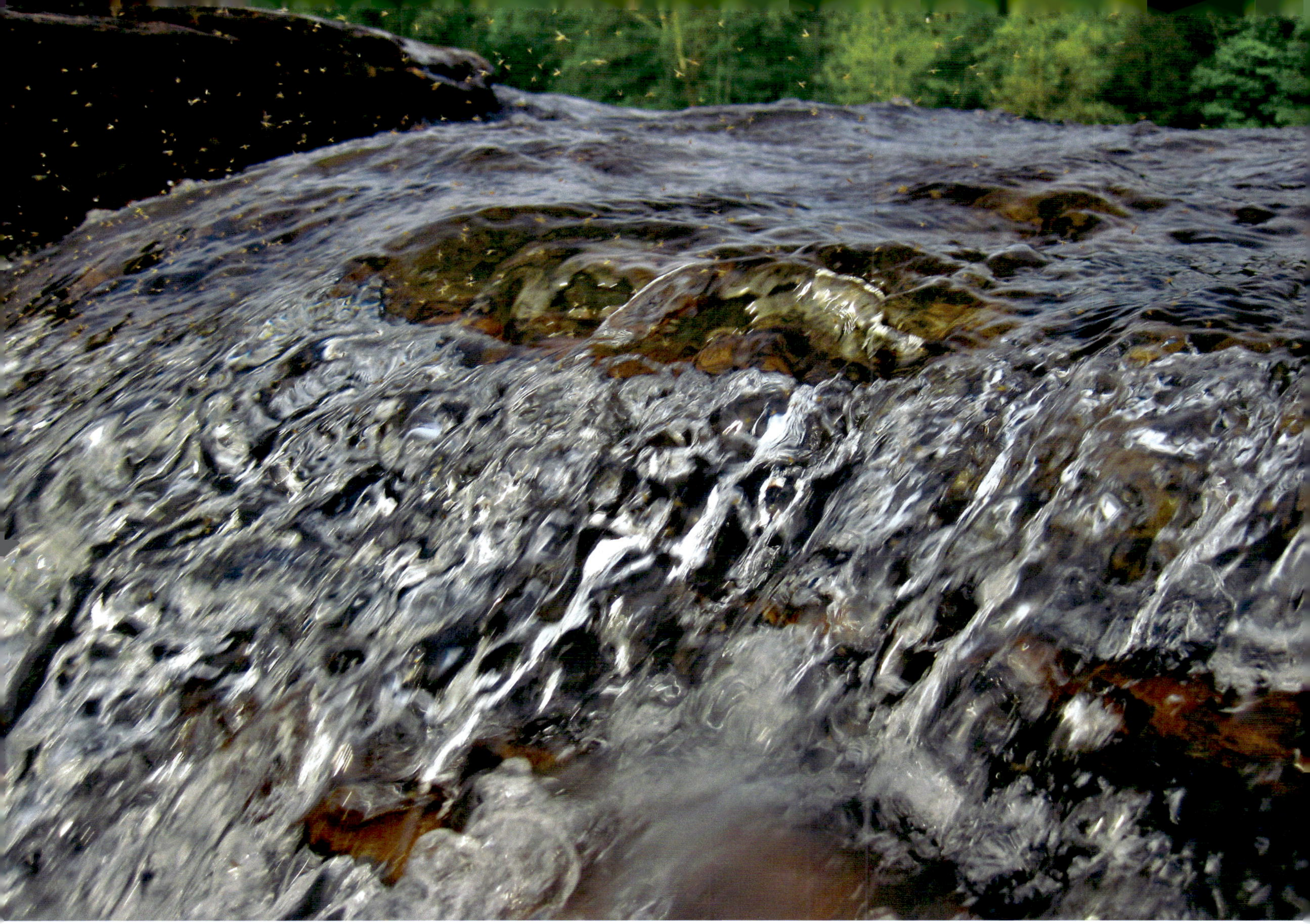

The river looks different at every new level of runoff. Eight hundred cubic feet per second (cfs), which equals about 2 feet on the gauge at the Ohiopyle State Park Visitor Center, run transparently clear in midsummer, rapids distinctly outlined in white, eddies spinning at shorelines, pour-overs fizzing like champagne. However, after an intense storm, 3,000 cfs or more rage as an angry deluge—muddy, brown, impressive with raw power that you can even smell, earthy with dirt, pushy with indifference to any wishes we might have for a flow that's relaxing, calming, or even swimmable.

Meadow Run tumbles over ledges of bedrock while tiny, rarely noticed midges hover at the surface of the flow. These insects are excellent feed for trout and sometimes constitute half their diet.

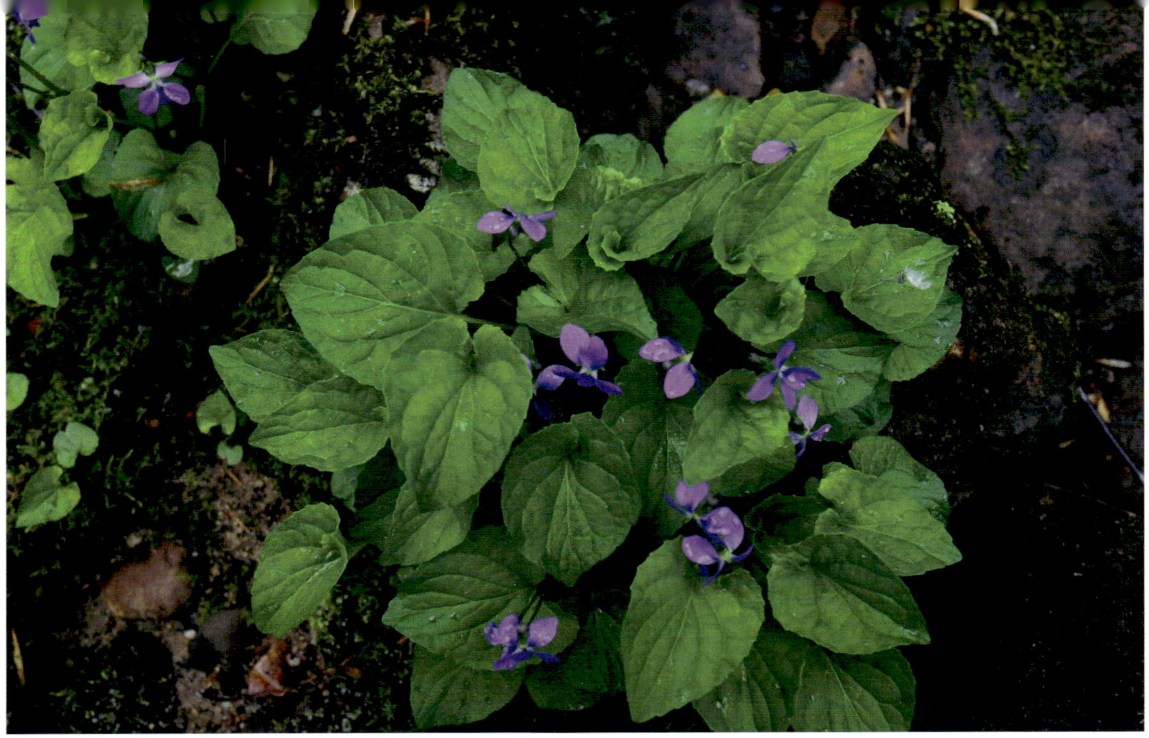

Violets bloom along the river and in its woodlands.

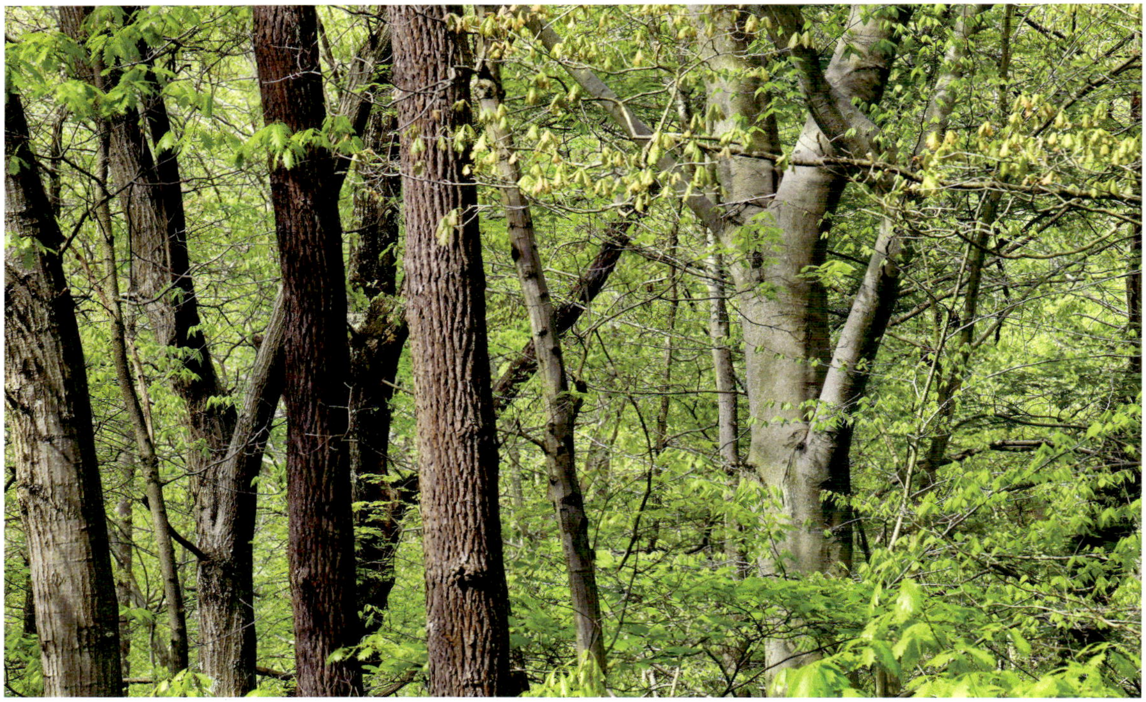

Yellow greens of early May brighten the Appalachian forest here with smooth gray bark of a beech tree on the right along with white and red oaks at the Middle Youghiogheny.

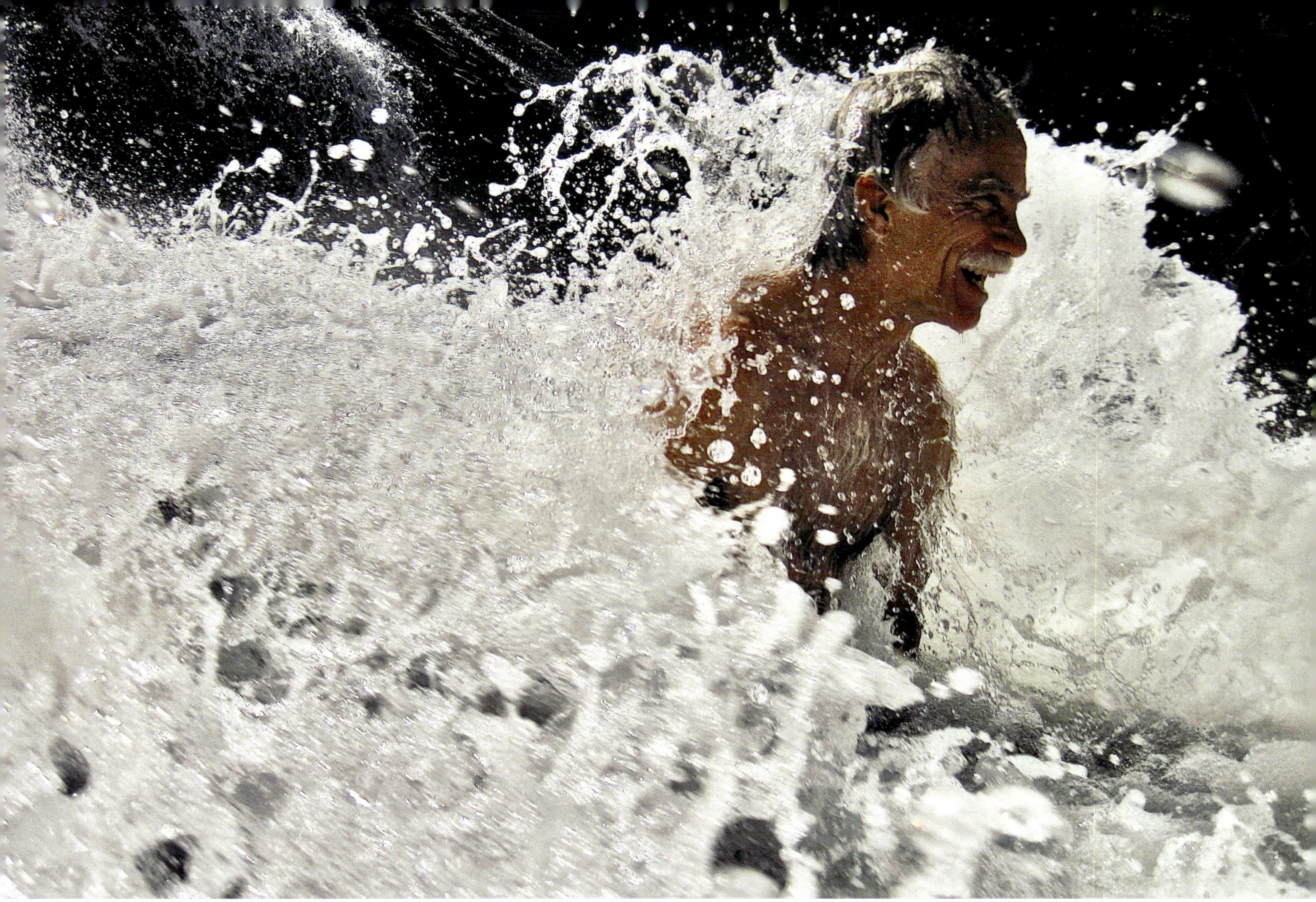

Then, too, each season brings to us a new world, custom painted for the calendar. Leaves bud out like yellow-green stars in April or May, unfurling to become the saturated green of June and July. October's yellow birches, red maples, and orange oaks reflect in mirroring pools, followed by winter's stark gray woodlands whitened by occasional blankets of snow. Ice in the dark depths of January lacquers shoreline rocks like an inhospitably harsh Arctic, never imagined during the playful days of summer. For my book, and for anyone who really wants to see the unabridged river, this is all worth a look.

The river changes by the season and, for that matter, by the hour. Anyone who gets up early might see a spectacular sunrise when light angles sharply

Summer brings warm days and refreshing dips in the Youghiogheny. With care and precautions for issues such as foot entrapment between sharp rocks, swimmers can enjoy the spray and foam of the upper river at Swallow Falls State Park.

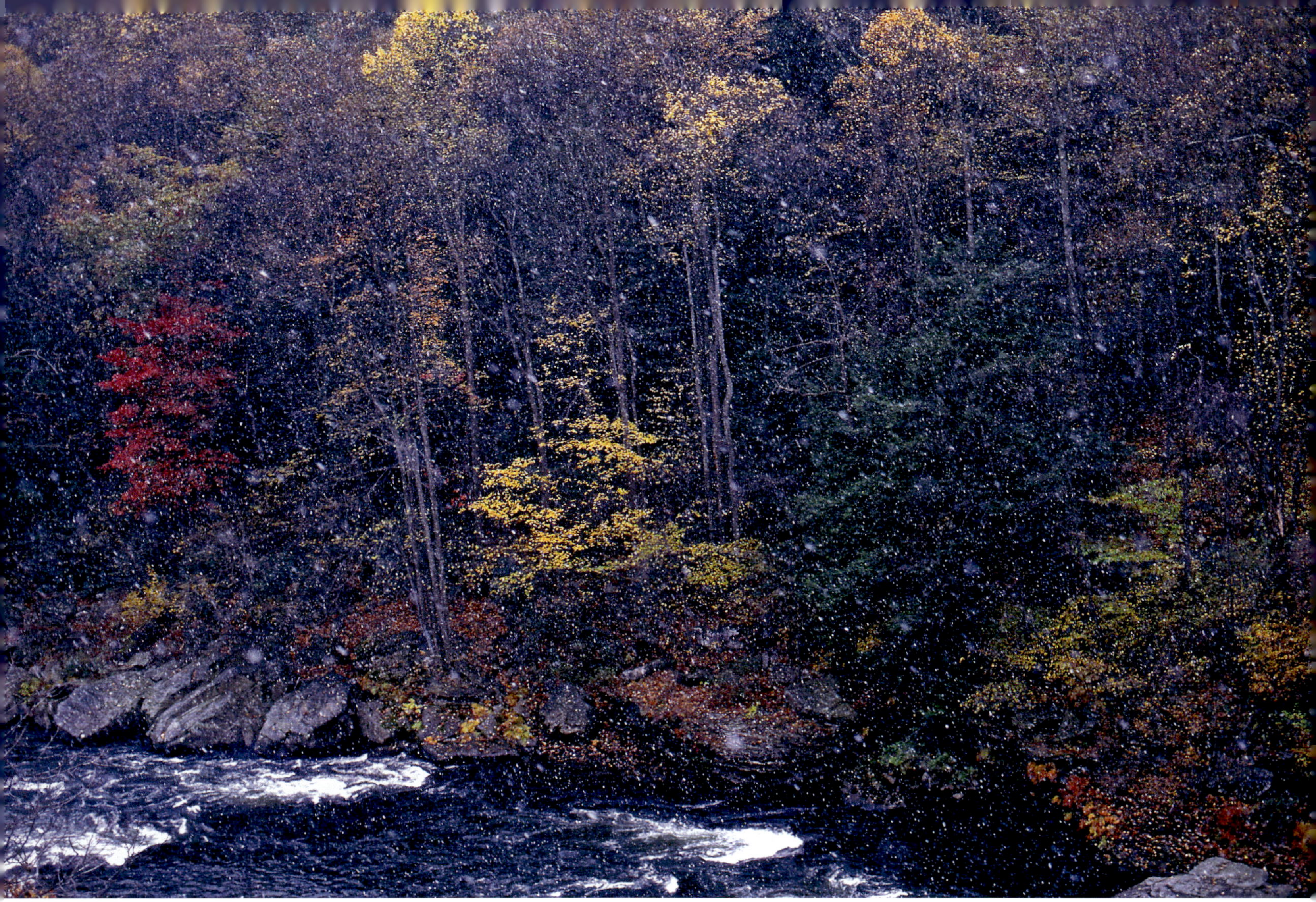

Autumn colors fade as the season's first snowstorm falls on the Appalachian forest, seen from the High Bridge downstream from Ohiopyle.

through shoreline hemlocks and onto bubbling rapids. If you stay, you'll see the day bloom with possibilities, as do our lives, out there when we're open to nature's inspirations. Afternoon sunshine makes for bright light with welcoming warmth on beaches where the water in summer begs for wading. Or maybe for a plunge, just for the sake of full immersion, in every sense of the phrase. Then the light peacefully fades as the earth completes its turn. Twilight darkens with its soothing tint of blue. Finally, if you're not afraid to miss dinner and stay out late, hundreds of stars fall into the river where they reflect on the surface of its quietest pools. Their pinpoints of light make the colorless mirror of the water seem even deeper than in daytime, even wilder than

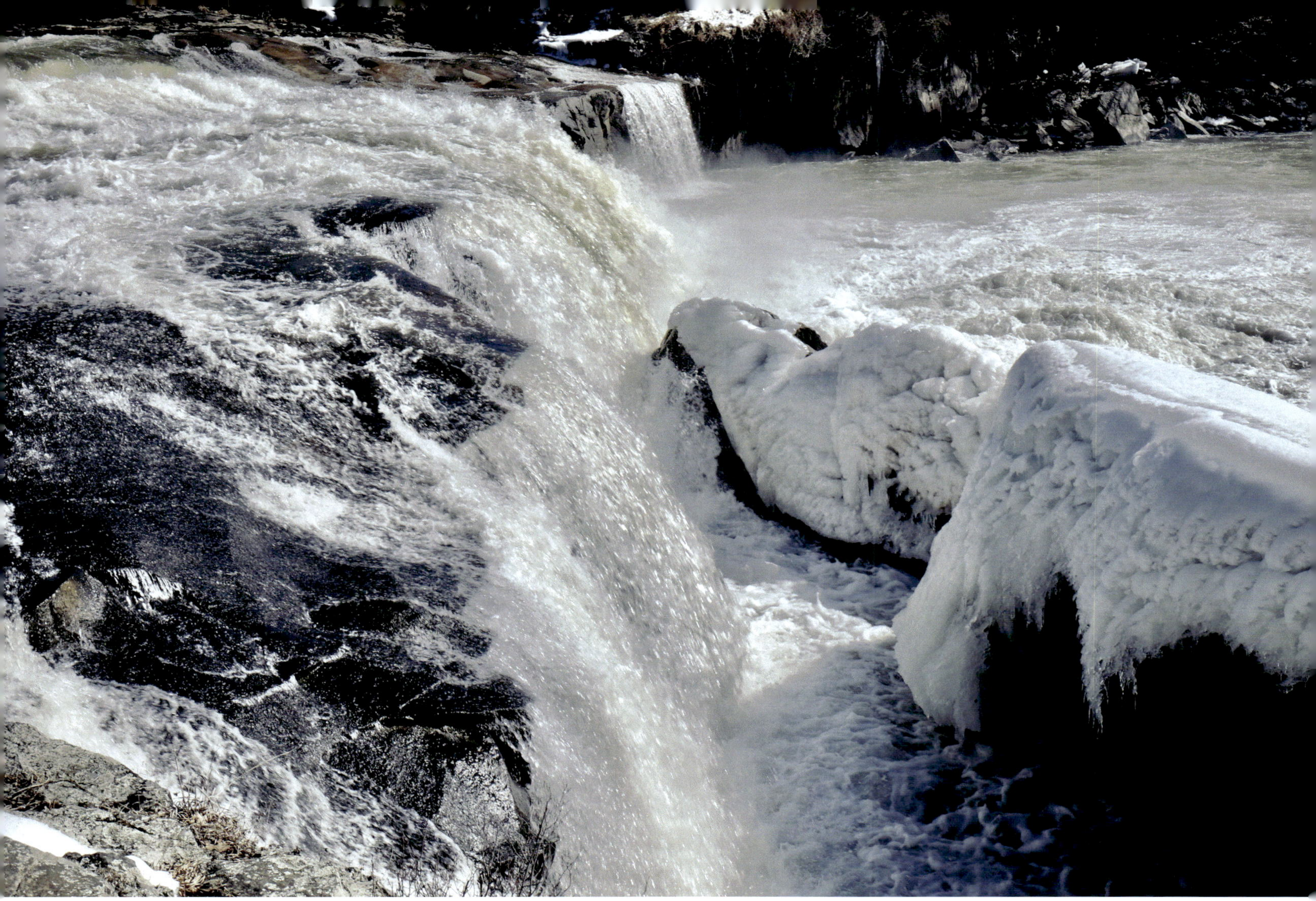

when we can see more clearly. The night appears wonderfully familiar yet mysteriously unknown in the current, which glimmers for the few of us who linger to see the remaining beauty of the Youghiogheny as darkness drops like the closing curtain of the day all around us.

Those who are patient might see the river in yet another way just by sitting in one spot and watching the water flow. It helps if you take some time to do this, and if you go alone. I happened to stumble unexpectedly into this meditative state long ago, almost but not quite a teenager, sitting on a rock at the Youghiogheny's edge. I had wandered downstream from the two-story, white-frame, big-porch Ohiopyle home of my grandmother, which sat directly upslope

In winter's grip, ice thickens in the mist zone at the base of Ohiopyle Falls.

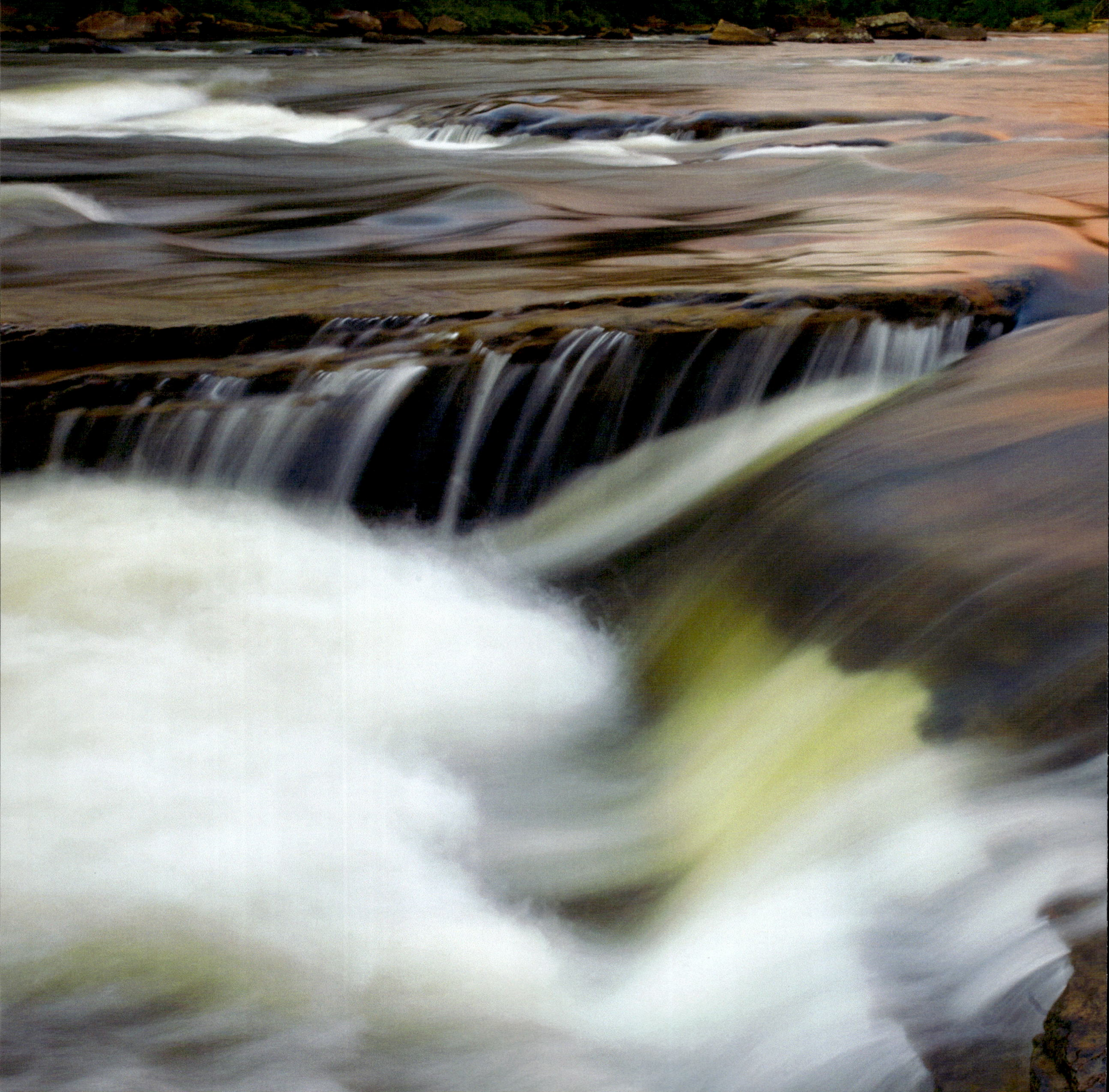

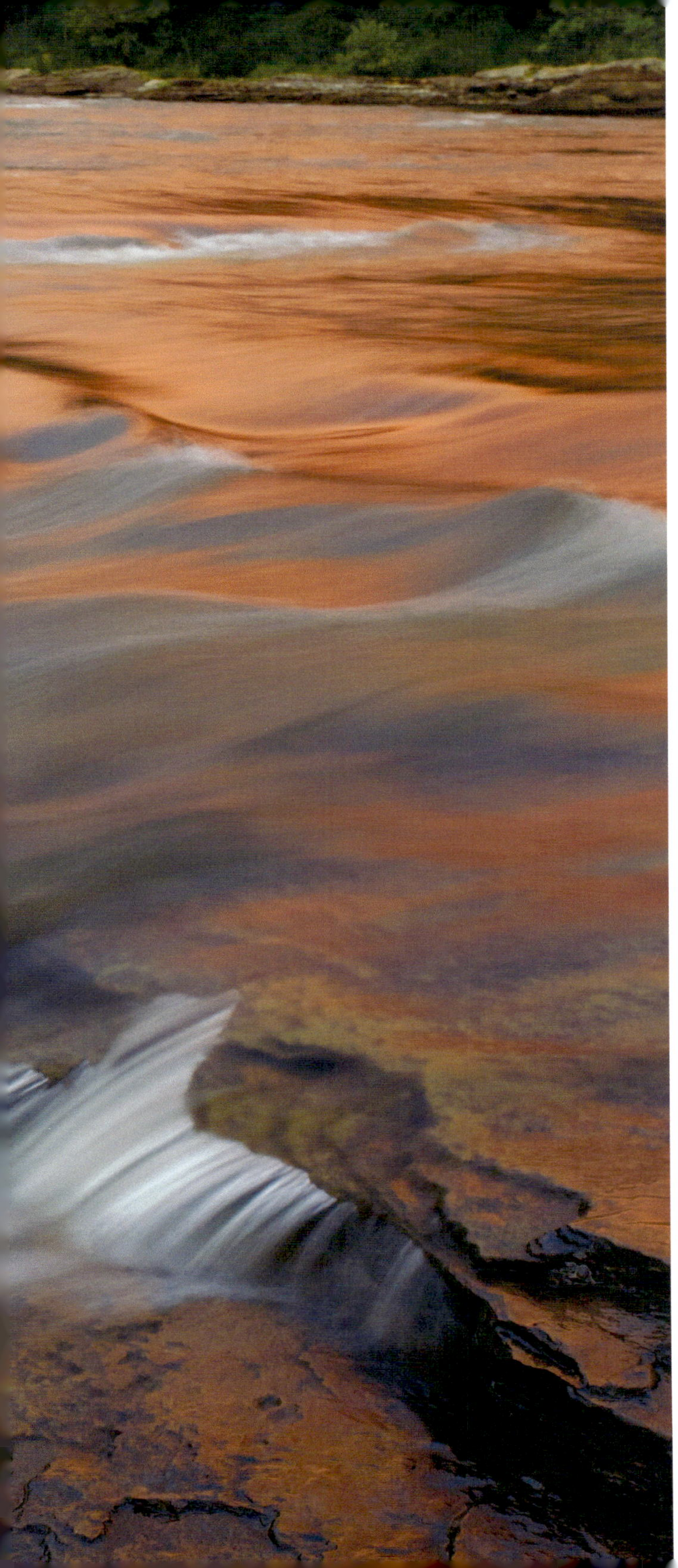

from where today's visitor center now stands. In summers I stayed there for weeks at a time with my father's mother, before the Ohiopyle State Park was created.

Alone, as usual on one of my evening explorations seeking some of the first wild adventures of my life, I wouldn't have heard Mildred Potter Palmer call even if she had tried to find me. The only sound came from the water rushing by. With my daily chores done at the family homestead where my grandmother had been raised after her birth in 1898 to Dalton and Goldie Potter, I had no promises to keep, no schedule to meet. I found myself staring into the current, not a thought on my mind. The rest of the world had disappeared. It could have been a hundred miles away. While the land all around lay stone-still, the water's smooth motion held me enraptured. For long moments the Youghiogheny became my entire everything, my whole life. Then, as if in a trance, I felt as if I were in the flow, magically moving with it, truly part of a world that was so much larger than I was. More important, I realized—even as a twelve-year-old—that the place was perfect. Remarkably, I knew why. It was perfect for the simple reason that it had been left alone. It was natural.

Seeing it all with such clarity eventually made me want to know more—more about how this extraordinary place came to be, more about how it so vitally supports life, more about everything around it, everything touching it, everything dependent on it. My own personal journeys of many kinds along the Youghiogheny began on that warm summer night, at a river rapid that, to my way of thinking, has been perfectly named. It's called Entrance.

On a warm summer evening the river elegantly reflects sunset at Entrance Rapid.

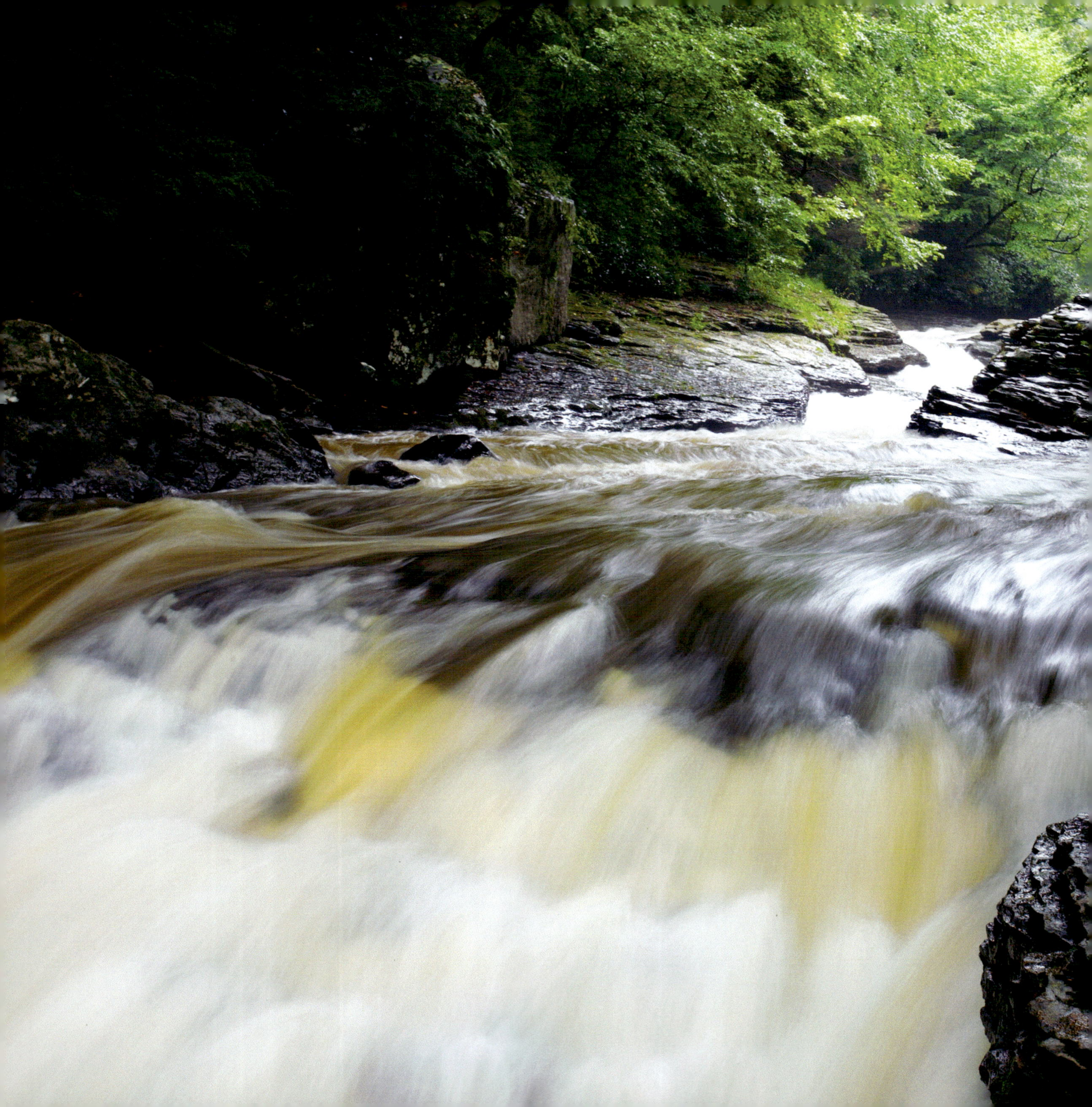

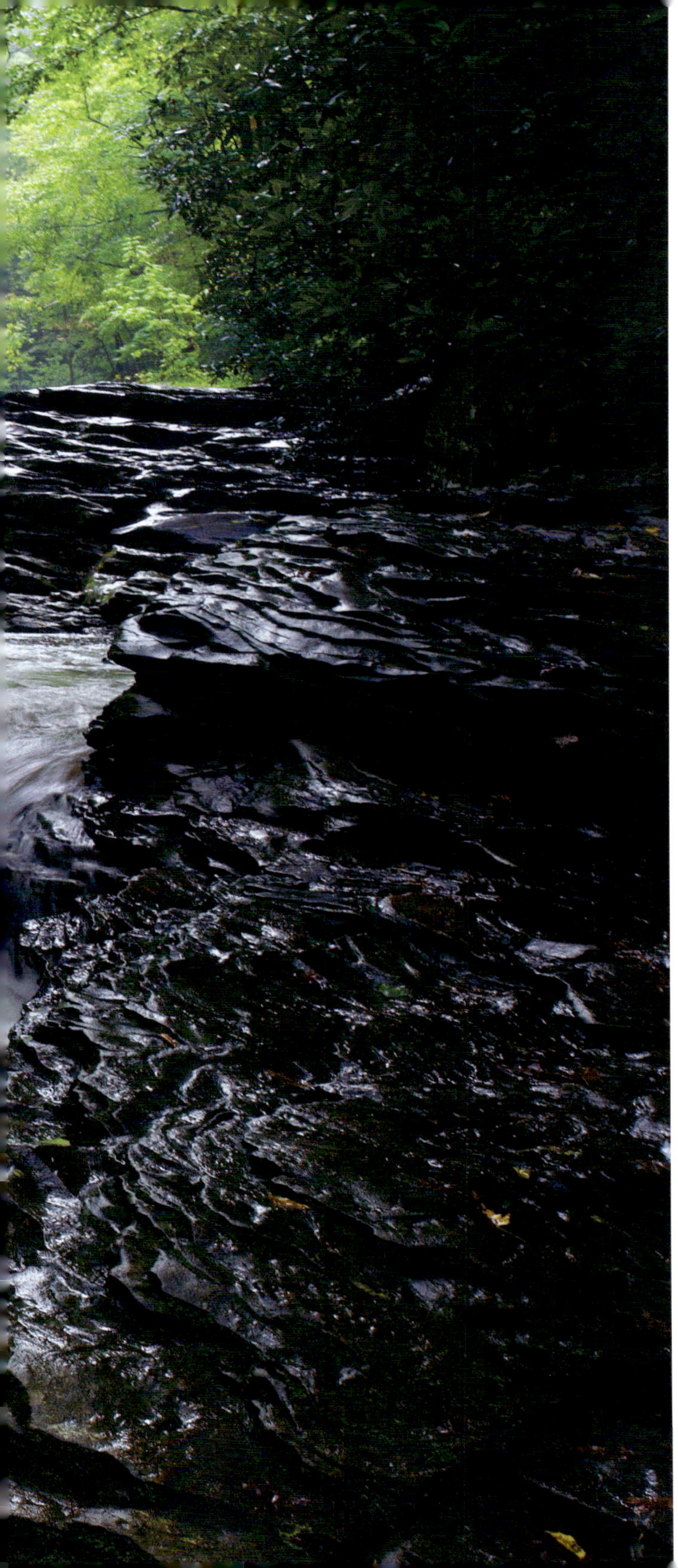

Chapter Three

Seeing and Knowing the Natural World

The beginning of a story can always be started earlier, and earlier, but perhaps it's enough here along the Youghiogheny to go back three hundred million years. That's when the earth now underfoot emerged as mountains from an older earth with prospects of becoming the magnificent landscape we see today.

It's those mountains along with the rain falling on them that have made the Youghiogheny what we now know. So consider all that makes this place what it is: the geology that explains the land's creation, the climate that brings the essential rain and snow, the hydrology that governs the way the river flows, the fish that animate the river and need it in order to live, the forests that link inseparably to both land and water, plus the various forms of wildlife that depend on the fantastic composite of everything else. This all leads to our own personal and community use of a precious place. The backstory behind the river can add clarity, understanding, and insight to everything we see.

The Mountains Came First

According to geologists, the North American Plate (a tectonic plate containing most of North America) separated from a

Meadow Run—the largest Youghiogheny tributary in Ohiopyle State Park—has carved narrow chutes through ancient bedrock.

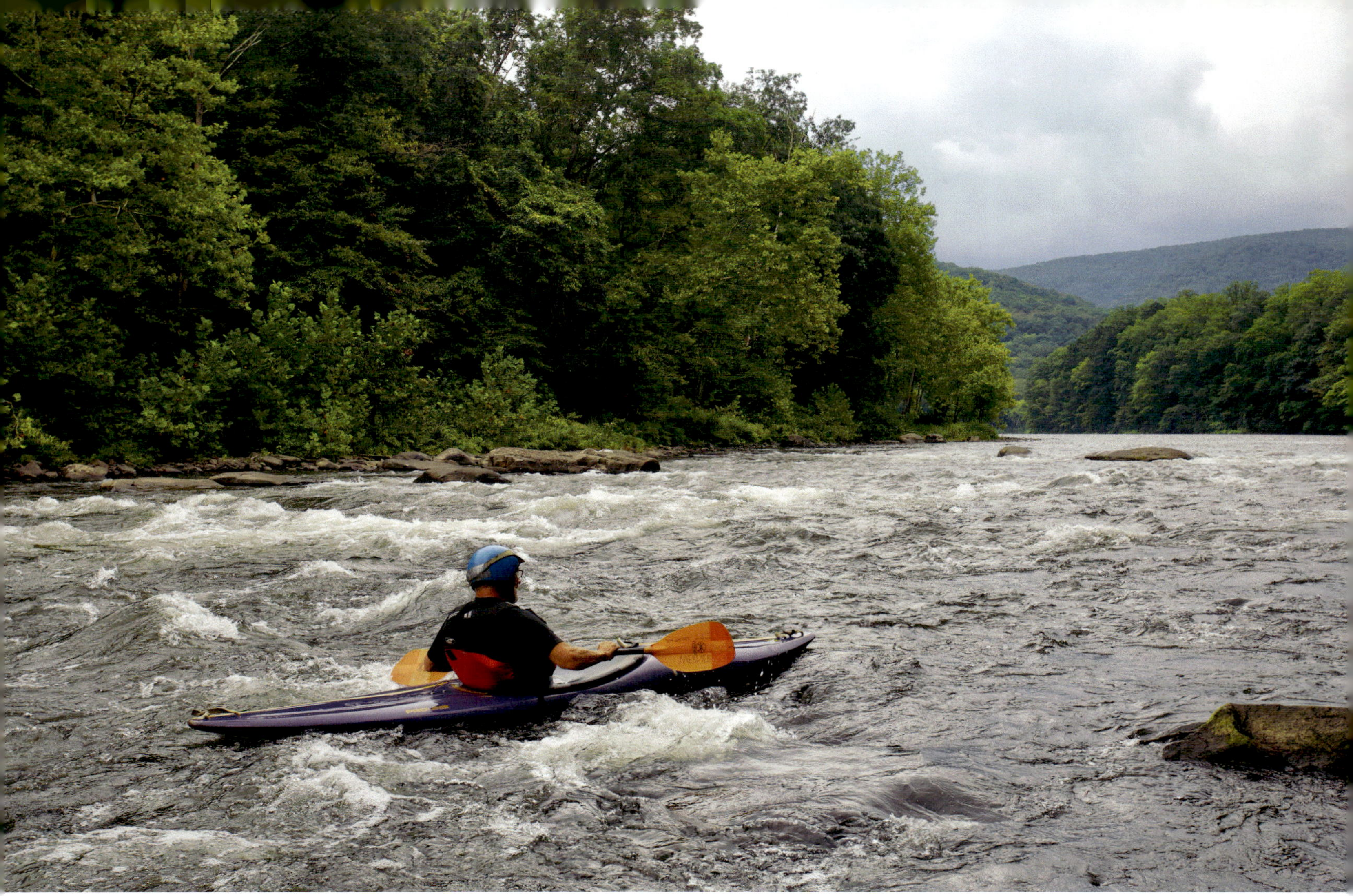

The Middle Youghiogheny between Confluence and Ohiopyle flows through the heart of Laurel Hill, whose ridgeline rises in the background.

contiguous worldwide continent called Pangaea, then drifted inch by inch northwestward. In the process, earthquakes wrinkled up multiple ridges of mountains, much like an untacked carpet would do if you pushed on one side, causing it to buckle up in the middle. This topography makes up what's called the Ridge and Valley Province—a subset of the Appalachian Mountains that now appears in strikingly parallel ridgelines with intervening valleys angling northeast to southwest through central Pennsylvania and beyond. These ridges start far east of the Youghiogheny with Maryland's Catoctin Mountain and continue in a washboard sequence of fifteen or so named uplifts that we have to climb to cross when driving interstate Highways 68 or 70 westward from Washington or Baltimore to Morgantown or Pittsburgh. The last

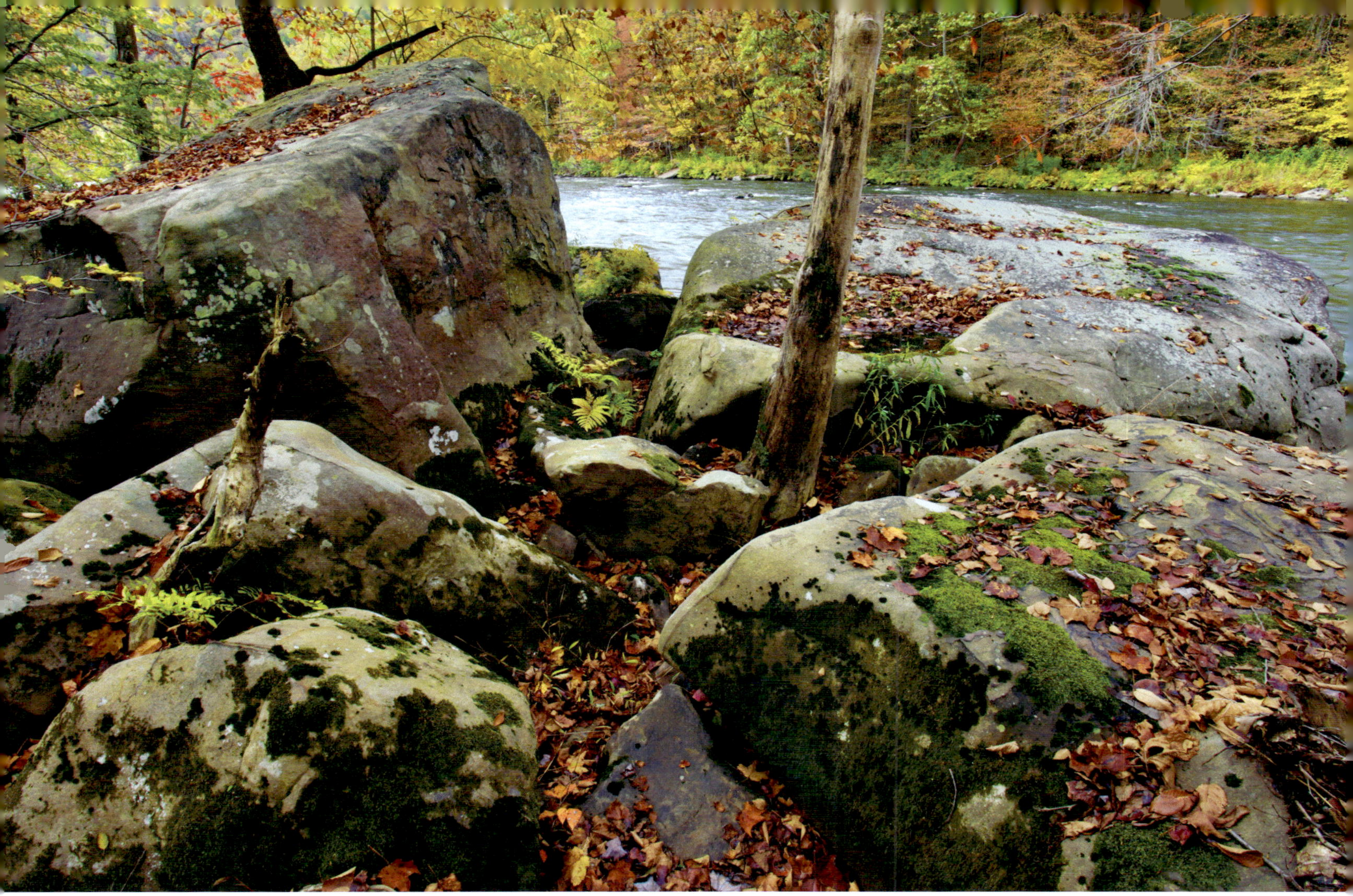

mountain, where the Ridge and Valley's distinctly sculpted crests have merged with an incised uplift called the Allegheny Plateau, is the 80-mile-long Chestnut Ridge, which the Youghiogheny transects downstream from Ohiopyle in a winding, rapid-riddled path east of Pittsburgh.

Tectonic forces pushed these mountains up while rainstorms sent water flowing across the land to accumulate in rivers that eventually eroded their channels like a hydraulic saw, cutting narrow gaps, gorges, and valleys through the ridges at roughly the same downward pace as the land was rising.

A mountain descriptively named Backbone had reared up east of today's town of Oakland, Maryland, marking the lofty Atlantic Continental Divide. From its eastern face the Potomac River flows to Chesapeake

Boulders resist erosion on the shores and in the channel of the Middle Youghiogheny below Ramcat Run.

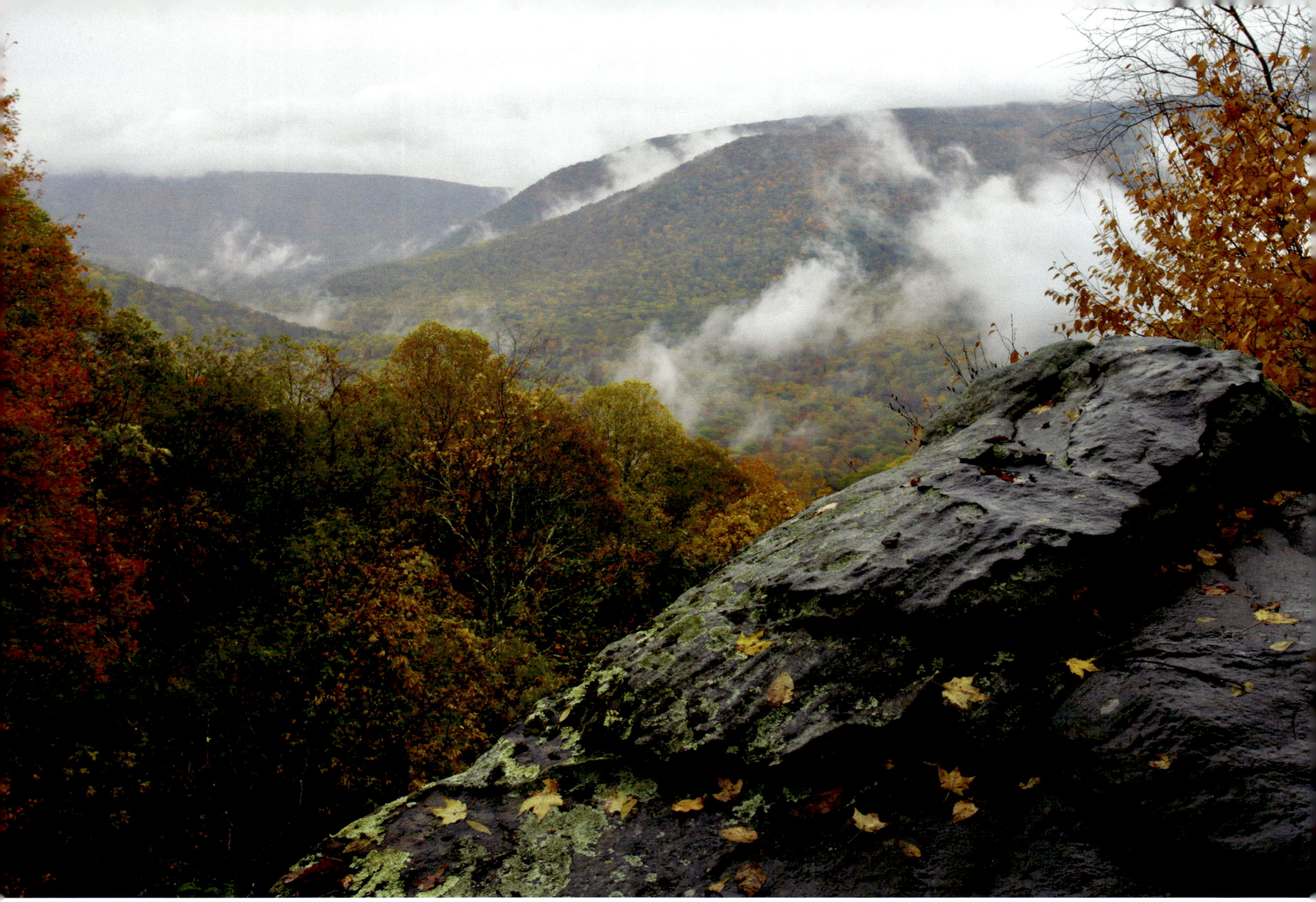

Baughman Rocks edge the Laurel Hill Gorge along the Sugarloaf Road east from Ohiopyle. Erosion-resistant sandstone endures weathering for millions of years as rimrock at mountain heights and also as rocks and ledges creating rapids in the river.

Bay. From its western side, Youghiogheny headwaters dug into Backbone's uplift, then sliced scenic gaps across Meadow and Negro Mountains and continued onward to the north and west. The river runs through Swallow Falls State Park, followed by the town of Friendsville. Below there, nearing the Maryland–Pennsylvania line, the Yough ("Yock") eddies into Youghiogheny Reservoir, which fills a 16-mile-long gap between mountains.

Below the reservoir's formidable dam, the topographic extravaganza escalates with the river's sharply cleaved valley through the 1,200-foot rise of Laurel Hill. By comparison, the famous Grand Canyon of Pennsylvania—carved by a sizable Pine Creek in the

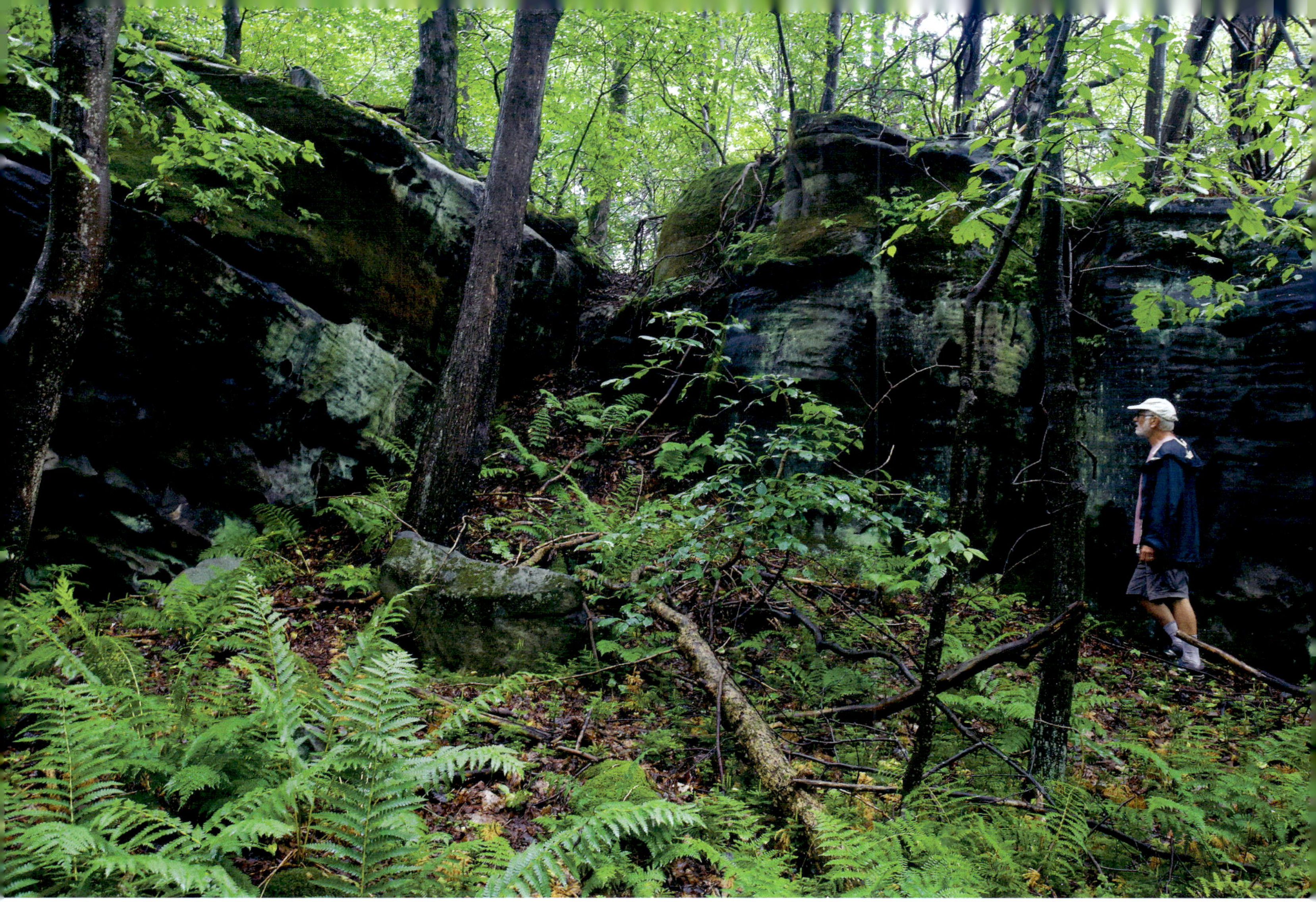

state's northcentral Susquehanna Highlands—showcases slopes rising 800 vertical feet, though Pine Creek's scenic topography veers dramatically steeper than does that of the Youghiogheny.

The "Middle" Yough, from Confluence to Ohiopyle, is followed by an even more notable gorge within the uplift modestly called Chestnut Ridge, which is in fact unquestionably a mountain, topping out at 2,400 feet above sea level, and severed in two by the river winding 17 miles through it, from Ohiopyle to Connellsville. Most of this distance highlights Ohiopyle State Park and is called the "Lower" Yough, even though the bottom half of the stream's length is yet to come. Downstream from Connellsville, the lowest reach of the

The McCune Trail loops at the summit of Laurel Hill just south of the Youghiogheny's ancient cut through the wide mountain massif. Slabs and walls of sandstone cap the wooded ridgetop.

132-mile-long river drifts through the western foothills of America's oldest mountain range.

Sandstone is one of the Appalachians' hardest types of rock, and the river confronts thick veins of it while slicing into the various gorges. Named for Pottsville in eastern Pennsylvania where the durable stratum was identified, this layer of intensely compacted and glued-together sand resists erosion and forms the showy cataract of Swallow Falls in Maryland, rocky rapids below Confluence, the vertical face of the Ohiopyle Falls, and the stair-stepping whitewater below it.

Over time the river has cut ever deeper channels into bedrock, and it continues to do so even as we speak, especially during big flows and floods. Geologists evaluating erosion estimate that one million years ago the water ran at the highest elevation now seen on Ohiopyle's Ferncliff Peninsula. Since that time the river has detoured clockwise, lengthening into the Ohiopyle Loop dropping 60 feet. Today this 1.7-mile-long hydraulic escalator downward is the second-longest section of the Youghiogheny lacking paralleling roads, or railroads, or blockages by a dam. The longest similarly wild section runs 18 miles, from Oakland to Friendsville, though with four bridge crossings.

The river's durable bedrock that we now see at the globe's 39th parallel of latitude was once located south of the equator where a sultry climate supported jungle-like vegetation. Imagine the Amazon. For eons, carbon-rich trees grew, died, and piled up on the ground, eventually morphing into black coal, interspersed and hidden deep between thicker layers of sandstone and shale.

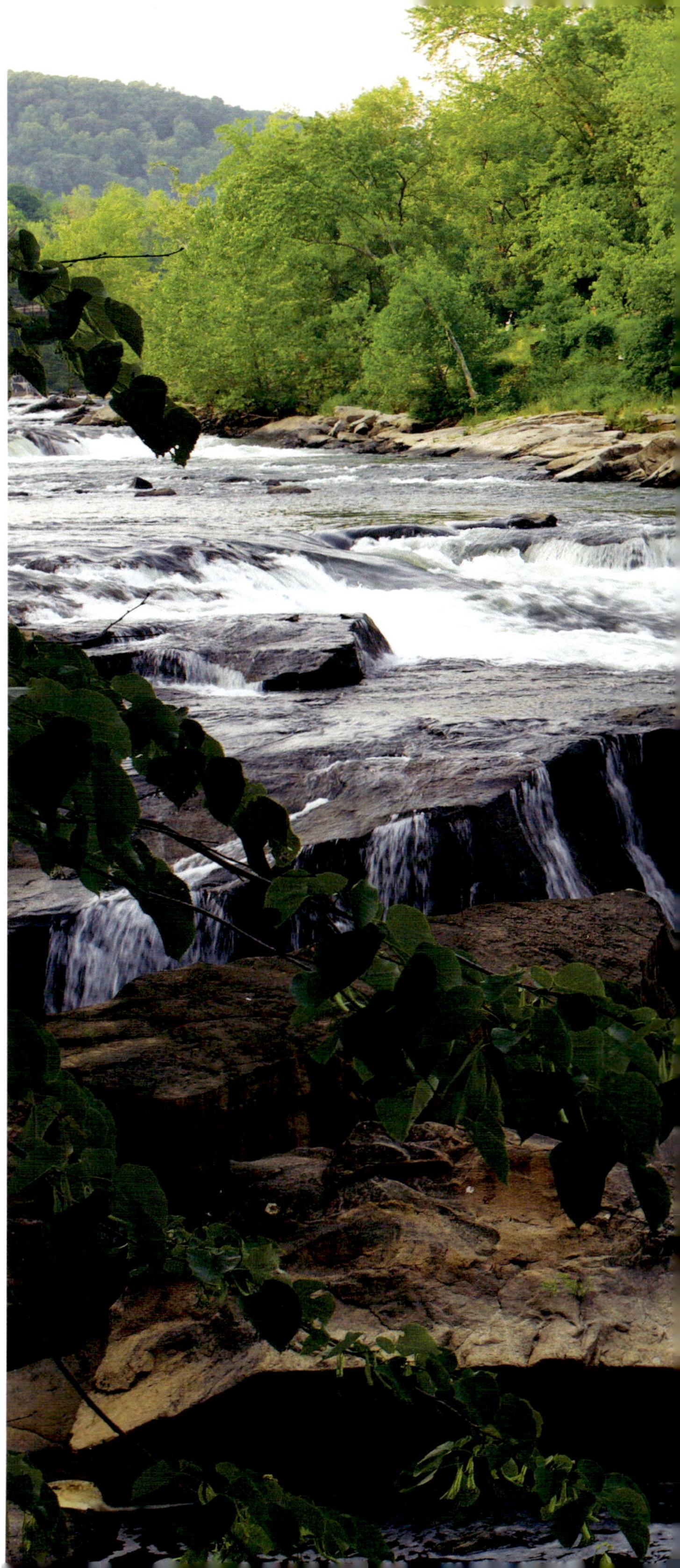

The Youghiogheny free-falls 18 feet over its extraordinary sandstone outcrop at Ohiopyle.

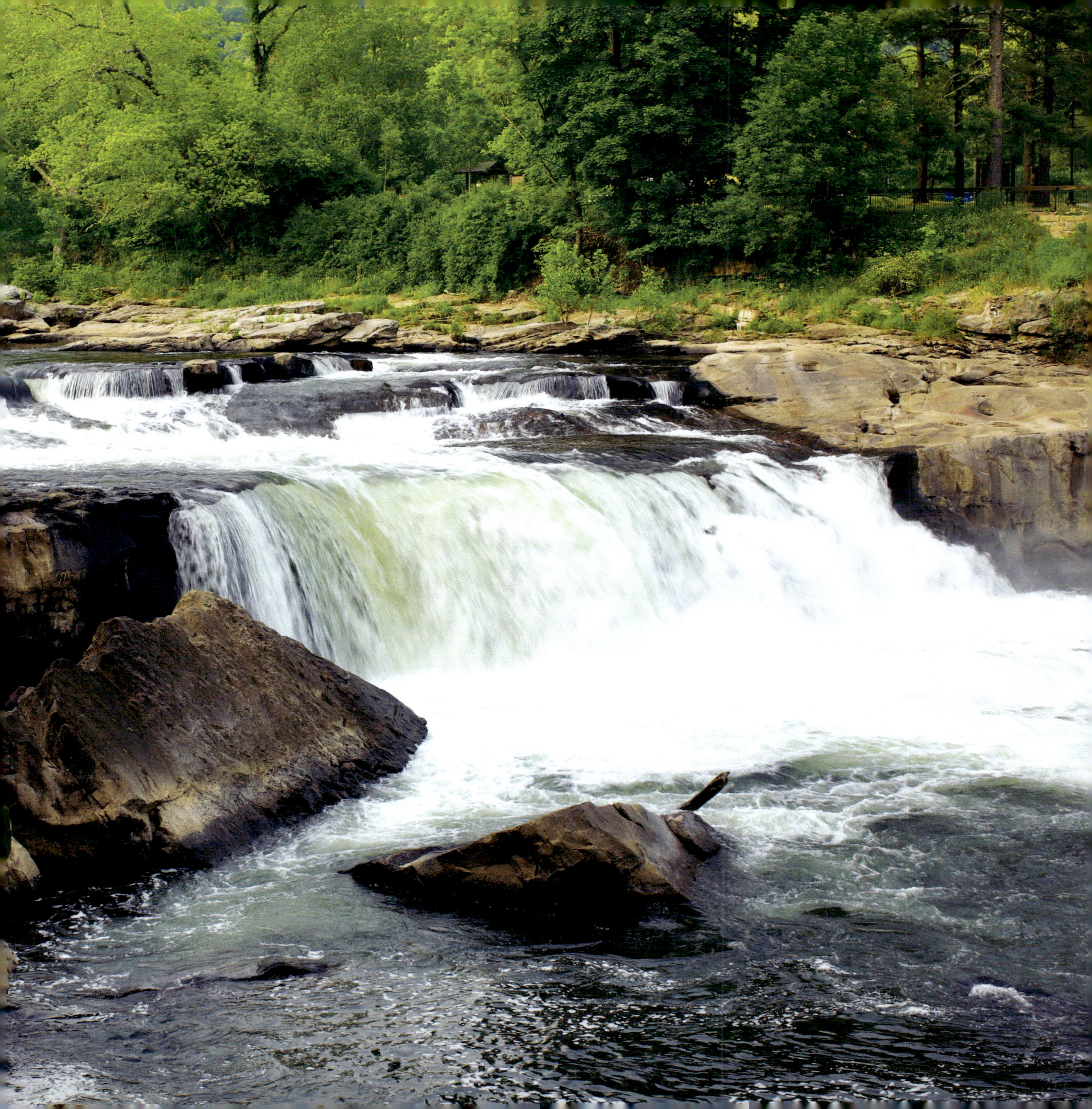

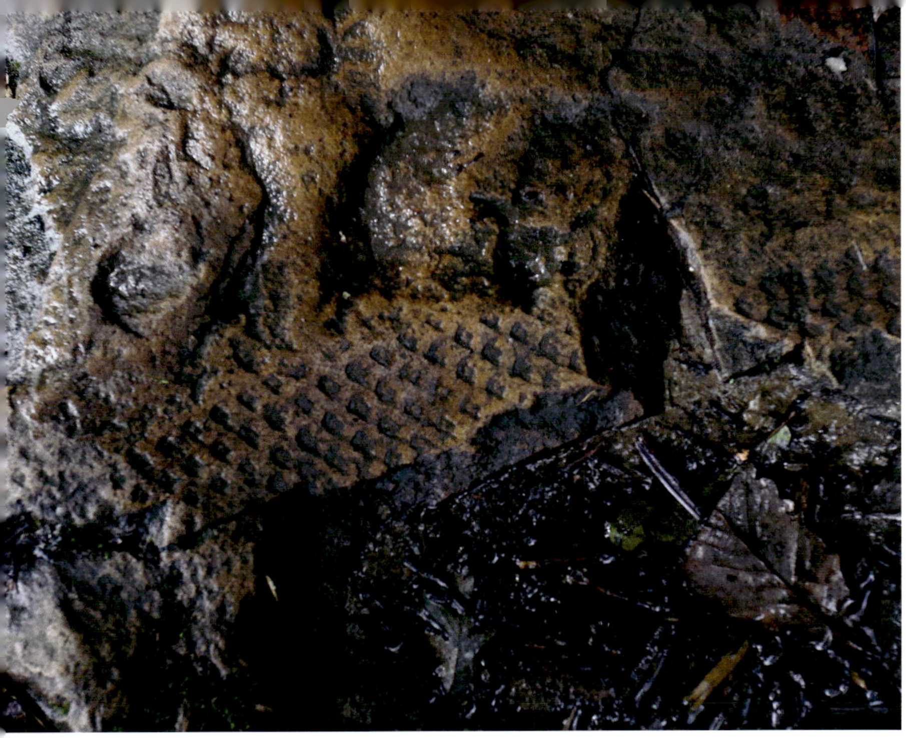

Fossils of Carboniferous Era *Lepidodendron* trees appear in sandstone outcrops of the Youghiogheny. This one lies above Ohiopyle Falls on the right side of the river when facing downstream.

OPPOSITE PAGE
In the Pre-Illinoian Glacial Epoch, the Youghiogheny's inlet to Lake Monongahela lay downstream from Lower Railroad Rapid, seen here beneath the High Bridge northwest of Ohiopyle.

Pressed into sandstone on the surface today are fossils called *Lepidodendron*—tall ancient club moss relatives called scale trees dating to the same Carboniferous Period of coal formation. Imprints of bark, textured like the rough skin of a pineapple, can be seen in bedrock slabs if you watch the footpath on river-right (meaning the right side of the river when facing downstream) on the trail's approach to Ohiopyle Falls. While evidence of a fallen hemlock or even a grandfatherly oak disappears within decades or, at most a few centuries, here along Youghiogheny shores we can spot remnants of a tree that lived three hundred million years ago, exhibited at our feet for all to see in the river's real-life, real-time museum of natural history.

More recent than the fossils, but still in a distant time from 1.2 million to 350,000 years ago, the ancient flow of the Youghiogheny was complicated by the Pre-Illinoian Glacial Epoch. This occurred long before what we usually call the Ice Age, which typically means the latest of many glacial surges that extended south from the North Pole and almost to here. The most recent of those continental ice ages is called the Wisconsin, ending only twelve thousand years ago. The fascinating route of the much older Pre-Illinoian Youghiogheny had emptied into an early version of the Monongahela River. Then, 30 miles northwest of today's Pittsburgh, the flow aimed directly northward as the Old Allegheny River along the present southbound route of the Beaver River with its tributary, Shenango. This ancient river ended in what geomorphologists call the Pittsburgh River, which ran eastward, roughly where Lake Erie lies today. However, rearranging all that, a solid wall of ice, thousands of feet thick, marked the continental glacier's southernmost advance near today's town of Beaver and acted like a dam, forcing the Old Allegheny's northbound water to stop flowing and to rise as Lake Monongahela. This inland sea grew to a length of 120 miles before overtopping to the west, establishing the route of today's Ohio River, which flows not north, as the Old Allegheny did, but southwestward to the Mississippi River.

Lake Monongahela peaked out at 1,100 feet above sea level. This is the Youghiogheny's present elevation at the mouth of Jim Run, about half a mile below the High Bridge of the bike trail west of Ohiopyle. Imagine such a sea, half the size of Lake Erie, lapping at shores now crowded with homes in the hilly neighborhoods of Connellsville! Similar glacial dams in a landscape that might have looked like today's Iceland likely formed and drained in the lower Youghiogheny basin three times.

Geographic processes such as these created the template on which the modern Youghiogheny flows. However, there would be no river without rain.

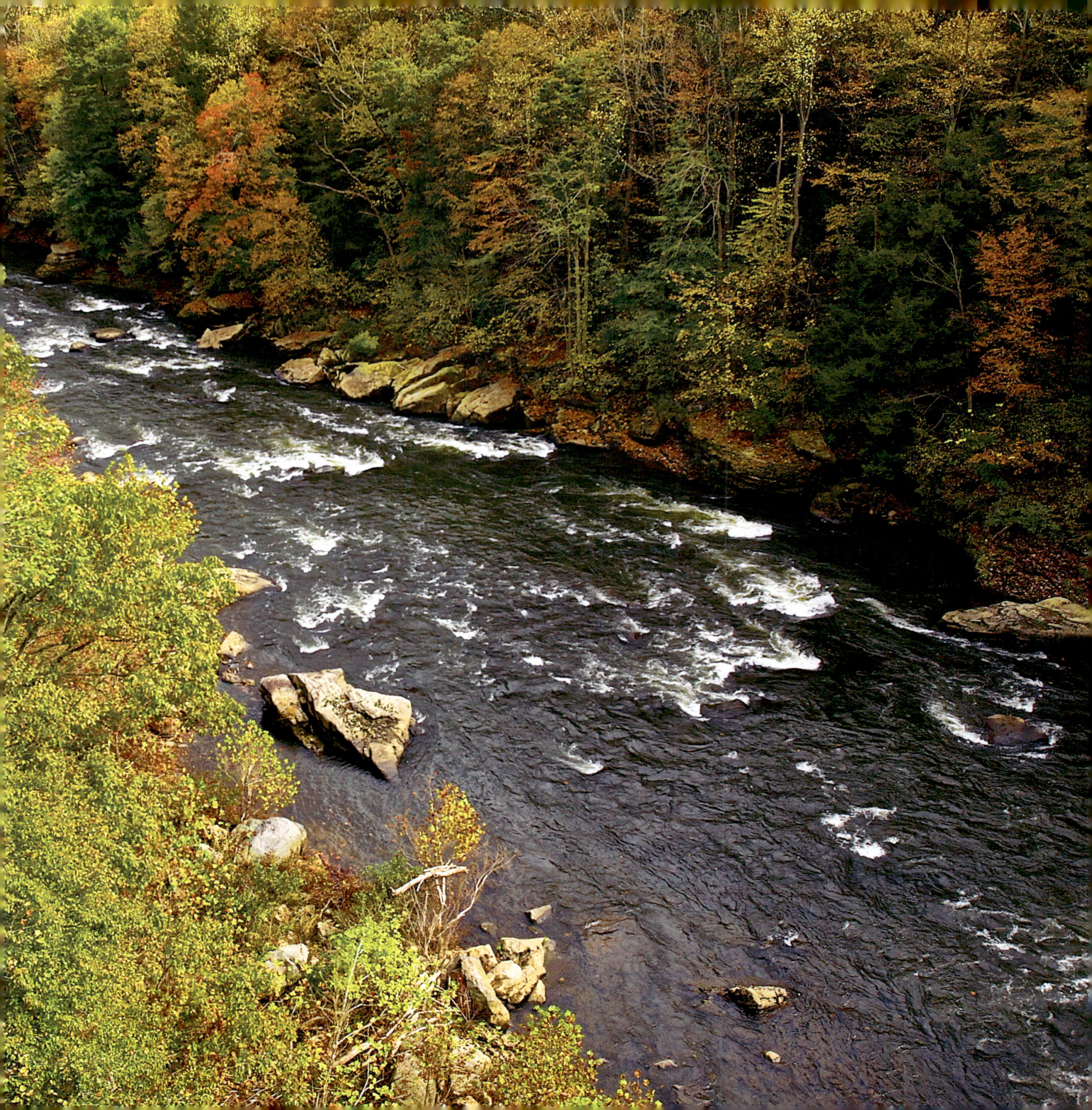

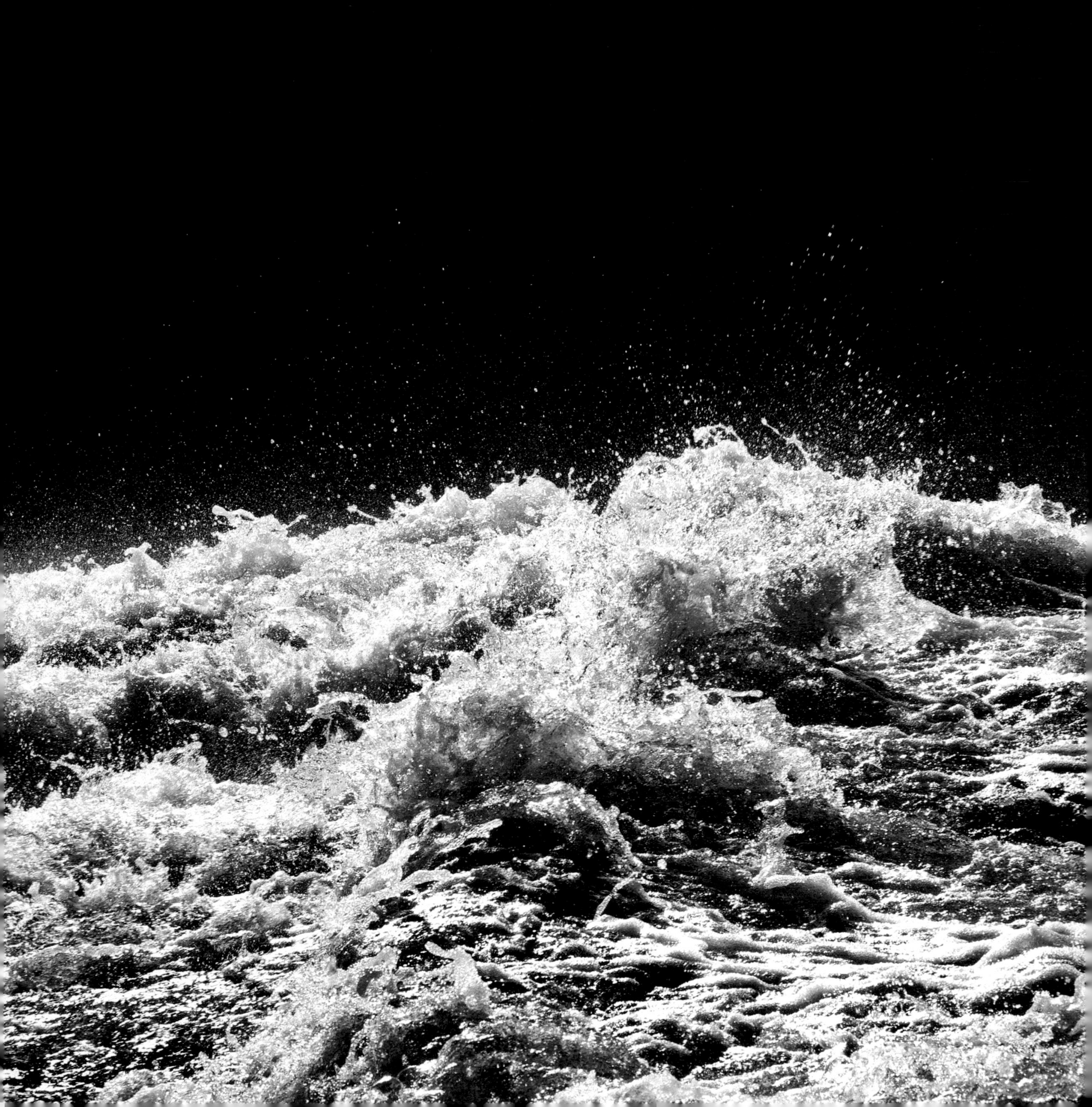

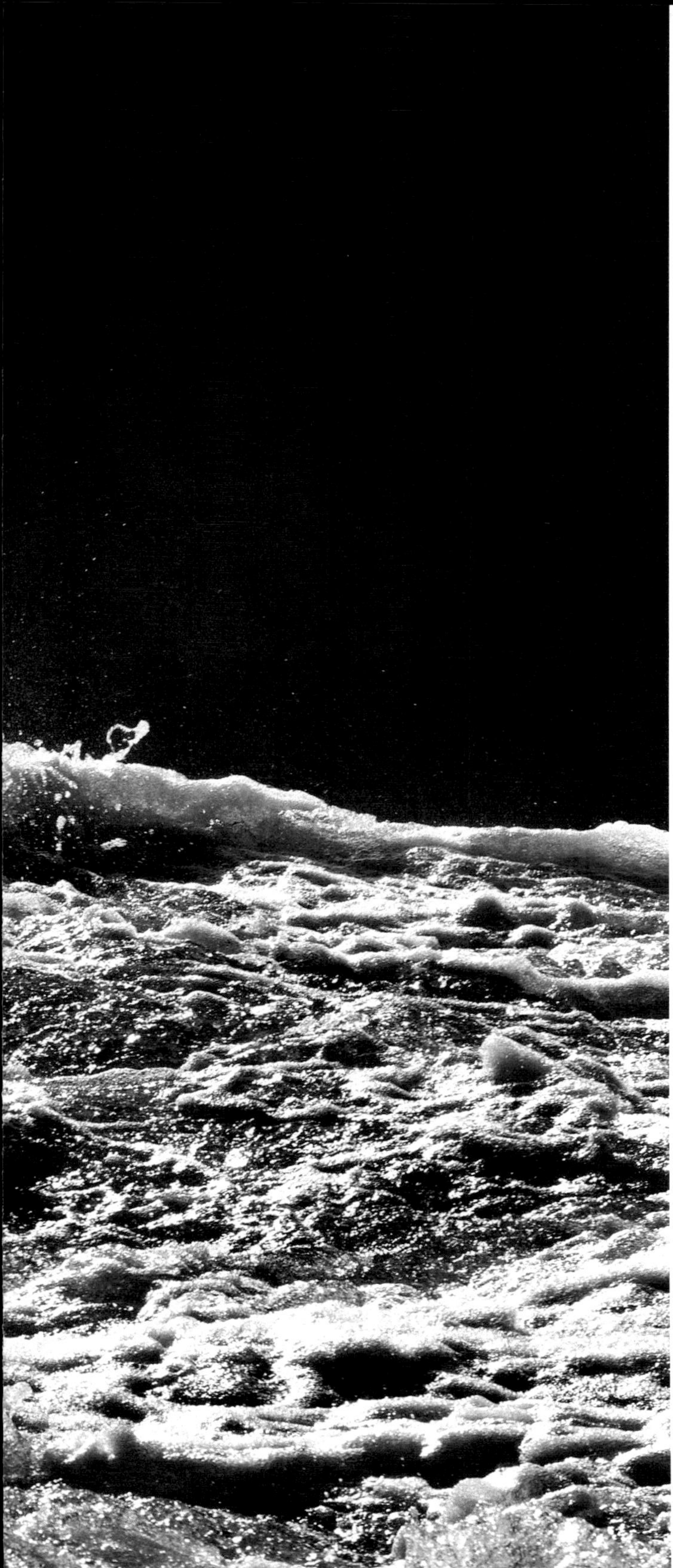

Rain and Runoff Make a River

Runoff from rain and snow has acted on the Youghiogheny's underlying geology and geography since the beginning of time, so to speak, and the stream's hydrology, or characteristics of flow, are the direct result. The varying rise and fall of this river are the heartbeat of a region. As Ohiopyle State Park's river ranger Amos Ludwig among others has said, "All places—and all of us—are connected by water." So, what *about* all that water?

We've seen how the currents of the Youghiogheny patiently but decisively cut their way through mountainous rocky masses that have arisen in the river's path, resulting in waterfalls, rapids, and riffles. Other aspects of flow stem from the weather's long-term patterns, meaning climate. While the greater region averages 35 or so inches of precipitation, higher mountains get 48 inches owing to moist air being pushed by prevailing winds up over the summits. This cools the clouds, reducing their ability to hold moisture, and thereby causes more of it to fall as rain or snow.

The Youghiogheny's wettest season, May through July, peaks with thunderstorms on hot afternoons when orographic or upward spirals of warm moist air gain impressive strength. Most of us reasonably take shelter from those drenchings, but summer storms account for some of the most vivid scenes one might ever see along the river.

Let me confess to an unusual seduction: these downpours are among my favorite things about the summertime Youghiogheny. They enable me to see and

Sunrise light shines brilliantly on waves at the brink of Ohiopyle Falls, midway down the length of the river. Runoff of rain and snowmelt feeds the Youghiogheny's hundreds of tributaries, which consolidate to the main stem's average flow of 3,348 cfs at the mouth—Pennsylvania's tenth-largest river.

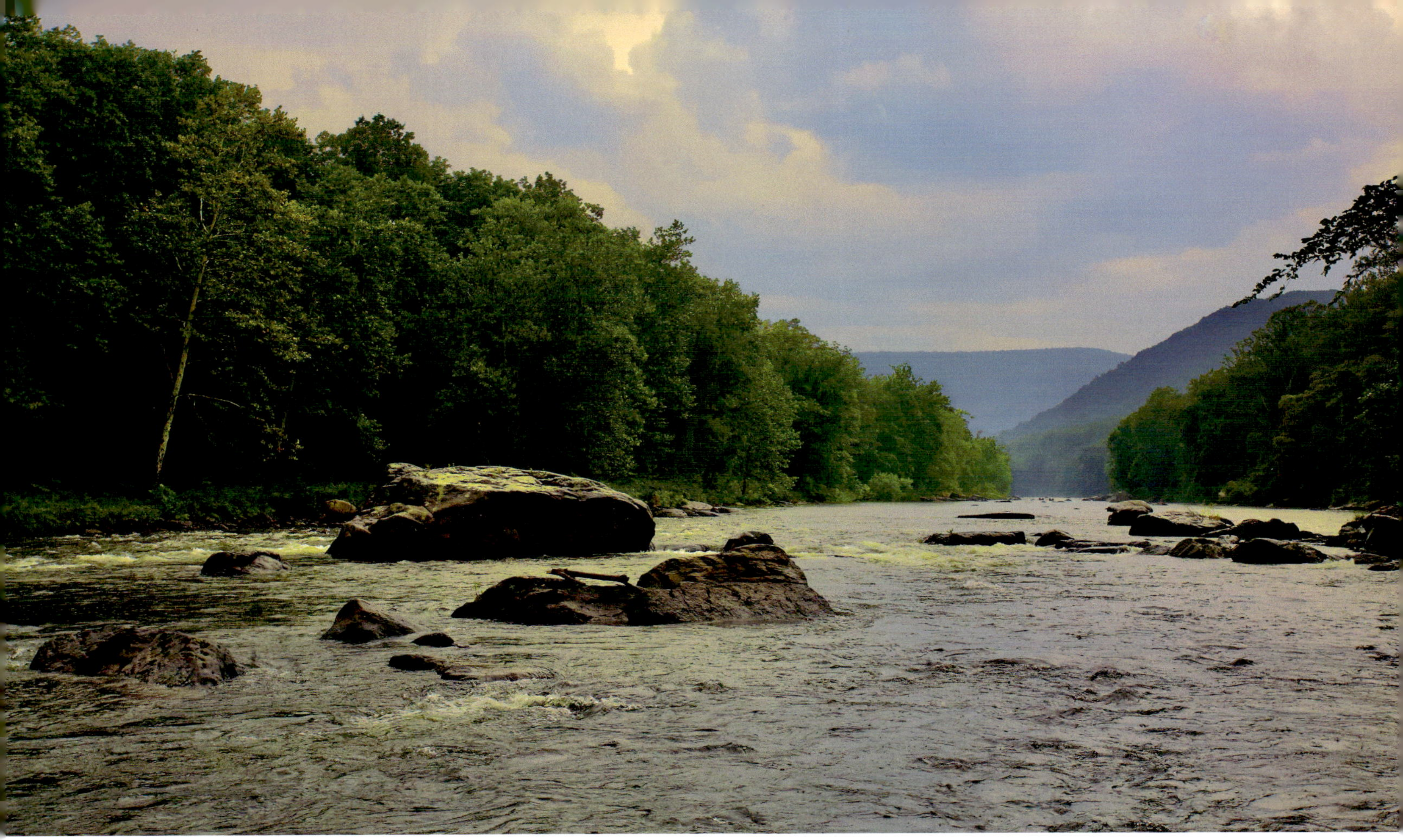

Thunderstorms brewing over the Middle Youghiogheny contribute to May, June, and July being the wettest months of the year.

literally feel the immediate, intimate connections of air to earth to river. The sky darkens. The wind gusts. The rain falls in curtains that fade with distance into flat gray overlays of atmosphere, the obscurity of details making everything look wilder. Anything but tame. So don't miss it! No need for soaking, discomfort, or misery: carry an umbrella, a raincoat, and maybe a sweater when thunderstorms threaten, and from your own private shelter enjoy these primal events. Check the weather report, but the afternoon summer storms usually don't last more than a half hour or so, if that.

Owing to both rain and snow, the Youghiogheny near its mouth averages a hefty 3,348 cfs, according to the US Geologic Survey's "gage adjusted" calculations. This is Pennsylvania's tenth-largest river in volume, the eighth-largest in length (counting out-of-state mileages of the Ohio, Allegheny, Delaware, and Monongahela), and also the eighth-largest in watershed acreage. Summer flows at Ohiopyle's halfway point typically run 800 cfs—which means good paddling on clear water without much problem of getting stuck on rocks or shoals.

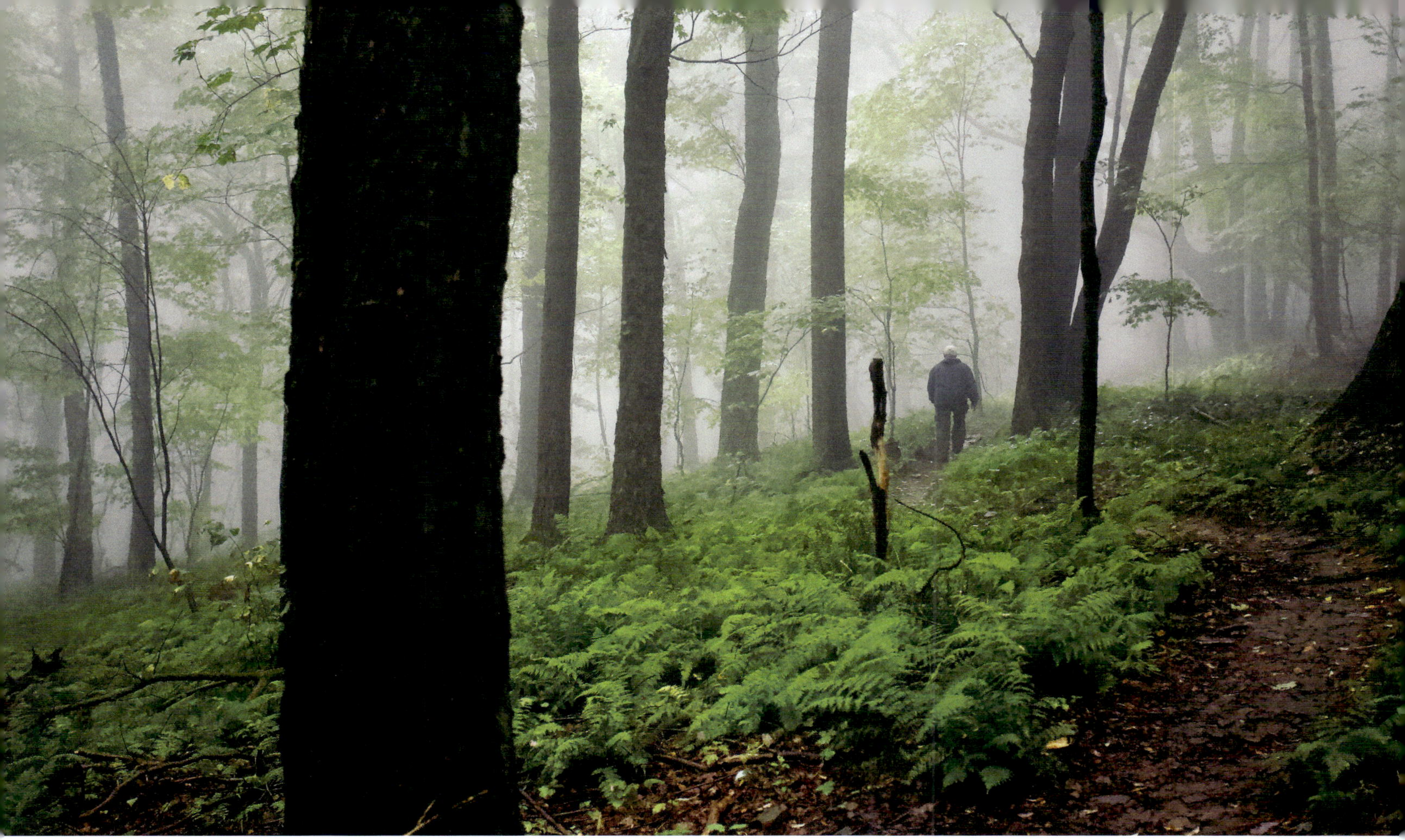

Pennsylvania's larger rivers in order of volume are the Ohio, Susquehanna, Allegheny, Delaware, Monongahela, West Branch Susquehanna, Juniata, and Beaver. In size, the Youghiogheny is just a tad smaller than the Schuylkill River of urban Philadelphia fame. The Yough's annual volume also matches the Cheat River—the next big waterway southward, in West Virginia, which peaks higher in spring but dips lower in summer.

The river's largest discharges of modern times occurred during the infamous flood of 1936, which happened before Youghiogheny Dam was built. Extremely big runoff came again in 1948 when the dam's record peak outflow clocked 13,700 cfs, only five years after the impoundment was filled, and in 1993 when the reservoir rose to 29 feet below the top of the dam. That seems like a safe margin, but it's only 7.2 vertical feet short of the spillway, whose overtopping would be a serious emergency releasing the full force of floodwaters on the valley below. This is likely to happen some day.

With greater atmospheric moisture in our global warming era and with the extreme climatological events expected by meteorologists, higher floods are

Continental winds push storm clouds over the mountains, cooling the air and causing rain to fall here along the McCune Trail at the Laurel Hill summit between Confluence and Ohiopyle.

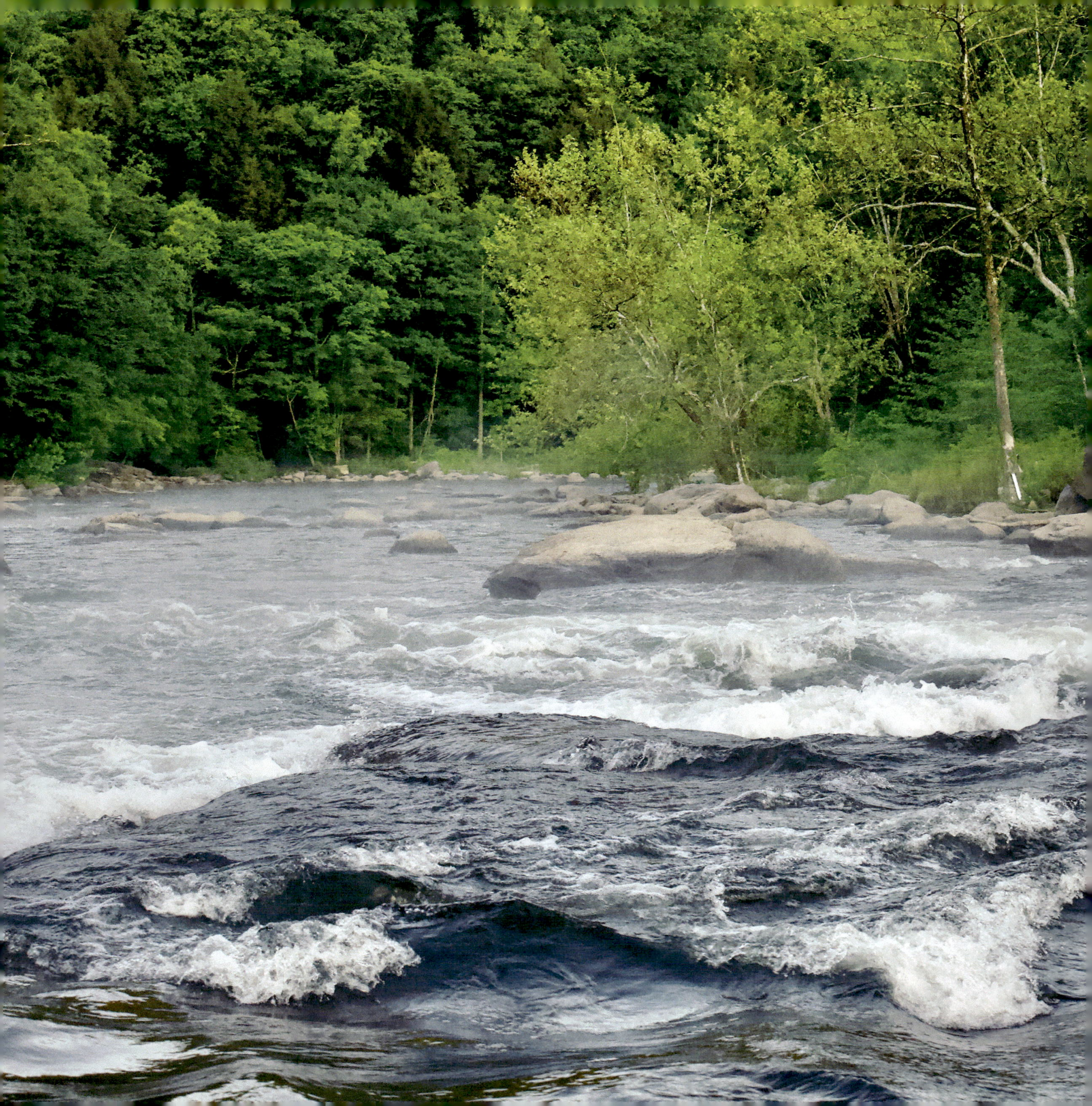

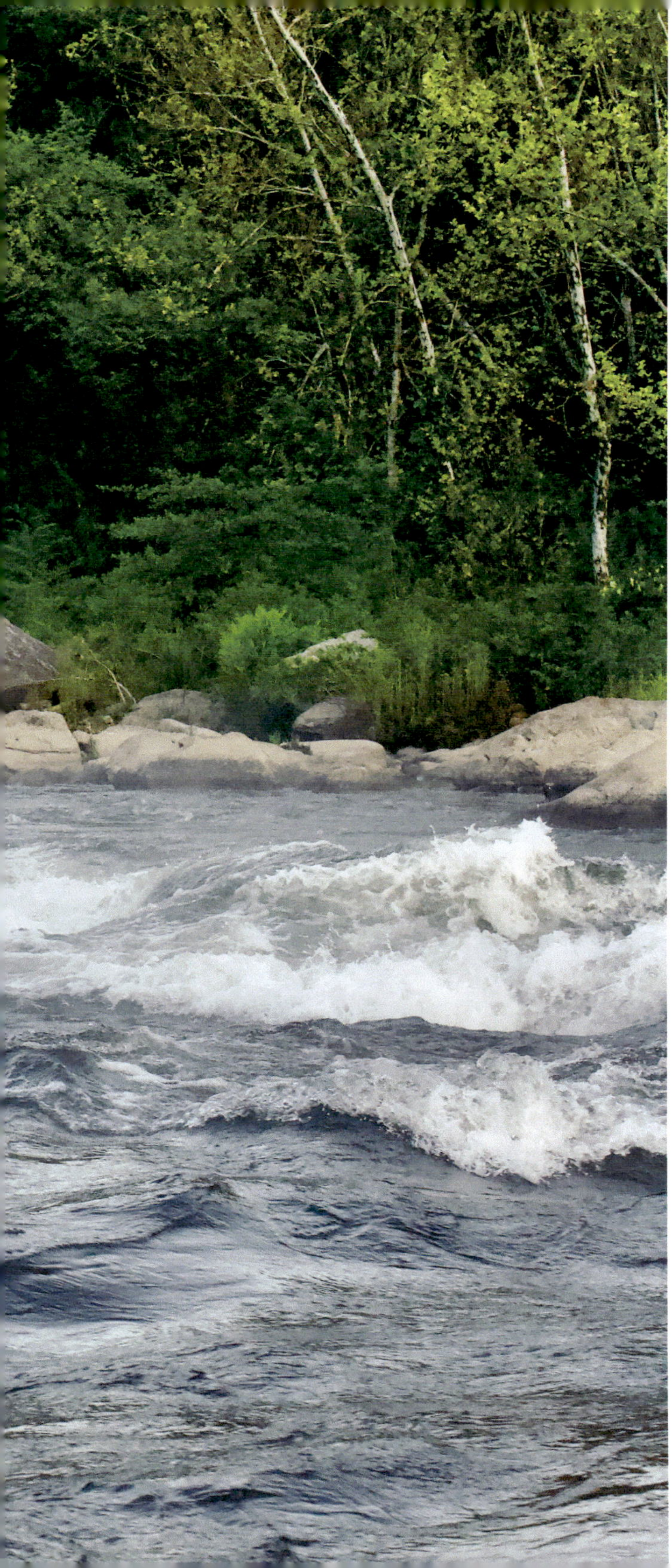

inevitable, reminding us that riverfront floodplains—even with the Army Corps of Engineers flood control dam upstream—should be left as open space. Seek higher ground if you want to build something! When that flood comes, the Ohiopyle Falls will be quite a sight to behold. Rather than the sharp vertical drop that we know nowadays, the Falls then will be more like a big nasty rapid—the most turbulent one imaginable.

The town directly below Youghiogheny Dam was named after a hydrologic term, Confluence, because it's a confluence or meeting place of not just two but three streams, an unusual phenomenon. The sizable Laurel Hill Creek joins the Casselman River immediately above its mouth at the Yough. From the air, the converging waterways resemble the three-prong toes of a turkey, hence the age-old nickname for the town of Confluence: Turkeyfoot.

The Casselman ranks as the Youghiogheny's largest tributary and suffers more than the main stem from mine drainage and its gritty discharges. With no dams on the Casselman, which otherwise cause silt to accumulate on the floors of reservoirs rather than flushing downstream, the muddy water that's seen at Ohiopyle following summer rainstorms is usually the result of pushy runoff in this heavily mined tributary basin. The turbid rush comes immediately after a downpour, often continuing several days before returning to typical summer clarity. So if you see a muddy Youghiogheny at Ohiopyle, a storm in the Casselman basin is often to blame. Be patient. Or come back after a few days or maybe a week without heavy rain, when the river will be clear.

Summer flows typically running 800 cfs at Ohiopyle create breaking waves, strong eddies, rocky pour-overs, and intricate rapids that appeal to whitewater paddlers and to others who appreciate the beauty of a wild river.

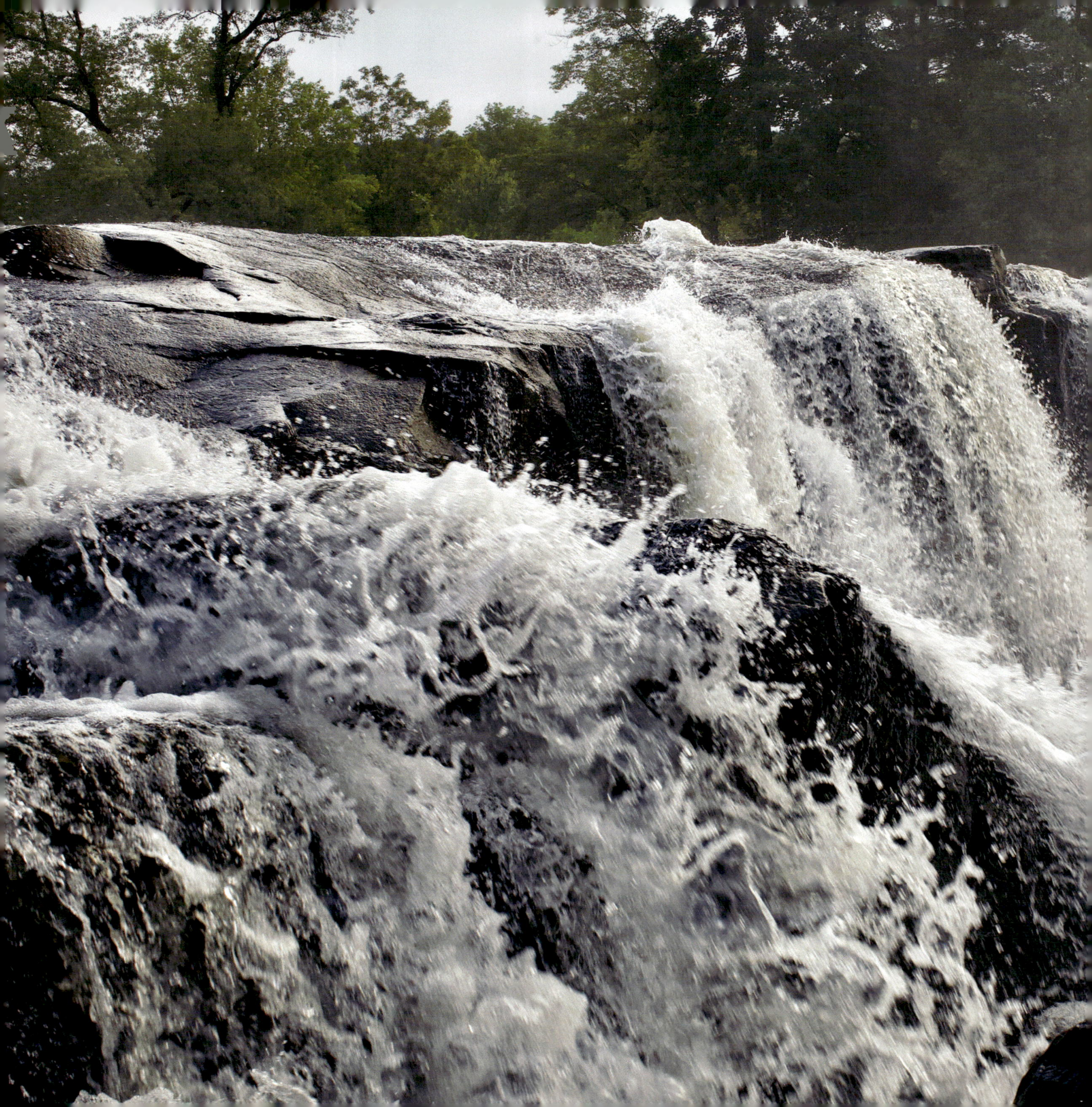

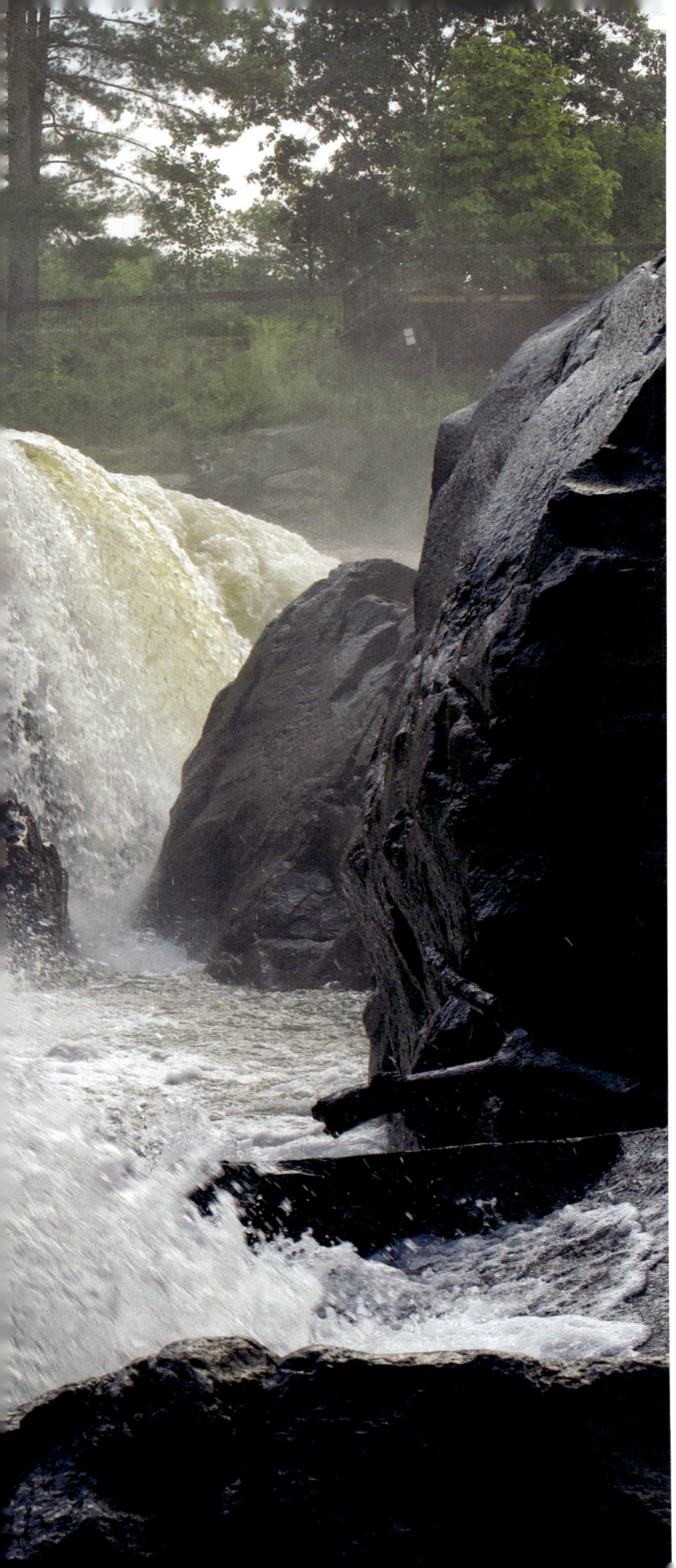

Below Connellsville, along the lower river, tributaries suffer far more erosion disturbance caused by development, roads, and mining. No surprise, the feeder streams there often run brown. Water quality has improved greatly since the 1980s, however, with less strip mining and with better treatment of municipal sewage, which at a number of places had oozed in stinking gray septic discharges, untreated, when I first canoed the entire lower section in 1982.

The river's geology combined with runoff have given us the Youghiogheny's signature feature: Ohiopyle Falls. This 18-foot-high sandstone ledge overlays a 2-to-4-foot-thick bed of softer shale. The force of water at the base of the Falls creates what hydrologists call a vortex helical flow. Picture a spinning eddy or a whirlpool—not on a horizontal plane as we see in a draining bathtub but on a vertical one with powerful force in the plunge pool at the base of the cataract scouring backward. Imagine a giant hydraulic grindstone spinning against the rock underlying the sandstone cliff. That vortex churns out the softer strata, exposing overlying slabs of harder sandstone, which ultimately lose foundational support, break off, and tumble into the current. This is especially evident on the river-right side of the Falls where angular boulders have tipped over in the outwash. As the bedrock further cracks and crumbles throughout million-year spans of geologic time, the sheer walls of sandstone that we see now forming the Falls have migrated upstream around the famous Loop that we see below Ohiopyle.

Here's another hydrologic and geologic claim to fame at this scenic bend in the Youghiogheny: the

Seen from its base on river-right, Ohiopyle Falls slowly but inevitably undermines softer shale beneath the sandstone outcrop, causing chunks of it to break off and collapse into the flow.

Ohiopyle Falls is the second-tallest vertical drop among large volume rivers in the entire Appalachian Range. Many small volume streams careen over higher waterfalls—much higher. For example, Whitewater Falls in North Carolina is 411 feet in its cascading drop. But among big rivers, only Cumberland Falls in Kentucky stands taller—at 68 feet, with flows averaging an impressive 3,600 cfs. Sandstone Falls on the New River in West Virginia may come to mind to those who have searched out its remote location, but it's mostly a lower and tumbling rather than sharply vertical pitch.

Outside the Appalachians, only a few big volume waterfalls in the entire East drop more than Ohiopyle. Niagara, of course, free-falls 160 feet with an utterly mind-blowing 100,000 cfs accumulated from the combined outflow from four of our five Great Lakes. The Potomac's Great Falls earns its name just outside Washington, DC. Letchworth Falls' big volume of New York's Genesee River plunges toward Lake Ontario's flat plain, and Hadley Falls of the Hudson, upstream from Albany, competes with the Yough for volume plus turbulence, if not sheerness. But even considering these watery landmarks beyond the Appalachians, Ohiopyle ranks among the most impressive waterfalls in all the East.

The erratic nature of natural stream flows in the Youghiogheny—from drought to raging flood—are largely evened out in the lower half of the river by the Army Corps of Engineers dam muscling up like an artificial mountain toward the sky within sight of Confluence, built for flood control in Pittsburgh while also providing for municipal water purveyors at Connellsville, McKeesport, and Westmoreland County. The

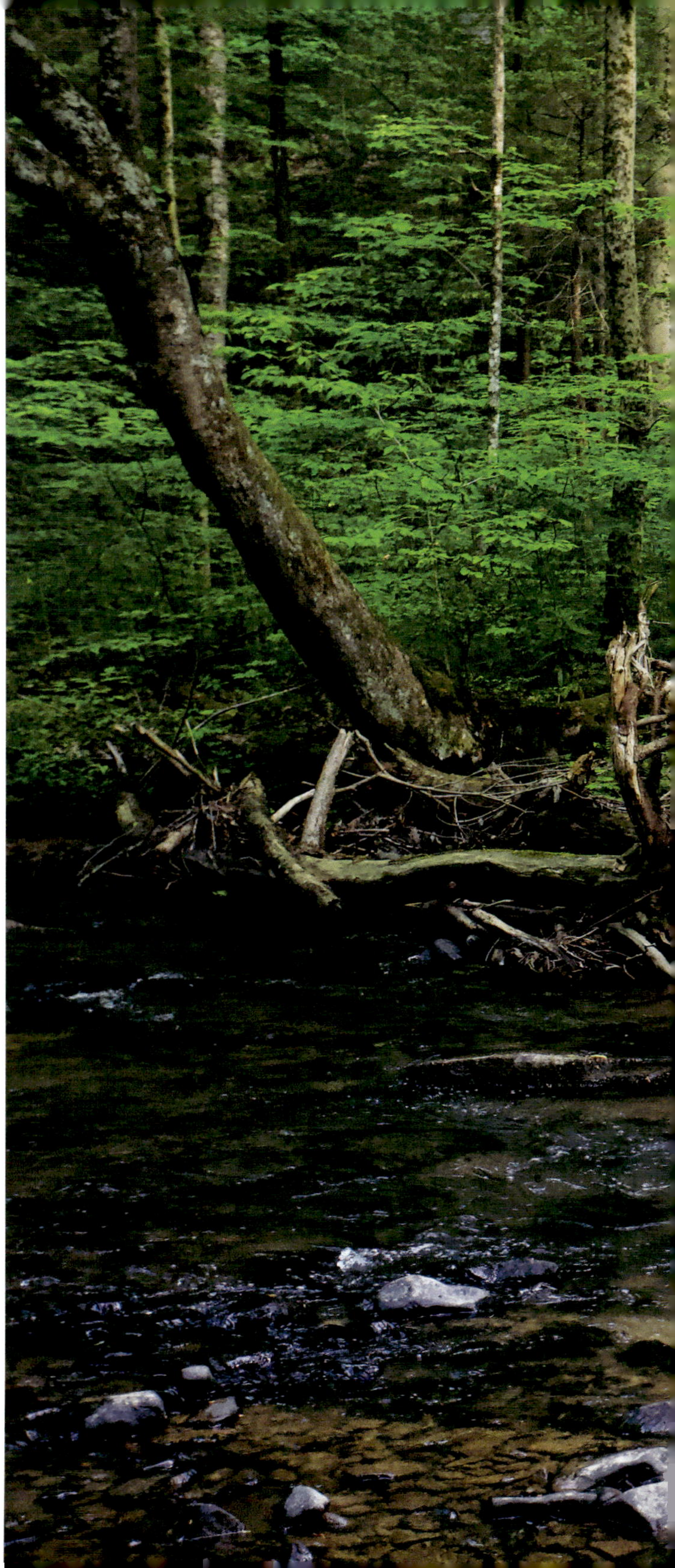

Once driven to local extinction by acid mine drainage, trout have recovered in some streams such as the Casselman tributary Whites Creek

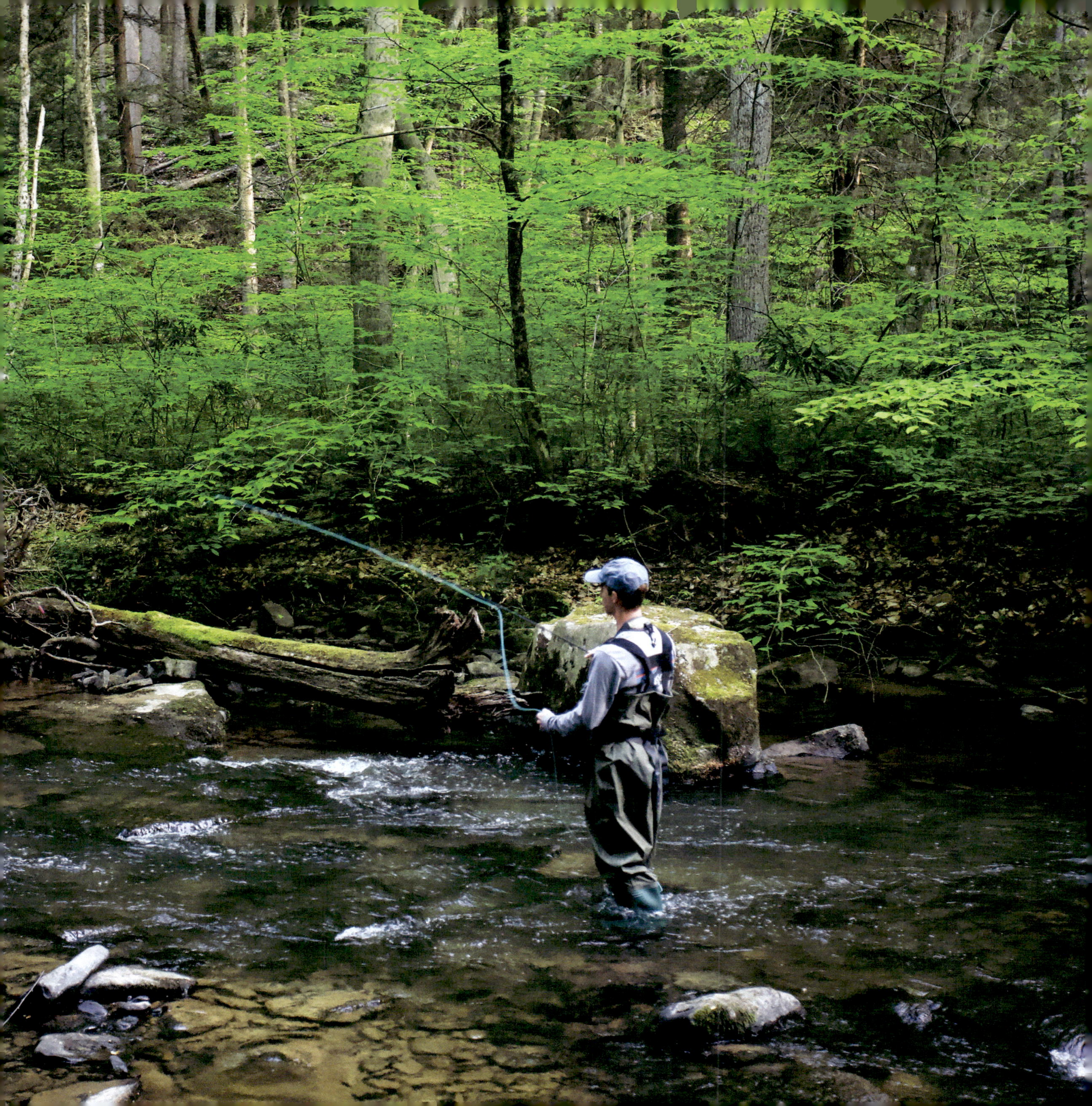

Corps's operation of the dam tends to stabilize flows for paddling below Confluence, an unanticipated by-product of the dam but now a staple ingredient in the local economy. Boyhood memories of my great-uncle Bill Holt included wading with ease across the river above the Ohiopyle Falls during late-summer flows predating the dam, something we wouldn't now survive without a long risky swim in swift current.

In contrast, today's Youghiogheny tops a short list of outstanding eastern whitewater rivers running with adequate flows for recreational paddling all summer, reliable every year. But for deeper historical perspective, going back before the dam, consider that, prior to the first round of clear-cutting the Appalachian forest, much of William Penn's paradisiac territory was blanketed with a thick, spongy, water-retaining layer of organic duff underfoot, full of absorption capacity with fallen leaves, the lattice works of soil-protecting roots, and thrumming communities of fungi, algae, and microbes by the billions in every shovelful of soil, all sheltered by a branching canopy towering a hundred feet up and higher, the entire organic package of the ages absorbing runoff and blunting sharp spikes and dips of rainfall and snowmelt. None of it, back then, was diminished by logging, mining, and related compactions, scrapings, burnings, and pavements destined to reshape entire drainage basins and rewrite the hydrographs of virtually every river in eastern America.

Those primeval rivers ran at far steadier levels than we see with the flashy, fluctuating patterns prevalent on today's Casselman, for example. In that sense, Youghiogheny Dam re-creates the steadier flow that typified many pristine rivers of the past, documented in writings of early travelers such as John James Audubon, who waxed eloquent about an unblemished Ohio River that cannot be imagined among the multicolored oil swirls and back-to-back dams stalling that big river's flow today. In this sense, the somewhat steady flows of the Youghiogheny at Ohiopyle offer a window to the river that people might have seen long ago before intricate balances of watershed health were upset.

Home for Fish

Speaking of changes: try to imagine how productive the pristine Youghiogheny must have been with fish finning upstream or down by the millions, all leaping to gulp nutritious insects buzzing at the water's surface. It's easier to realize—as many of us alive today remember so well—that fish had been virtually eradicated from the Youghiogheny during its booming era of mining with little regulation or responsibility before the 1970s. Fish that survived in isolated pockets of tributaries make for an untold story of survival by razor-thin margins, resurrection of entire species pushed to the brink of extinction, and ecosystem-wrecking replacement by exotic non-native fishes, from Asian carp to German brown trout. The recovery of fish in the Youghiogheny has been remarkable, but it remains notably incomplete for failing to re-create pristine or top-quality conditions that once existed or, for that matter, conditions that re-welcome native wild fish at all.

Professor David Argent of Pennsylvania Western University explained that the Youghiogheny now supports two distinct fish communities: those of colder mountain waters upstream from Connellsville and, in contrast, warm-water species below, down where the river's gradient flattens, temperature rises, and pollution worsens from sewage, abandoned mines, and leaking gas wells. Curiously, and counterintuitively, the more remote and wilder tributaries of the Allegheny Mountain

Section—meaning the upriver watersheds of Chestnut Ridge, Laurel Hill, and headwaters farther south—support fewer fish species and far fewer numbers of fish than does the lower Youghiogheny basin, draining what geographers call the Pittsburgh Low Plateaus geographic province downstream of Connellsville.

In the mountains, Argent along with biologist David Kimmel counted twenty-two species of fish, while in the basin below they found forty-one. Some species in both regions were non-native exotics, which often—no, let me say, virtually always—harm native populations. Mining plus logging have heavily affected both regions, but here's the surprising difference: the upper watersheds have far less naturally occurring limestone as bedrock. The lower basin is far more developed and industrialized, but the presence of limestone, dating to ancient sea deposits of decaying shellfish along with saltwater corals, makes those lower-basin waters more amenable to fish, though not always to species that thrive on clean water, not always to native fish, and not always to fish that anyone would want to catch.

In the Youghiogheny's two geographic regions the biologists found three species of trout (mostly in the upper watershed), nine species of bass and sunfish, four of suckers, eight of darters and perch, one drum, three catfish species, two kinds of sculpins, and seventeen minnows—meaning the *Cyprinidae* family, which is a genealogic group and not, as we might assume, just a name for small fish.

Here's another surprise that Dr. Argent illuminated, and an even larger disappointment: among all major tributaries to Pennsylvania's Youghiogheny above Connellsville, only three streams—Meadow Run, Indian Creek, and Hen Run—accounted for 77 percent of the species and 76 percent of the total number of fish counted. Dozens of tributaries appear to be depauperate of fish. For all the success in restoration of trout in the Youghiogheny, from Confluence to Connellsville, it's still a severely impoverished ecosystem compared to what it once was.

Possibly surviving, but not found, the longnose sucker and cheat minnow are "critically imperiled," according to Pennsylvania's *Natural Heritage Inventory*. These fish had thrived for the ages, so let's hope that missing-in-action here does not mean that they've gone extinct in the basin during our watch.

Native fish, meaning species that naturally lived here without human intervention, include brook trout—the poster child of wild fish Appalachian-wide. With flashing red undersides and showy spots of yellow, blue, orange, and maroon, for millennia these feisty swimmers excelled in cool waters from Maine to Georgia. But as in the wild country needed by grizzly bears, the conditions required by brook trout have become rare. These include clean water, cold water temperatures under 70 degrees, natural fluctuations of high-to-low flows, rich associations of native species including mayflies, and isolation from ravenously predatory or compulsively competitive non-native fish. Even expensive decades-long efforts to reclaim tributaries such as Indian Creek and the Casselman have failed to effectively reinstate wild brook trout. Consider that, for all the celebrated beauty in Cucumber Run's waterfall near Ohiopyle, samplers in 2009 found not one fish there. Of any kind.

Compared to watersheds outside the Appalachians' coal mining belt, the Youghiogheny remains poor in species richness and populations; of thirteen tributaries, five appeared to be fishless, and the greater basin scored a near total absence of brook trout. Native brookies, distinguished from stocked trout, were found only in the surprisingly low flows of Hen Run, just

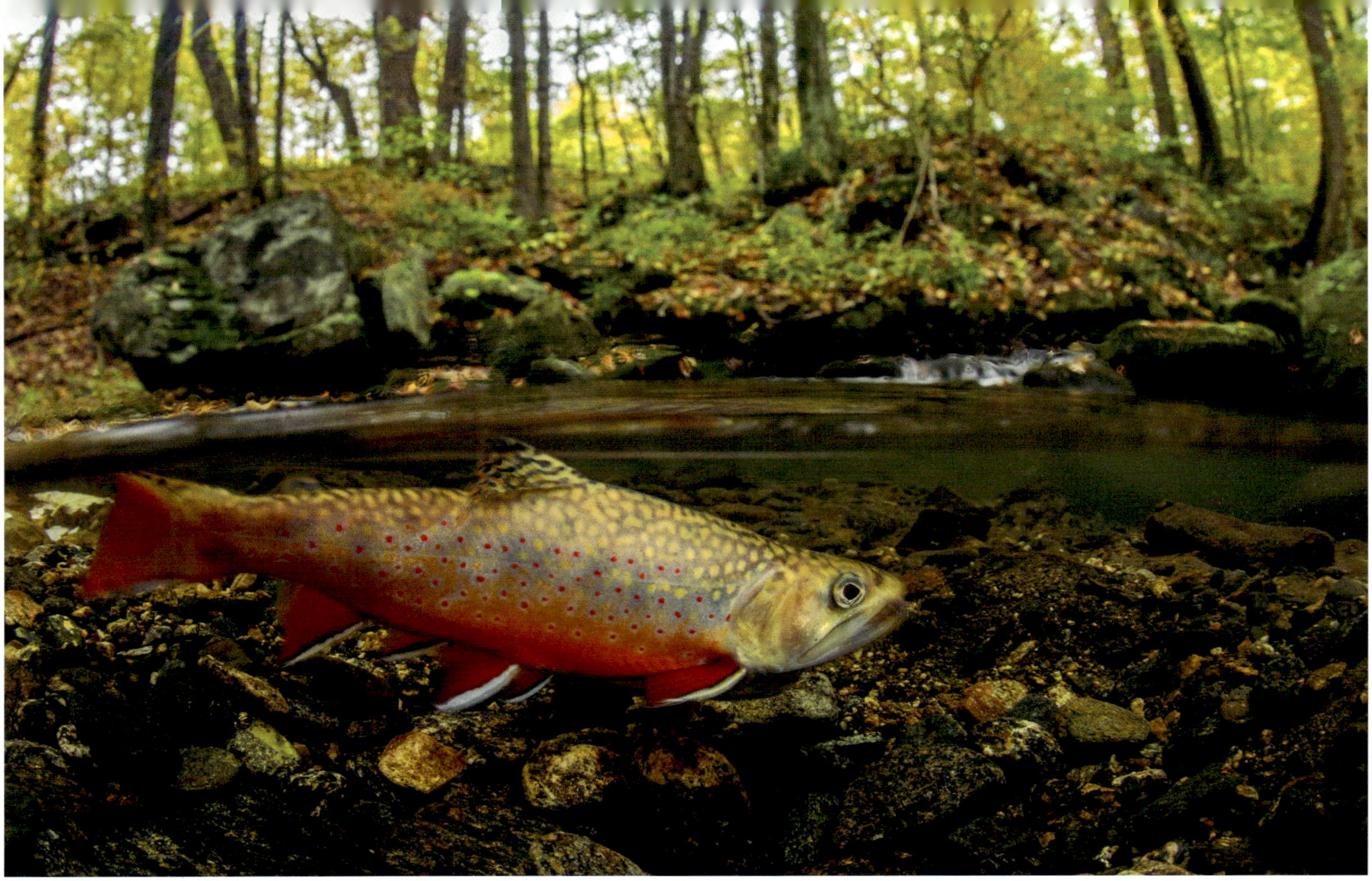

Colorful brook trout, once healthy here, have been thoughtlessly eradicated from nearly all their local streams, but continued watershed restoration efforts could bring wild populations back. This exquisite native brookie was photographed in a trout stream in the Great Smoky Mountains of Tennessee, but it could thrive in the Youghiogheny basin if restoration efforts continue. Photo by David Herasimtschuk of *Freshwaters Illustrated*.

upstream of Confluence. Slightly more hopeful, samplers later found a few brook trout in Ramcat Run below Confluence, and in Jonathan Run below Ohiopyle, which were likely of stocked lineage. Native blacknose dace and creek chubs were also spotted there.

Although a lot of work has been done to reclaim the Youghiogheny from the legacy of mining's costly pollution, Argent reminds us that we have a long way to go to reestablish a healthy network for native fish—once cherished, no doubt, before the original wildness was lost.

But here's the good news in what we might recognize as a shifting baseline of expectations: with less pollution from coal mines—severe by any account in historic times—trout fishing for stocked brown and rainbow trout has become a big draw to anglers from throughout the East. The Argent/Kimmel survey acknowledged that stream reclamation work has produced a "significant recreational fishery." Fly-fishing below the Army Corps dam for stocked trout has become a major attraction of the Laurel Highlands. Guided fishing trips by raft or in drift boats powered by oars—unimagined just several decades ago—now fuel local businesses. For anglers wanting more intimate waters, Meadow Run has become noted for stocked trout, with the lower 2 miles designated for artificial lures only, with catch-and-release required in months other than summer. Likewise, Whites Creek of

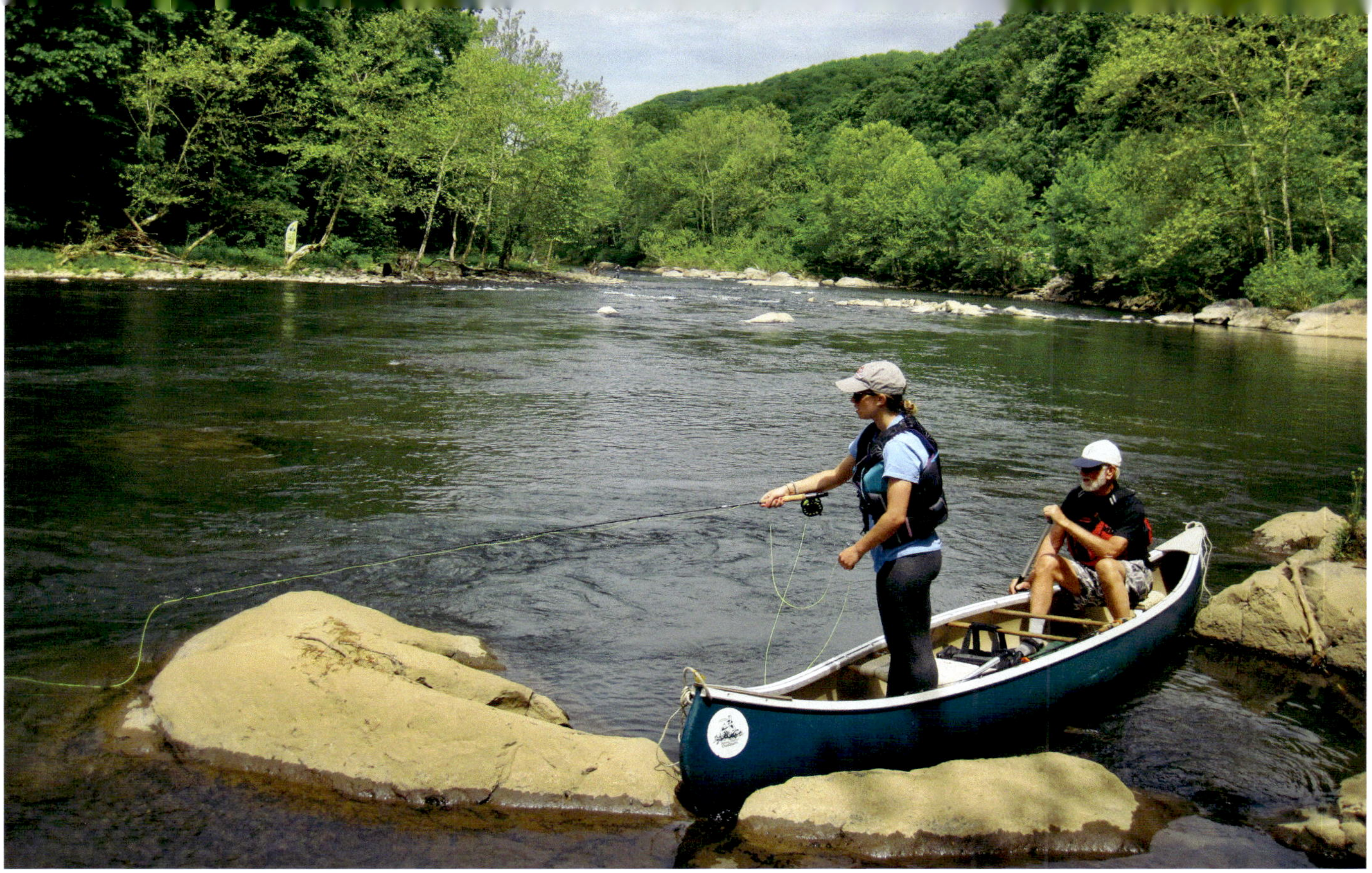

The Middle Youghiogheny, from Confluence to Ohiopyle, was devoid of fish in the mid-1900s but has recovered to become a popular angling river.

the Casselman basin is stocked by the Fish Commission and draws trout anglers to refreshingly shaded shores.

Today's reinstated populations of fish, such as they are, also mean good news at the top of the food chain, where river otters have reclaimed status as key predators. Long ago driven to local extinction by ruin of the clean water needed for their food supply, plus trapping for their lustrous fur with its mind boggling, almost unbelievable 150,000 to 400,000 hairs per square inch, otters have been translocated from Pennsylvania's Pocono Mountains to Ohiopyle State Park. Don't worry, anglers: otters eat principally crayfish or small slow swimmers such as suckers. The playful, charismatic furbearers have resettled throughout much of the river's length, with guest appearances as wildland emissaries even in McKeesport.

Return of the otters, not to mention trout of any kind and our fish-eating national emblem, the bald eagle, is good news for the Youghiogheny, but even beyond any lingering problems of acid, iron, aluminum, and silt caused by mining, most tributaries remain surprisingly crippled by the rail lines that were built on both sides of the river going back to 1883. Whenever the rail builders encountered a tributary, they crossed it typically not with expensive bridges giving streams room to flow, but with culverts that concentrate and shoot tributary runoff through a pipe, and some pipes have their downstream ends dangling in space, creating their own fake

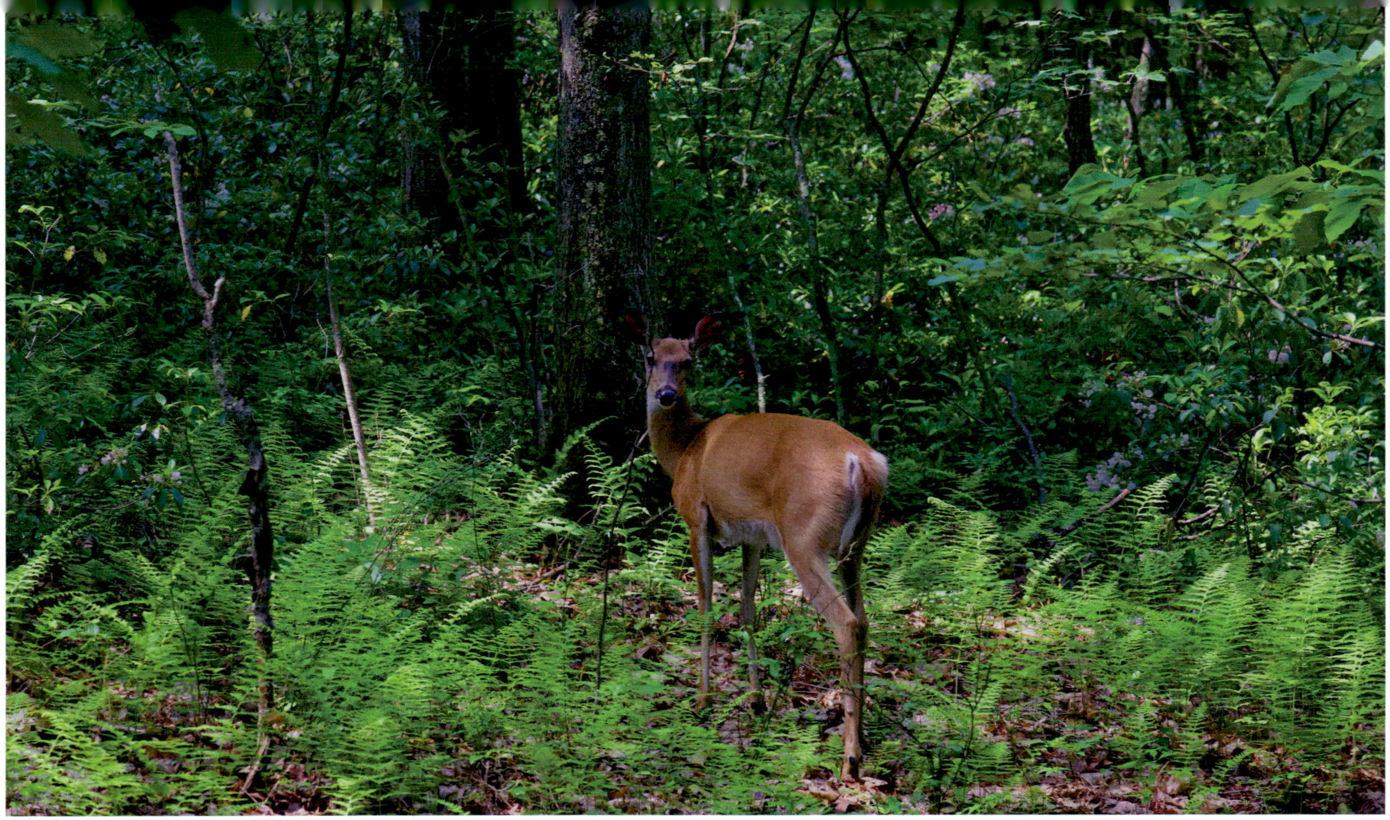

Wildlife of all kinds including this whitetail deer along upper reaches of the Youghiogheny depend on the river, its tributaries, and the land throughout the watershed for habitat, food, refuge, and migration.

waterfalls. These obstructions cannot be negotiated by fish. Even with the conversion of the railroad on the lower river's south side to a bicycle trail, the typical tributary culvert remains a formidable impediment, sharply limiting fish numbers and migration to and from the main stem for spawning, feeding, or seeking refuge in either smaller or larger currents. Who would think a facility as benign as a bicycle trail would have inherited such an intractable problem?

Strides in stream reclamation have been impressive during the past five decades, but regarding fish, and looking forward, having a healthy river goes beyond having stocked trout to catch. It involves native fish, non-game species of fish, and the complicated if not vexing interplay between water and our treatment of land. So let's look closer at that.

Reclaiming a Ruined River

Two centuries of coal mining have increased the severity of runoff extremes and caused water temperatures in much of the Youghiogheny along with its tributaries to climb beyond viable limits for trout or other cold-water fish. But more devastating, mining has exposed the region's acid bearing shales and coal seams—buried safely underground and sealed up by earthy overburden since the Carboniferous Era two hundred million or more years ago—to air, combined with water, together catalyzing a foul brew of modern-day chemical reactions. The main culprit, sulfuric acid, is augmented by deadly metal compounds of iron and aluminum—useful metals but not ones you want in your drinking water or your pan-fried fish.

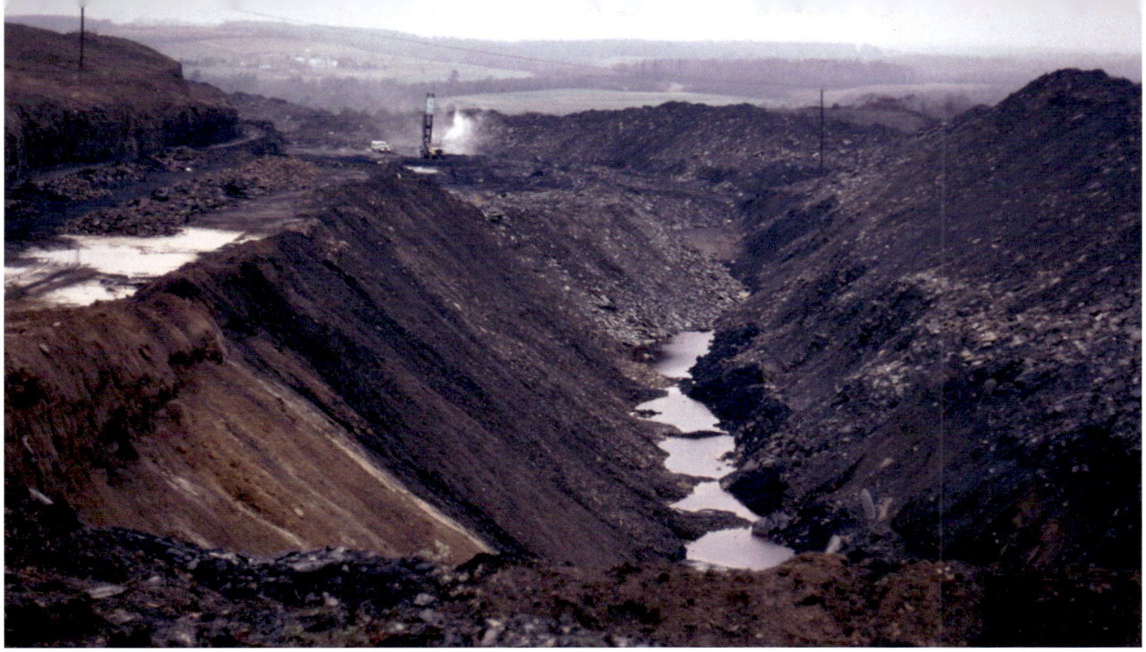

Strip mines for coal have degraded many tributaries and the main stem of the Youghiogheny. The number of mines and the damage they cause declined as the energy economy shifted and as pollution and reclamation rules became more effective, but coal is still being exported and new mines are still being proposed.

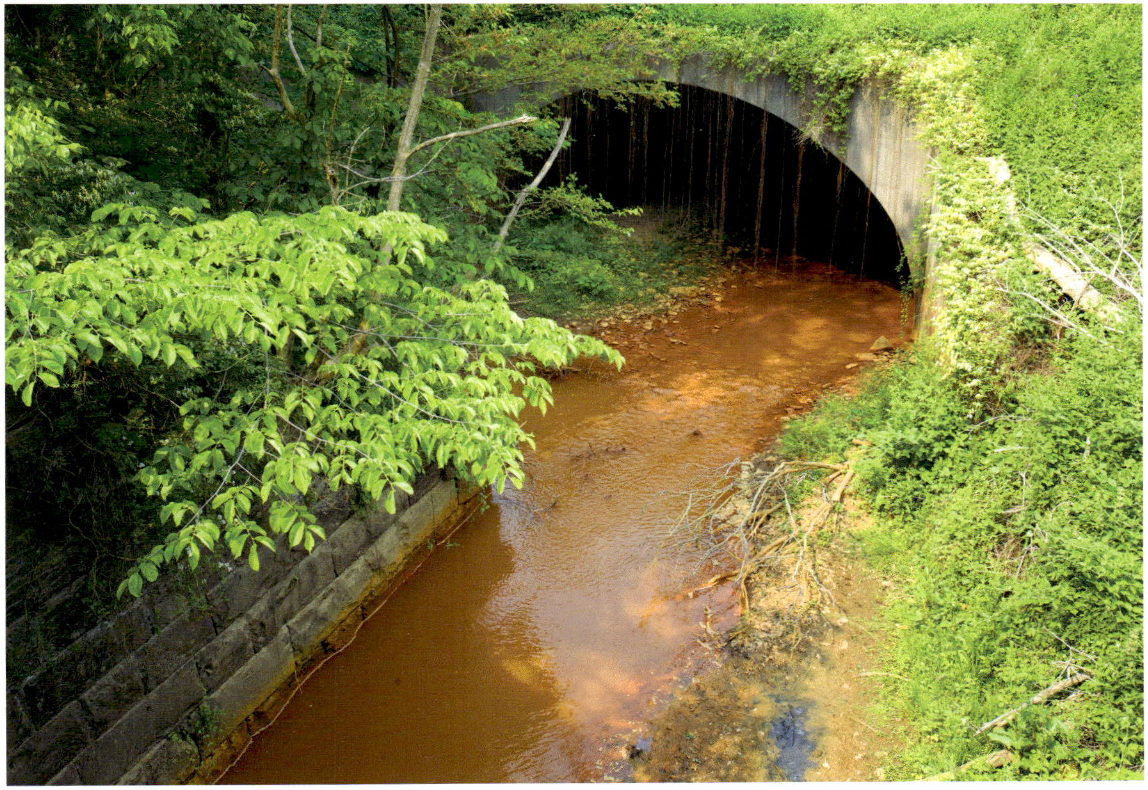

Drainage from abandoned mines and their iron, aluminum, and other toxins continue to plague tributaries such as Gillespie Creek downstream from West Newton. The Mountain Watershed Association strives to address acid mine drainage wherever it occurs in the river basin, including discharges from sources that have been polluting for more than one hundred years.

The problem goes way back, even to the Youghiogheny's early settlers, including my own ancestors who, for household use in cookstoves and furnaces, dug their own "groundhog" mines in their back forty, if not their backyard, along streams such as Meadow Run (in my case). Later, industrial mining generated pollution beyond the pale, at massive scales and uncontrolled for decades, from shafts tunneled deep below the surface. After the coal was removed, those cavities filled with water and then overflowed from mines that had been abandoned in the dig-and-run pattern of the day. Then, from the 1950s onward, strip mines caused even more trouble where the soil and rocky subsurface—whole mountains of it—was blasted, bulldozed, and scooped up by draglines in order to extract the coal. Some of the black seams were as thin as a foot, but the mining of them sterilized entire creeks. One of my grandmother's cousins, Shelby Mitchell, had owned and managed small mines in a bygone era before World War II, and when I asked him, many years ago, how they had regarded the pollution that plagued the region as a hidden if not so hidden cost of coal, he responded with a regretful shake of his head, explaining, "Back then you just didn't think about that."

The mining process killed off virtually all aquatic life for decades. Orange stains in the lower Casselman appear less frequently today than they did even in 1990 and far less than in 1970, but acid mine drainage persists in less noticed tributaries. See, for

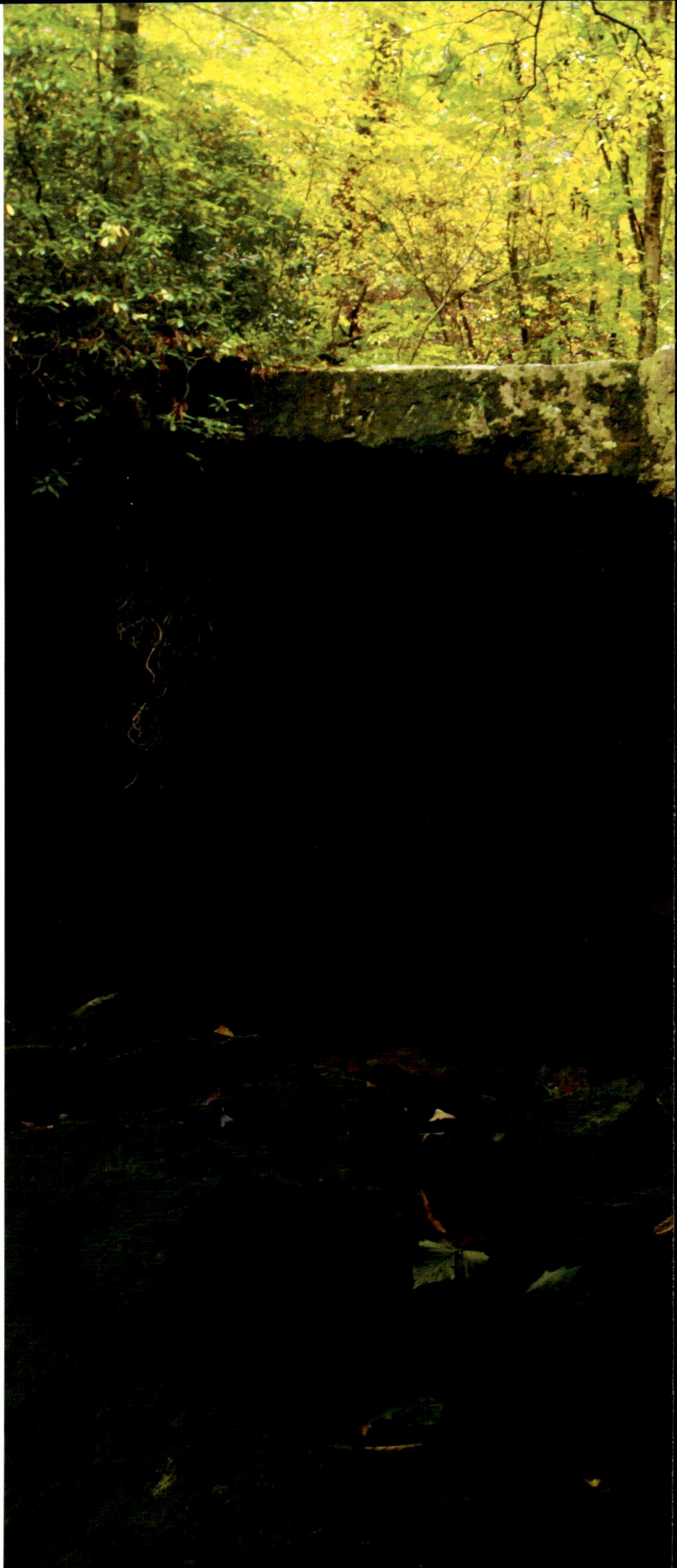

At the iconic waterfall of Cucumber Run, mine drainage once stained bedrock as orange as a pumpkin. Remediation efforts require continuous treatment many decades after the damage occurred—all at continuing public expense for coal that was mined for private profit a century ago.

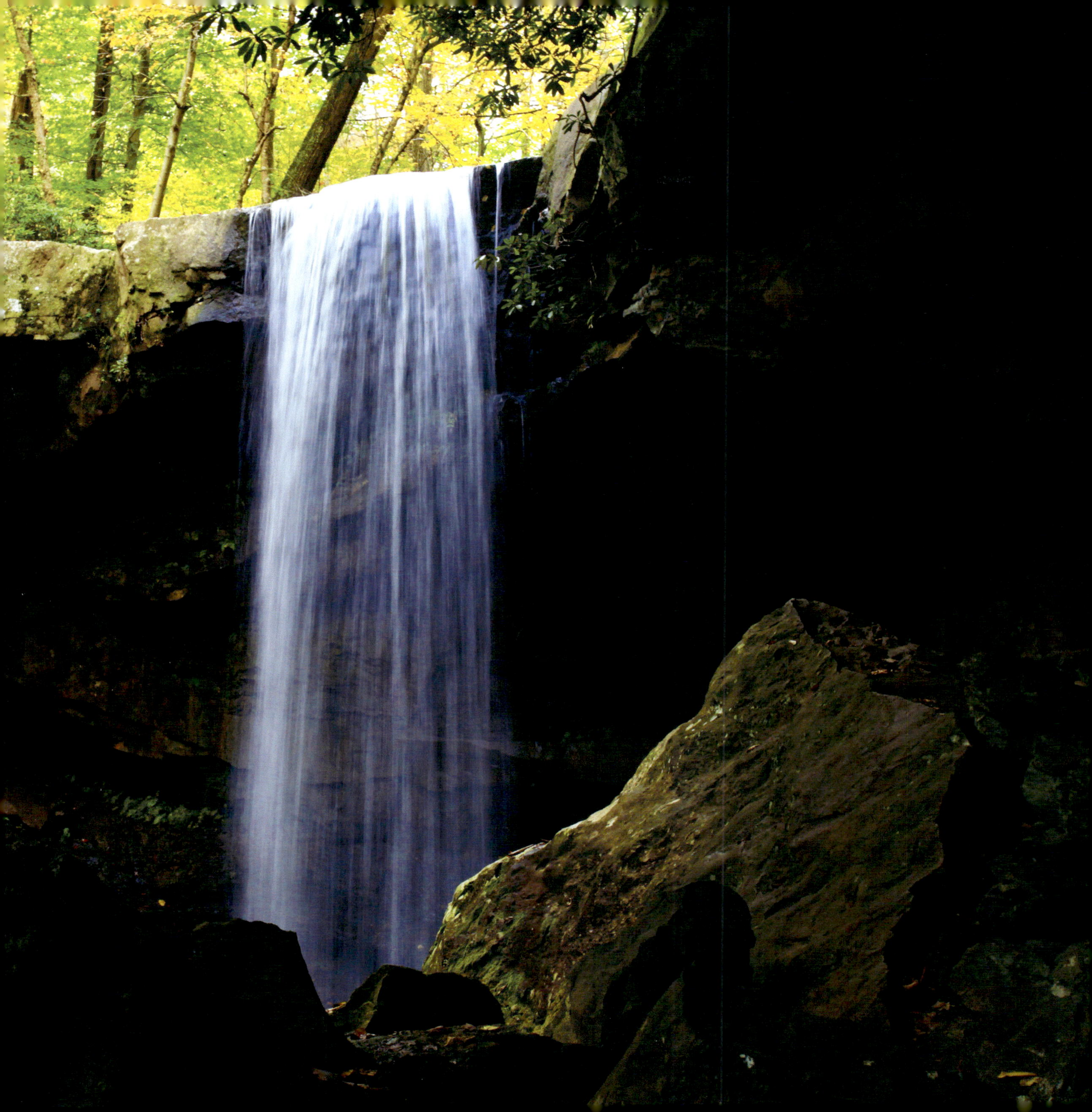

example, the streams crossing the Great Allegheny Passage bike trail a few miles above Connellsville and at Gillespie Run downstream from West Newton. Profits were gleaned a century ago, but the problems were left to the public and to generations still unborn today. Although controls have certainly improved, the Pennsylvania Department of Environmental Protection continues to grant permits for new mines in acid-polluting strata.

Thankfully, payoffs for mine reclamation can be seen at the Youghiogheny below Confluence, now transformed from a corrosively acidic waterway of the 1960s to a fishery drawing anglers casting for trout the whole way down to Connellsville. Cucumber Falls—glaring orange with disgusting iron precipitate during my youth—now delights Ohiopyle State Park visitors as one of the most elegant waterfalls in Pennsylvania, its reclaimed color perfectly natural, even though no fish can be found.

Non-native, stocked trout have thrived in parts of the river, but even they, with their subsidized maternity wards (hatcheries), socialized medical care (bacterial inoculations for all that need it in the hatcheries), and taxpayer funded life insurance policies (public money for replacement when fish kills occur) cannot be taken for granted. Important, too: remediation of mine acid does not typically eliminate the problem but, rather, treats it with facilities and limestone additives that amount to life support for the fish, requiring continuing care, trucking in supplies, and ongoing maintenance along the Casselman, Cucumber Run, and others.

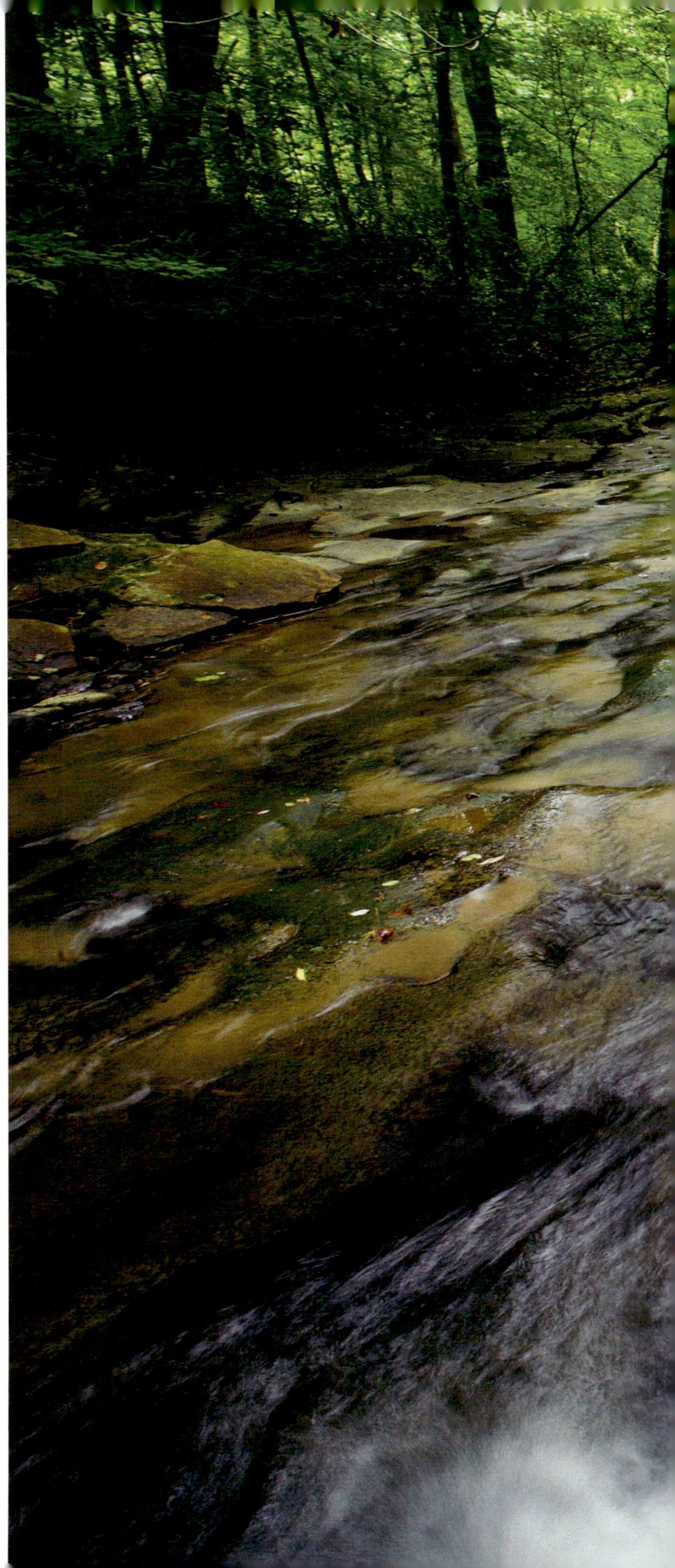

Reached by a scenic woodland trail on the west side of Ohiopyle State Park, Jonathan Run has been polluted by mine drainage in the past. However, reclamation efforts greatly reduced damage to this tributary, which now appears pristine in lower reaches.

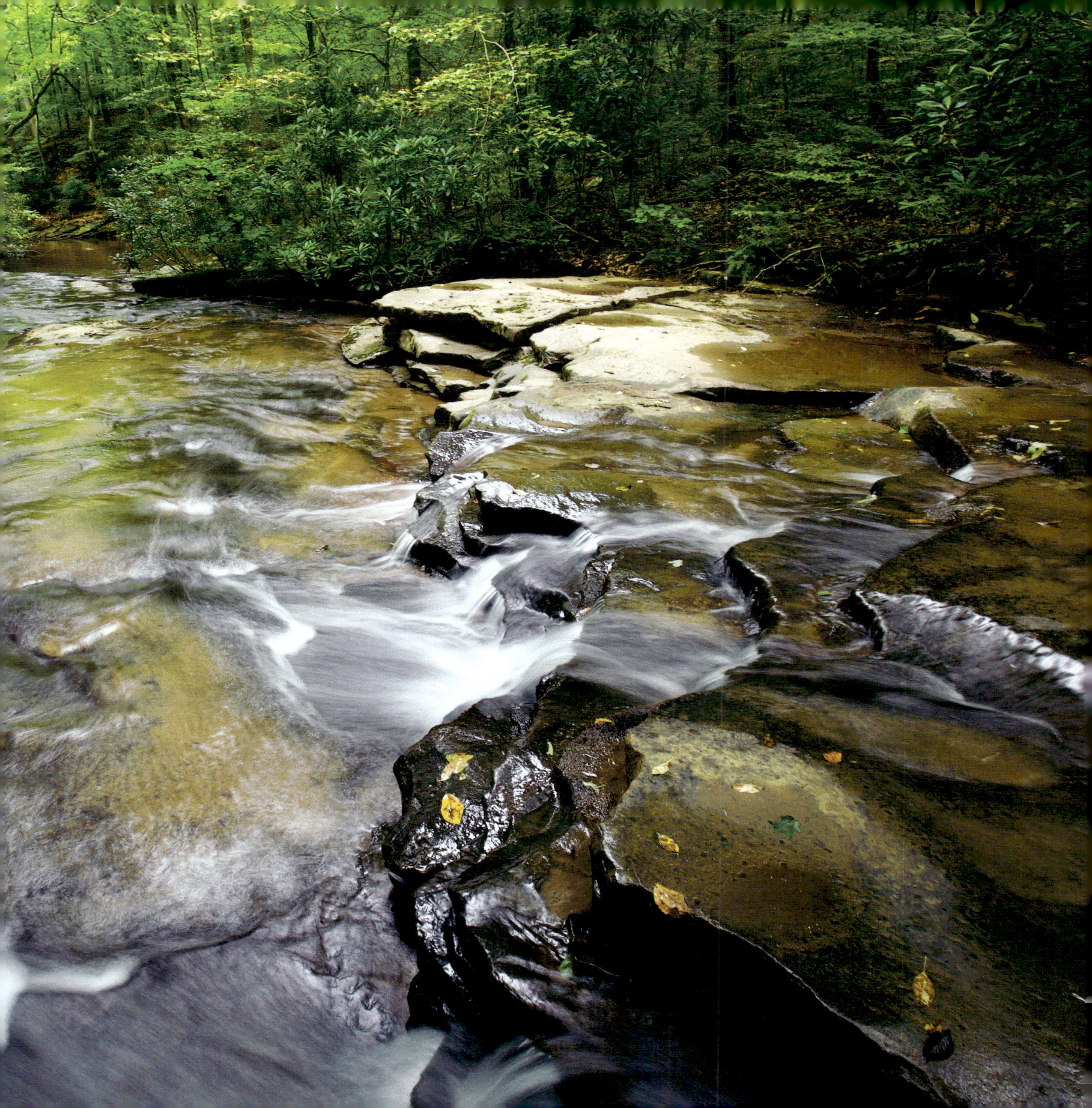

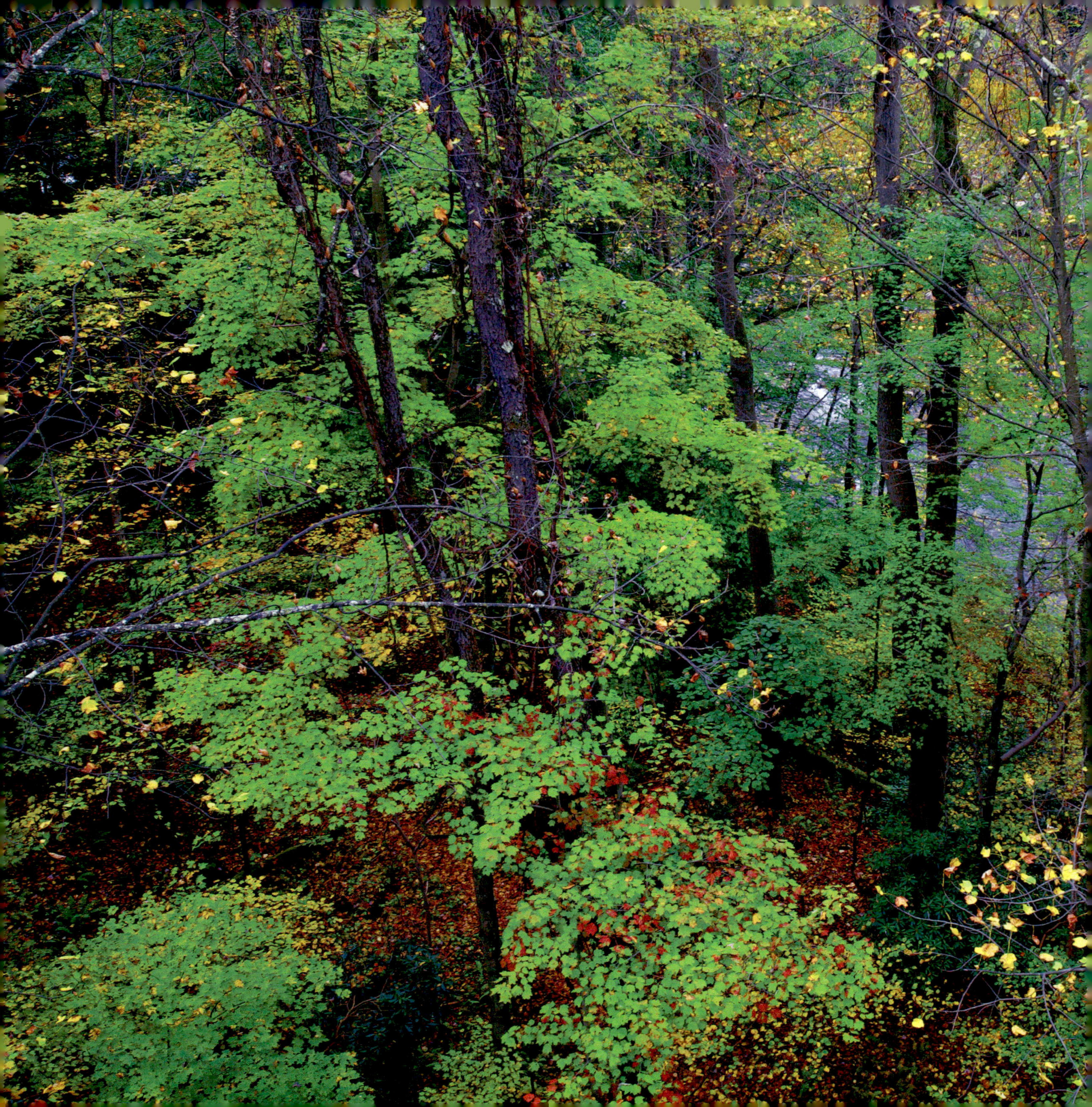

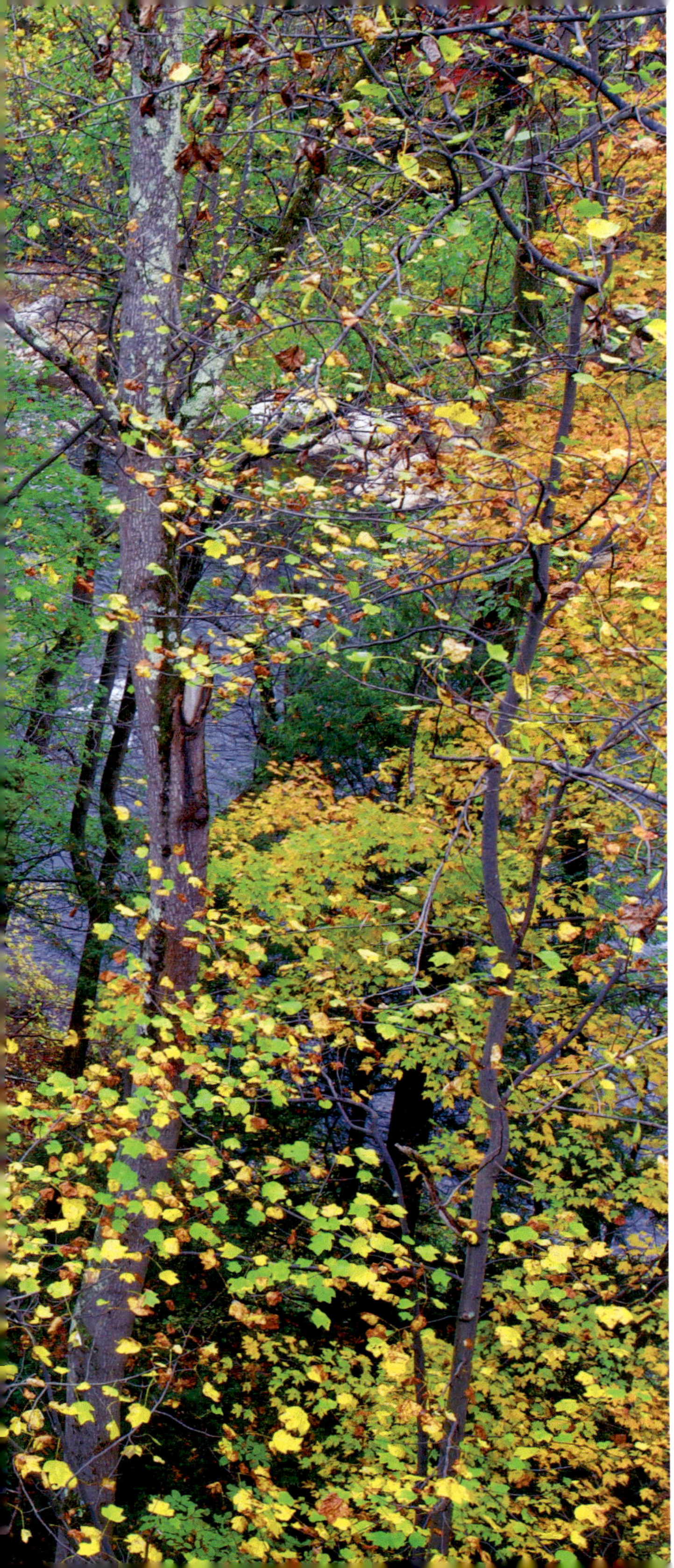

We pay for this through taxes and also private contributions—just one among many hidden costs of coal. Without due diligence, the acid will reoccur. As biologist David Argent noted, "The periodic discharges from tributaries serve as a reminder of the fragility of recovery." Even with acid reduction that makes trout fishing possible, Meadow Run, for example, still suffers fish-stunting concentrations of iron laced with aluminum.

Jonathan Run, joining the Youghiogheny in its spectacular gorge downstream from Ohiopyle, now supports stocked brook trout at least intermittently, as does Glade Run, a road-free tributary of Dunbar Creek upstream from Connellsville. These and other recoveries have been waged courageously by the Western Pennsylvania Conservancy, local chapters of Trout Unlimited, the Mountain Watershed Association, Youghiogheny Riverkeeper, and State Fish Commission along with other public agencies and organizations.

Coal mining's persistent problems and their multi-billion-dollar taxpayer burdens might be regarded as the legacy of neglect from a time when people knew no better. But after suffering the consequences of water pollution, then paying dearly to restore degraded waters to nominal standards for fish, swimming, and public health, one might assume that crucial lessons would have been learned by one and all. But the early 2000s rush to extract underground fossil gas makes one wonder: did we learn *anything* from the boom-and-bust cycle of coal with its legacy of bankrupting cost and heartbreaking loss in everything from belly-up

From the High Bridge over the Youghiogheny west of Ohiopyle, walkers and bicyclists enjoy a rare opportunity to look down into the canopy of red maples, chestnut oaks, and tuliptrees.

fish to black lung disease sending our own ancestors to early graves?

Miners dug tunnels or stripped mountaintops to get to the coal lying hundreds of feet underground, but now the drillers for "natural" gas bore down not just hundreds but thousands of feet to inject huge quantities of water and sand laced with toxic chemicals through a process called hydraulic fracturing or "fracking." Like miniature earthquakes, the high pressure infusion of water, sand, and lubricating additives cracks open layers of rock and exposes groundwater to the injected toxins. The gas is then pumped up to pipelines while the wastewater is pushed back into subterranean cavities, if it will go. Optionally, it has been trucked to municipal sewage plants, which are incapable of accommodating such volumes of poisons whose names are withheld secretly from the public under "proprietary" laws favoring the gas drillers, even though private water wells may be at risk and even though the wastewater has been dumped into public treatment plants. Residential ratepayers for that treatment have been the victims, not to mention the streams below the overstressed sewage plant discharges. Or the chemically laced waste can simply be dumped into surface streams where Department of Environmental Protection staff cannot effectively police or even monitor what's happening day and night statewide at—get this—hundreds of thousands of gas wells.

More than 210,000 fracking and conventional gas wells scattered across Pennsylvania have been issued 57,000 violations of regulations between 2008 and 2023 alone, according to state data via the nonprofit watchdog organization FracTracker. And the violations continue. Beyond pollution of both groundwater and surface waters, streams and springs in the Appalachians' fracking belt have been dried up to feed the fracking process, and increasingly so, as the industry turns to methods consuming more and more water to extract less and less gas. At one small stream, 1.5 million gallons of water per day were proposed to supply a single methane well, according to *Inside Climate News*, September 17, 2023. The effects go far beyond fish and wildlife and extend to rural development that depends on drinkable groundwater.

Economic deposits of frackable gas have thus far not been exploited in Ohiopyle State Park, though the park's *Management Recommendations* note that upper slopes including Sugarloaf Mountain and Laurel Hill could be threatened by drilling.

Surprisingly, the state does not own mineral rights to most of the park—nor for most state forest and park land statewide. For that acreage, Pennsylvania law allows miners to build roads, strip mine for coal, and drill for fossil gas. The miners may "reasonably" use surface rights in order to extract the underground rights. As the court rulings have come down, "reasonable" use typically means there are few limits on what may be done to surface values, provided restoration laws—however inadequate—are followed. In other words, Pennsylvania courts rule that you can preempt a property owner's surface rights, in plain view since the time of William Penn, if a gas driller happens to own underground rights that lie thousands of feet down and wants to invest as little as a year or so seeking whatever profits there might be from drilling, followed by the minimum in state requirements for restoration, such as removal of the drilling equipment and backfilling earth that was removed from a now ruined landscape.

The energy corporations' advantages date back to the 1800s bias for laws that blatantly favored rapid and unfettered development of seemingly unlimited resources. But while the pick-and-shovel mines of the Civil War era have given way to draglines with booms thirty stories tall and to explosives capable of leveling a mountaintop in a day, the laws governing mineral rights have scarcely evolved at all. One might as well assume that this year's transportation legislation were based on the use of nothing but horses and buggies.

Most fracking in the Youghiogheny basin has thus far occurred in lower-river watersheds such as Sewickley Creek. It remains to be seen if fracking's legacy of pollution, water consumption, and incursions on public and private property will worsen as the price of gas escalates or if they'll diminish as our economy slowly transforms from fossil fuels to renewable energy.

Forests Guard the Waters

The geology beneath us, the climate above, and the flow of streams all around set the stage for plants and wildlife to thrive, and the interplay between forests and waters is intimate, extensive, and constant. It's not complicated. For example, trees shade streams, which keeps them from heating up. This is not only essential for trout but also for lesser known coldwater sculpins, stonerollers, and dace, which deserve their own share of life even though no one wants to take them home in their creel for dinner.

In shaded ravines, Pennsylvania's official state tree—the eastern hemlock—is especially important in the forest-to-stream connection because the conifer's dense needle masses block out 90 percent of the otherwise warming sunlight. Meanwhile, carpet-like roots of riparian or riverfront plants protect streambanks from erosion, safeguarding water from silt that otherwise clouds clean flows. Nutrients in the form of fallen leaves and plant-associated insects beneficially feed aquatic life. Trees fall into the channels where the logs become essential habitat for fish along with other aquatic creatures. Logjams, the biologists tell us, are not the forest's perverse way of snagging fishing lines but instead are crucial to a whole chain of aquatic life.

The unexpected bottom line here is that fish need trees and rivers need forests. It's all an integrated ecosystem. So consider this crossover world involving both terrestrial and aquatic life, and specifically the forests with related plant communities of the Youghiogheny. We can't live without rivers, and our rivers would not be the lifelines they are without trees.

Botanists in Ohiopyle State Park have counted 458 species of vascular plants—trees, shrubs, flowers, and grasses with veins in their leaves for transporting water. Native species account for 390 while 68 were introduced, either intentionally or carelessly, from elsewhere, principally other continents. Botanists warn that 20 of those exotics are highly invasive and harmful to native plants and wildlife. Some have the potential to devastate whole Appalachian ecosystems and, in fact, are in the process of doing so. Furthermore, whole economies depend on those ecosystems. More on this unexpected threat later.

Understanding, appreciating, and caring for the forest take us back to the geography and climate that underpin everything. The river's unusual path northward from Maryland to Pennsylvania allows seeds of southern species to be carried north. Meanwhile, the relative warmth of the Youghiogheny's valley bottom provides an open door for that migration of plants and animals via water, wind, and wildlife. This ecological element of the river grows more important as the climate warms and as needs increase for native life to relocate northward. All along the way, the topographic reach from warm valley bottom to cool ridgetop directly above also makes the Youghiogheny basin an outdoor museum of plant diversity.

Owing to these factors, plus the basin's central position in the 2,000-mile-long Appalachian Range, from Alabama to Canada, some plant species typical of the South reach their northern limits at the Youghiogheny. The range of the pawpaw, for example, centers in Kentucky but extends to here. With distinctively oblong leaves in the understory, it produces one of the few fruits that can be picked from wild Appalachian trees, provided you get there before the squirrels do, which you usually won't. With a fruity scent—imagine a deluxe apple-banana-pineapple smoothie—pawpaws can be baked into a sweet desert. The umbrella magnolia also reaches its northern limits here.

Coming from the opposite direction, Northern Hardwoods include hemlocks, sugar maples, and sweet (black) birches, each as New England as the Pilgrims but flourishing here, intermingled with temperate climate hickories and oaks of several species each. Knowing all this, Paul Wiegman, for three decades the Western Pennsylvania Conservancy's chief naturalist and still the reigning dean of the region's natural history, noted, "The Youghiogheny has become a hotspot of native species diversity, which biologists regard as an essential feature of a healthy environment."

Hiding in the understory, slender blue iris and autumn willow find welcoming habitat though they normally reside south of the Mason-Dixon Line.

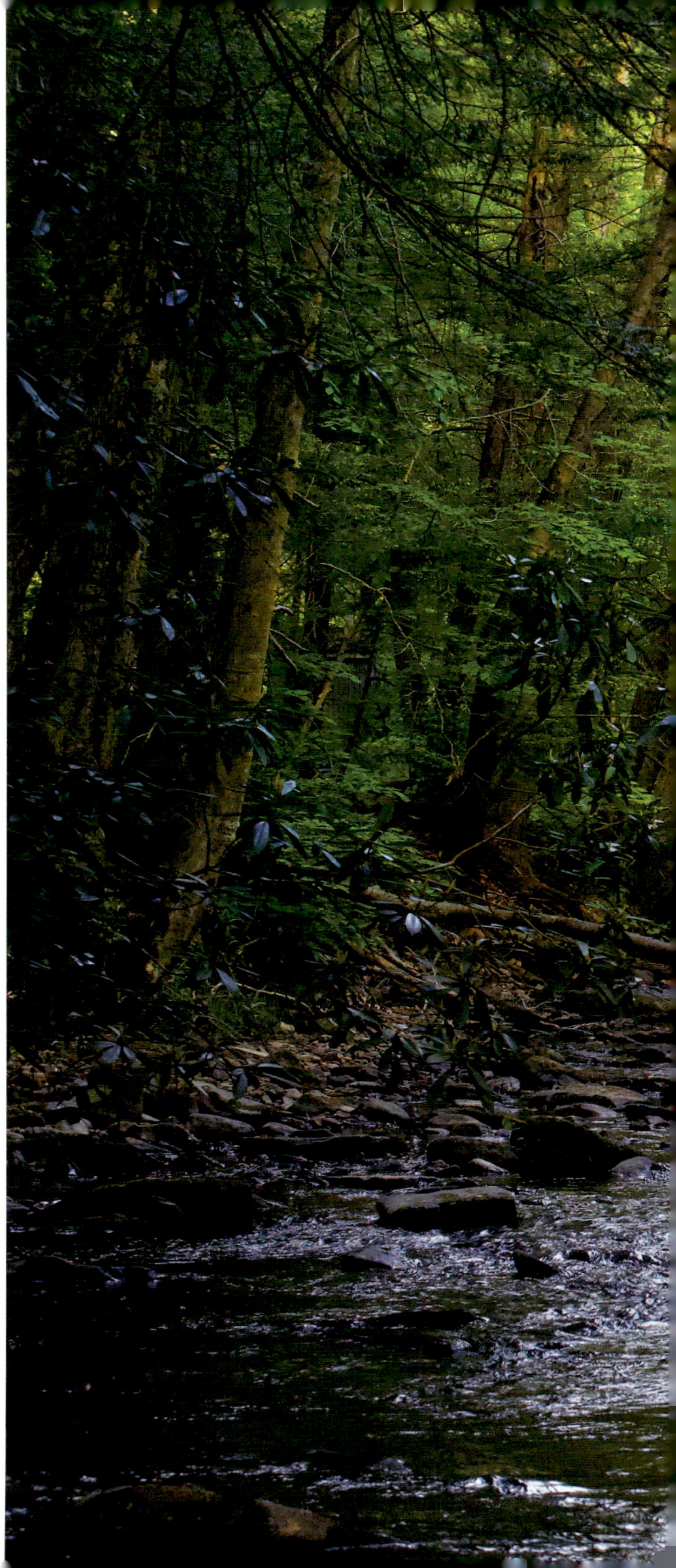

Trees along streams cast essential shade needed to keep the water cool enough for brook trout and other cold-water life. This protection grows increasingly important in the age of global warming.

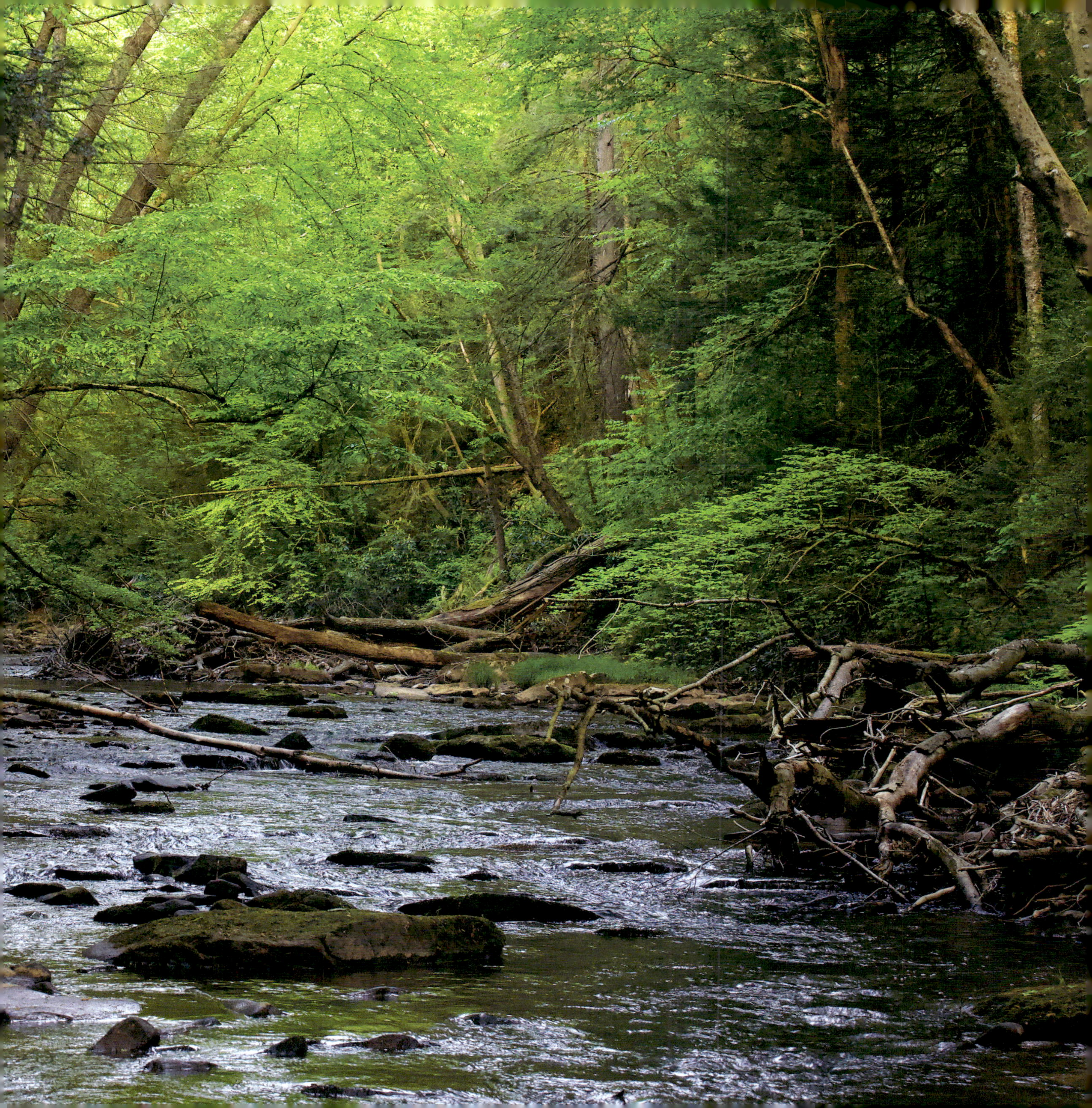

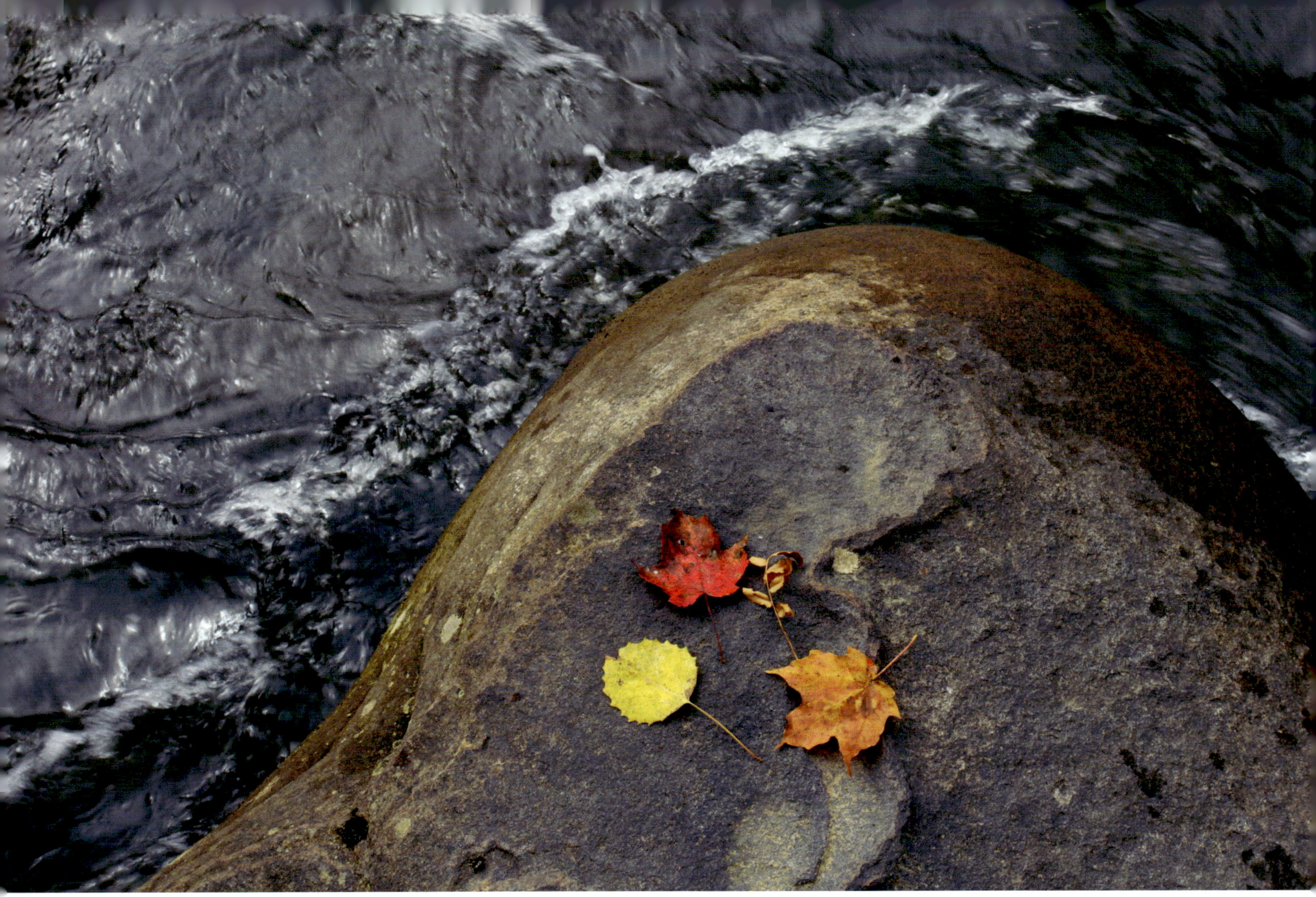

Leaves of a bigtooth aspen, red maple, royal fern, and sugar maple have all landed on one boulder at the edge of the Middle Youghiogheny near Drakes Run.

Uncommon throughout the mountains, a shrub called buffalo nut persists on Ferncliff Peninsula. Carolina tassel-rue sprouts nowhere else in Pennsylvania. Noting other plants in the same boat, which we might call a real-life Noah's Ark, the Ohiopyle State Park *Resources Inventory* reports that the Youghiogheny's arrangements of plant communities are "not found in other parts of the state." The river is a critical player in this diversity game.

Down at the water level, for example, floods periodically wash across sandstone slabs where several plant species evolved specifically in rocky cracks near Ohiopyle Falls. This distinctive habitat, "riverside ice scour," is maintained by episodic flows

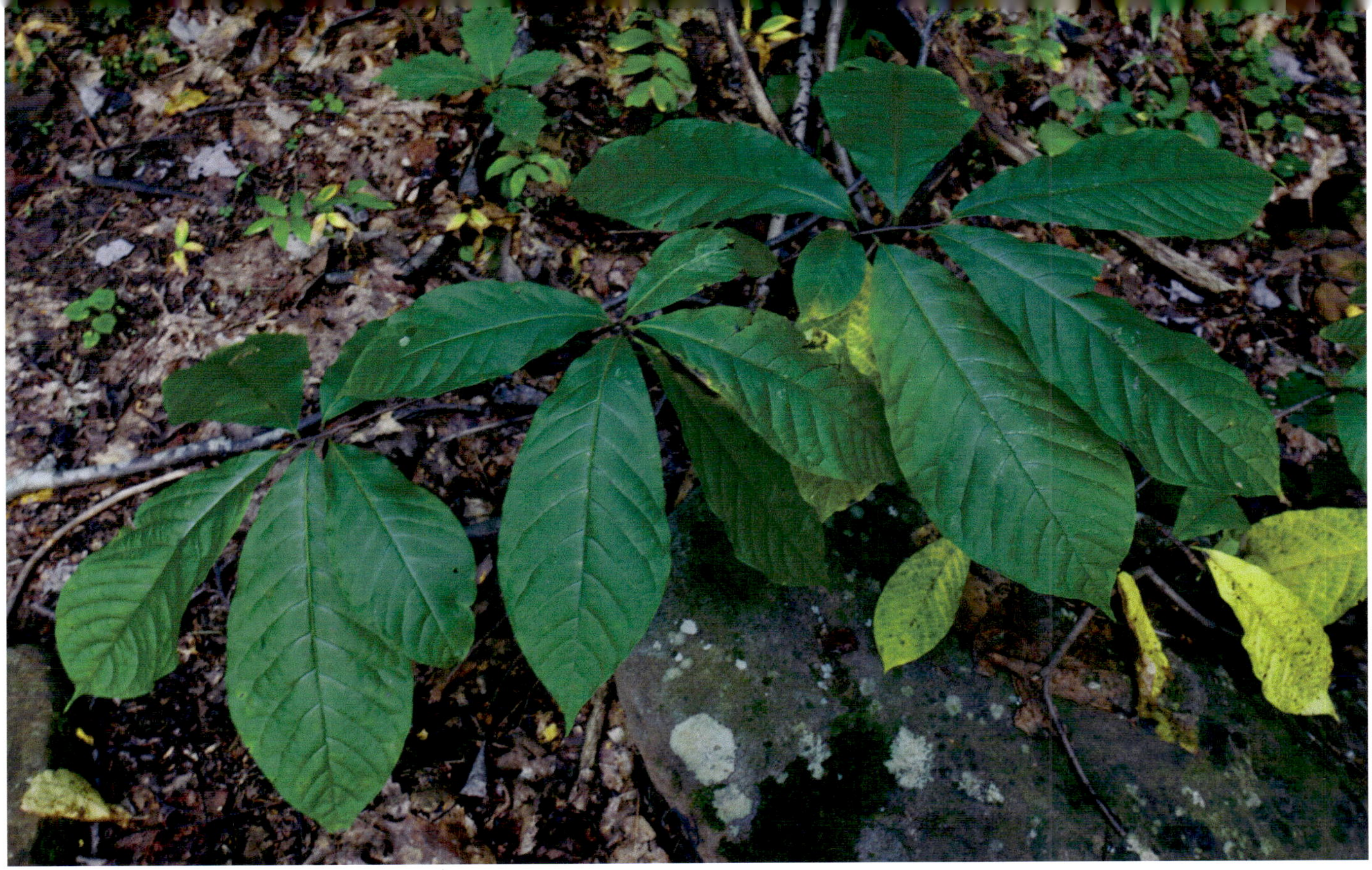

Along the Laurel Highlands Trail this fruit-bearing pawpaw tree grows at the northern end of the species's range.

of ice pushed by peak runoff floating miniature ice bergs that scrape against shoreline bedrock. A globally significant, imperiled community, it includes the flowered Barbara's buttons. Always scarce, these plants that thrive in the aftermath of river-borne ice have dwindled owing to milder winters of the global warming era. Pointing to the rocky riverfront, the state park management plan notes that "these communities represent the most ecologically significant areas in the park."

Going beyond the growing scarcity of ice on this increasingly hot planet, the Youghiogheny's history of logging and mining has caused many plant species to suffer or disappear entirely. The Pennsylvania Natural Heritage Program declares the survival of seventy-five species here to be of "special concern," though the loss of just one would be enough for alarm in the essential task of ecosystem repair. As the famed ecologist Aldo Leopold cautioned, "The first rule of intelligent tinkering is to keep all the pieces." Yet we're at the verge of losing many of them.

The Ohiopyle Loop of the Youghiogheny plays prominently in botanical narratives. Beginning in 1916, an internationally known Pittsburgh scientist, O. E. Jennings, boasted of a "botanist's paradise" of eighty-seven woody species on Ferncliff Peninsula. This owed principally to the rocky aspects of the gorge not being logged. Much of the rest had been cut

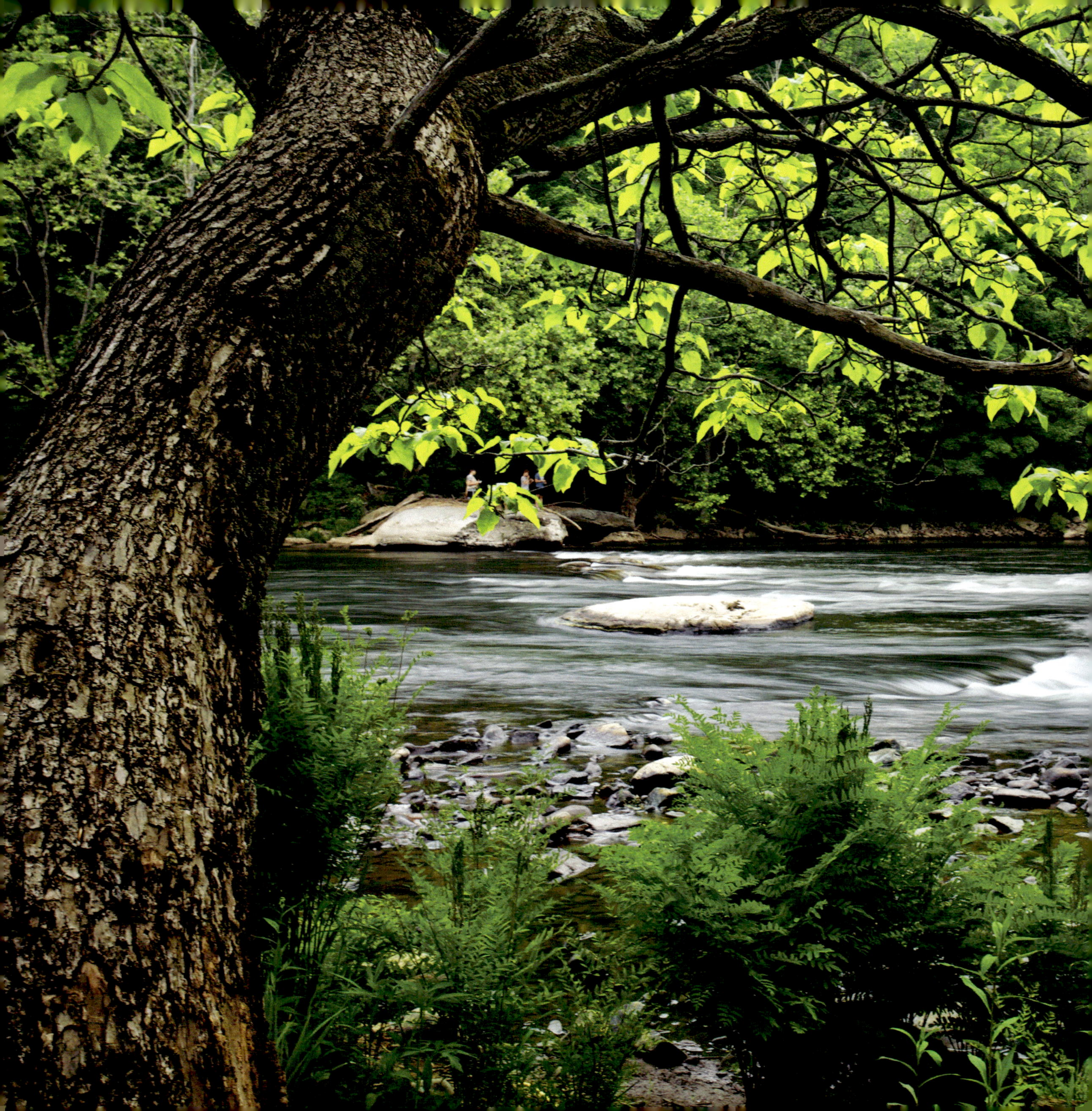

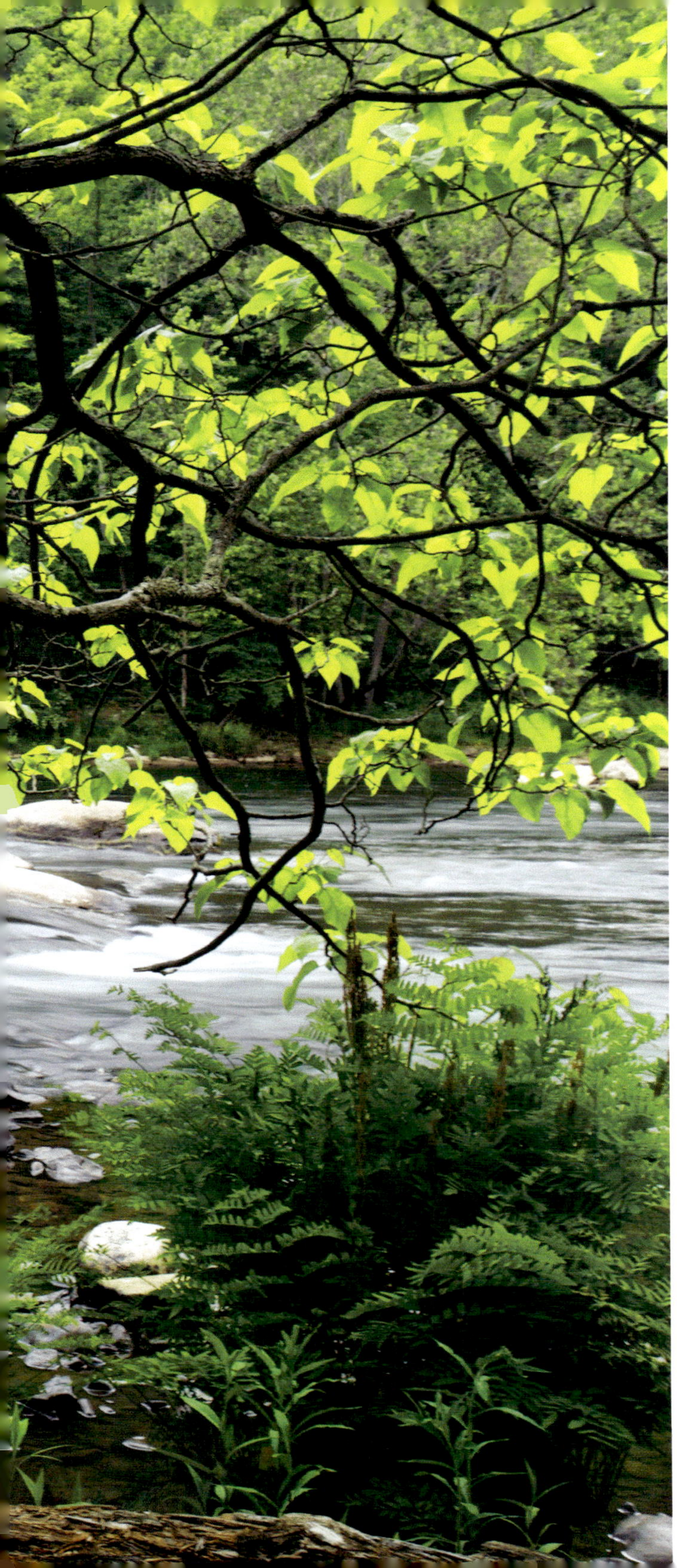

over in 1911 and was selectively logged again in 1947. In time, a robust suite of tuliptrees, oaks, beeches, basswoods, white pines, and hemlocks regrew across the entire Ferncliff tract, which survived undeveloped. The only house nearby was that of my great grandmother's sister, Cora Mitchell, tucked north of the Western Maryland railroad tracks. The Ohiopyle State Park later leased that house as a youth hostel, and it was subsequently torn down in 2004.

Now, throughout the Ferncliff Peninsula, including Aunt Cora's former front yard, a fabulous canopy again arcs overhead—the closest thing we have to forest primeval for many miles around, though none of this was preordained. In fact, Ferncliff's fate was an extremely close call. Jennings's Paradise was saved from a plan to sell it for board feet yet again, then for an amusement park—think Kennywood's roller coasters southeast of Pittsburgh—with clients arriving by rail, all of it topped off with a 337-lot subdivision of new houses crowding the riverfront of the Loop. Lillian McCahan of Ohiopyle saw the value of the forest and, by way of others who were socially connected, engaged Edgar J. Kaufmann, who owned Kaufmann's Department Store, Pittsburgh's classy shopping place to buy anything. He generously deeded the entire acreage to the Western Pennsylvania Conservancy for safekeeping as one of that organization's formative land preservation projects. Later rolled into the 20,600 acres that became Ohiopyle State Park, Ferncliff endures as a botanical highlight of Pennsylvania and the Appalachians.

An umbrella magnolia, normally thriving farther south in the Appalachians, beautifully overhangs the Middle Youghiogheny at Paddler's Lane campground and resort downstream from Confluence.

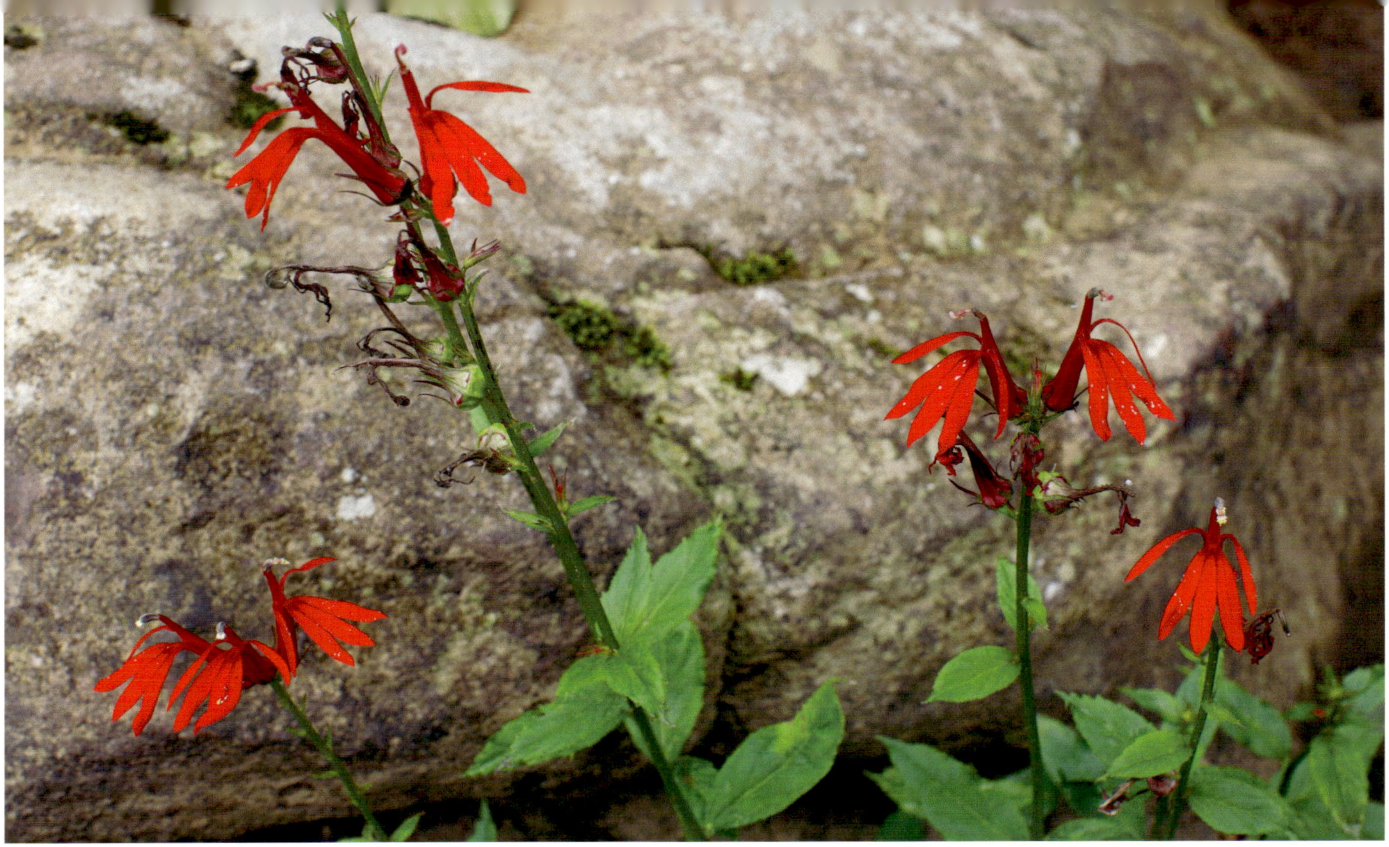

In summer bloom, cardinal flowers brighten Appalachian streamsides and become favorites of hummingbirds and long-tongued butterflies.

Nearly all the rest of the Youghiogheny watershed and, for that matter, virtually the entire states of Pennsylvania, Maryland, and West Virginia, were logged, then clear-cut again a varying number of times. Almost every acre of what became Ohiopyle State Park had been cut at least once. Seventeen percent of the park was cobbled together from mountain farms, some of them continuously cleared for crops, mowed for hay, or grazed as rocky pasture back to the 1700s.

After Ohiopyle State Park acquisition beginning in 1964, that land began to resprout trees. Unexpectedly, the whole park became an unofficial laboratory of forest succession and recovery. Given enough time, light-loving sumacs and locusts that dot, then fill in abandoned fields, graduate to red maples with colonies of white pines, followed by tuliptrees after the first wave of pioneering forest has crested. Increasingly dense canopies benefit successive species that thrive in shade that grows progressively cooler with time: oaks, hickories, and basswoods. Ultimately this natural succession of deepening shade favors sugar maples and hemlocks, which germinate, grow, and reproduce better than any other species beneath their own thickening canopies that filter out nine-tenths of the sunlight.

All of this constitutes what botanists call the Central Appalachian's Mixed Mesophytic Forest, meaning trees of many kinds in a wet climate. The most diverse temperate forest worldwide, it nourishes up to forty tree species per acre. To compare, only twelve native tree species grow in all of Wyoming, which takes eight hours to drive across on an interstate highway.

Most impressive of all is the commanding presence of tuliptrees. Their odd name refers to faint similarities between the tree's showy flowers and the popular Easter Day bloomer, though the tree's leaves are also oddly reminiscent, in shape, to that of a tulip flower. But the similarities end there. These woodland giants are deciduous, not evergreen, but for their size and the likelihood of inspiring awe, I regard them as redwoods of the East. Fast-growing, mature tuliptrees can bulk up to 6-foot diameters. They can tower an impressive 190 feet skyward. Picture the stairs of a nineteen-story building that you'd have to climb just to carry a drink of water to the foliage at the top. That fluid, of course, is impressively pumped up to the sky through nothing but osmosis in the conductive living tissues of these ultimate botanical wonders.

Thriving in sunlight where mountainsides had been cut over but then allowed to grow again, tuliptrees fill a gaping hole in the ecosystem left by the loss of chestnut trees. That formerly keystone species, accounting for one-third of the forest in parts of Appalachia, benefited myriad life forms with habitat and food, all eliminated a century ago by a deadly fungus haplessly introduced by boat from Asia. Chestnut trees remain, but only in sapling form because the persistent air-borne pathogen infects the bark as soon as cracks in fattening stems begin to appear. The chestnut's demise left big shoes to fill in eastern forests, but at least we have tuliptrees in their place. Fingers crossed for the future.

No less important, and maybe more so, oaks of the red, white, and chestnut varieties thrive as mainstays in Youghiogheny forests. Producing dense wood that is of legendary value for furniture, tool handles, barrel staves, plus any number of hardware essentials, oaks are more fundamentally a prerequisite home for squirrels and a food staple for deer along with other nut and acorn eaters. Oaks require longer than tuliptrees to mature to great height and girth, but they cope with hostile elements exceptionally well. They regerminate in shadier conditions, which allows them to reign as a climax species, meaning that young trees can grow in the shade of their elders. This produces successive generations, if they are spared fire, windthrow, disease, and most critically, logging.

Champions of this climax forest, accommodating woodland shadows best, hemlocks rise in riparian strips and chilled pockets totaling a widely beloved and disproportionately important four hundred acres of Ohiopyle State Park. Individual trees stand for several hundred years and up to eight hundred, which accounts for only the recent past. Hemlock groves continuously linking multiple generations backward in time can boast of family ancestors dating to millennia on land roamed by mastodons weighing ten thousand pounds each, while the Wisconsin Continental Glacier, just to the north, began to drip away into our region's streams. Much of the Youghiogheny's hemlock acreage lies in the Great Gorge where the river charts its Loop below Ohiopyle, and on Ferncliff Peninsula at the inside of that big bend in the river, and within nearby Meadow Run's scenic hall of nature. Superb groves likewise survive in Swallow Falls State Park of Maryland and in West Virginia's Cathedral State Park, located along a tributary to the uppermost reach of the river.

Unlike people along with all our animal kin, trees grow throughout their entire lives, and because

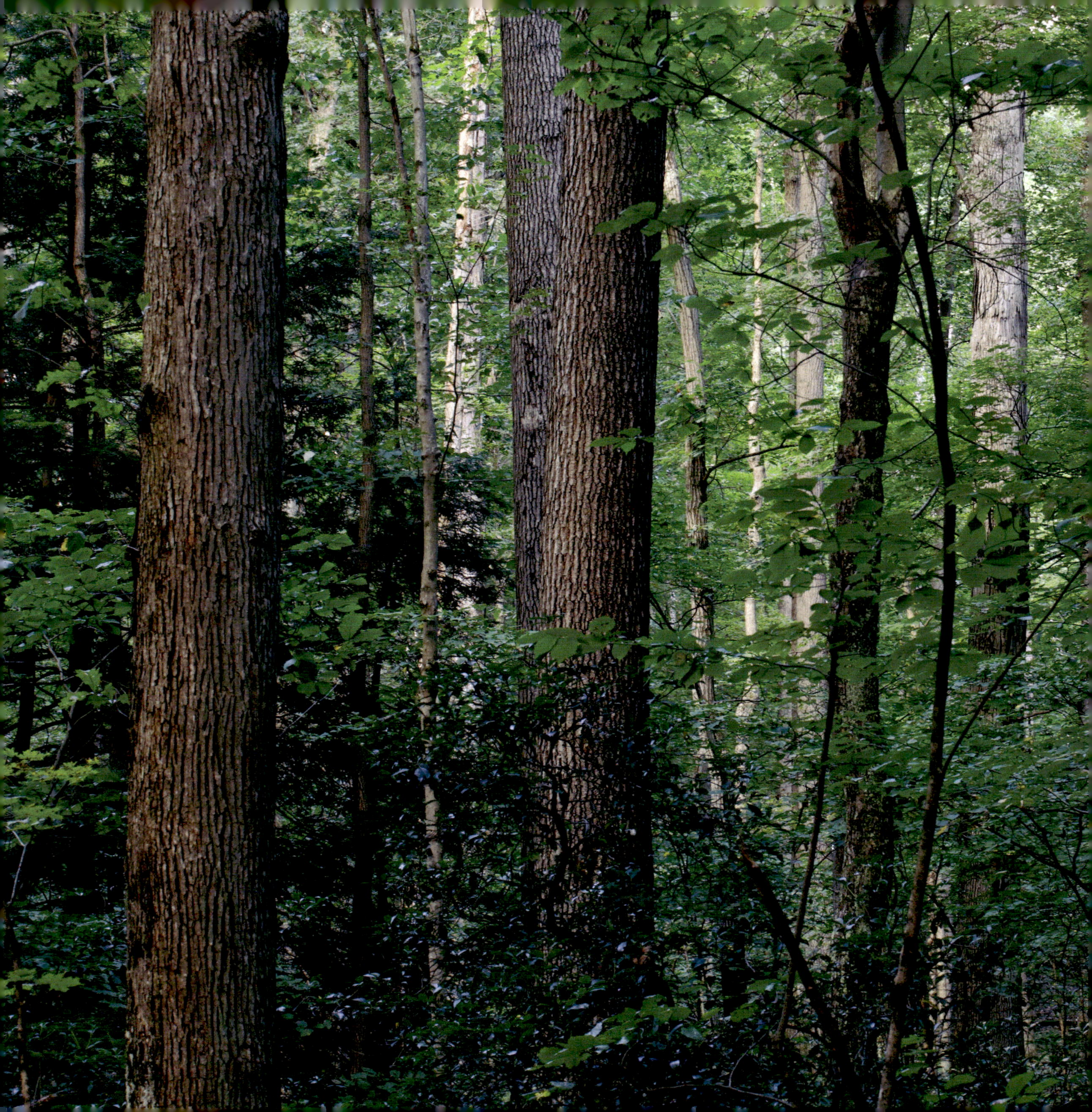

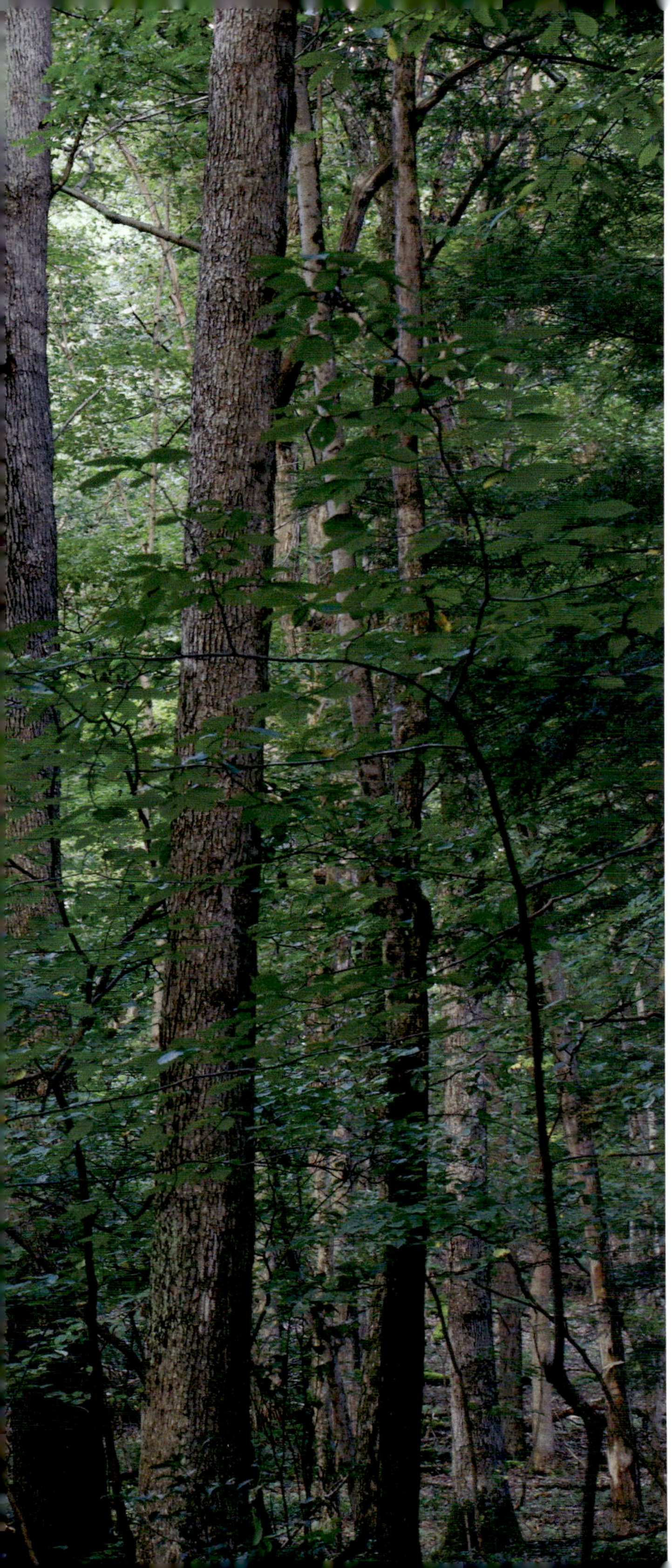

Ohiopyle State Park is now mostly protected from development and logging, its trees get larger until leveled by trauma: windthrow, fire, drought, invasion of disabling insects, exotic fungal blights, road building, and logging. Now recovering from two centuries of unrestrained cutting, much of the forest within Park boundaries has aged in just fifty years from struggling saplings to impressive towers of cellulose linking soil to sky as nothing else can do. Trunks of tuliptrees can fatten a half inch or more yearly; unexceptional trees of my own youth now stand tall with diameters from 20 to 30 inches greater than when I first laid eyes on them. That's what I consider progress.

The forests within the basin's three state parks now resemble the Appalachians' once widespread ancient forest far more than does acreage anywhere else. These tracts increasingly support a whole ecosystem of forgotten bounty in the groves of big trees, all creating a haunting beauty that has become exceedingly rare. In fact, almost entirely gone.

Classic characteristics of old-growth forests include not just cathedral-like columns that lend a sublime if not sacred effect for anyone willing to notice but also a rich mix of aged snags, which hold nesting sites for birds, plus cavities serving as fully furnished homes for tree-dwelling mammals ranging from ubiquitous raccoons to reclusive fishers—the sleek, seldom seen second-largest member of the weasel family. Underfoot, roots protectively enmesh whole hillsides and riverfronts, which otherwise are victims to

Tuliptrees sprout in sunlight and can grow for hundreds of years to become monarchs of eastern woodlands. Taller than any other Appalachian tree, they can exceed diameters of 6 feet. This magnificent forest thrives on Ferncliff Peninsula.

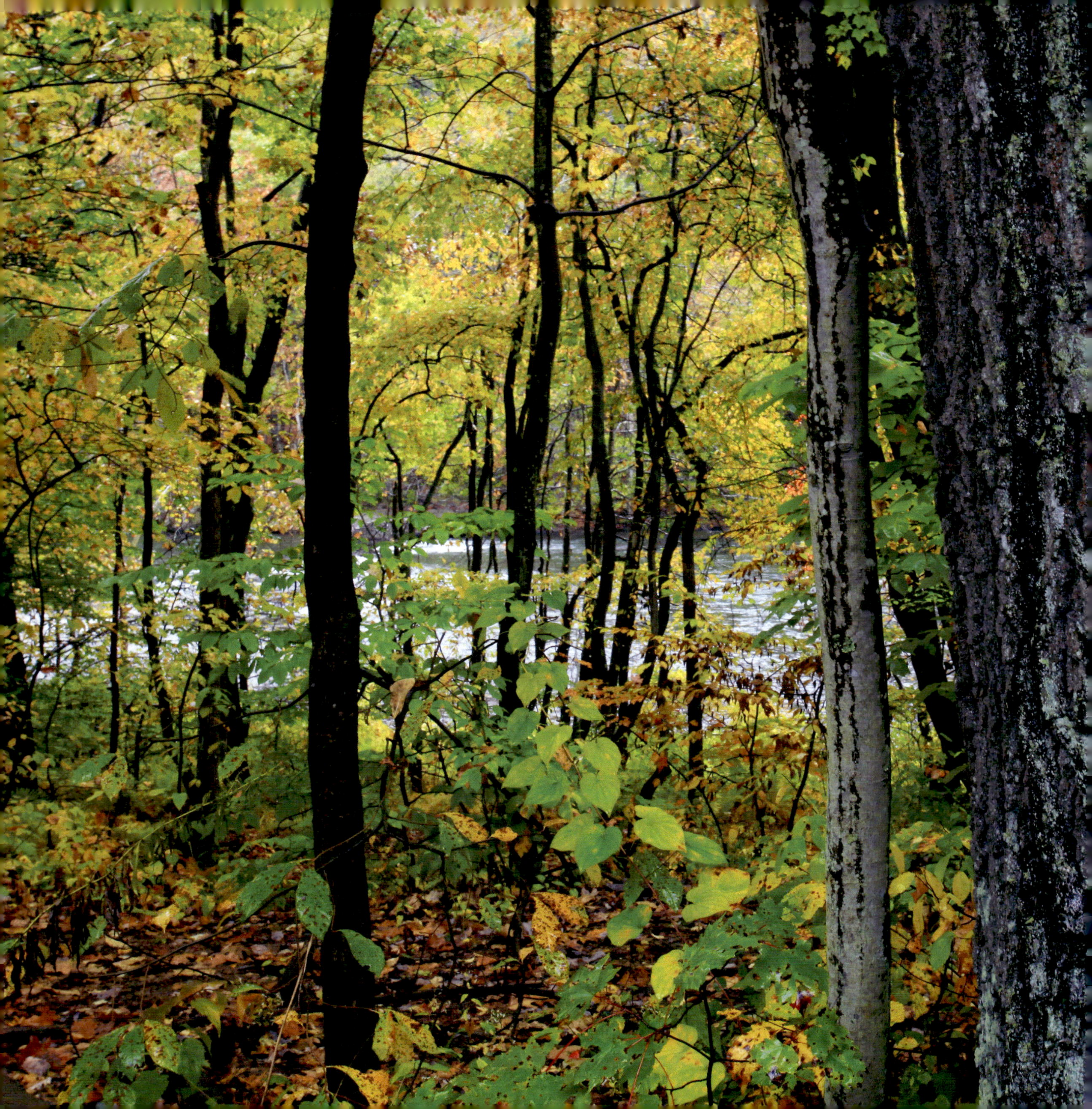

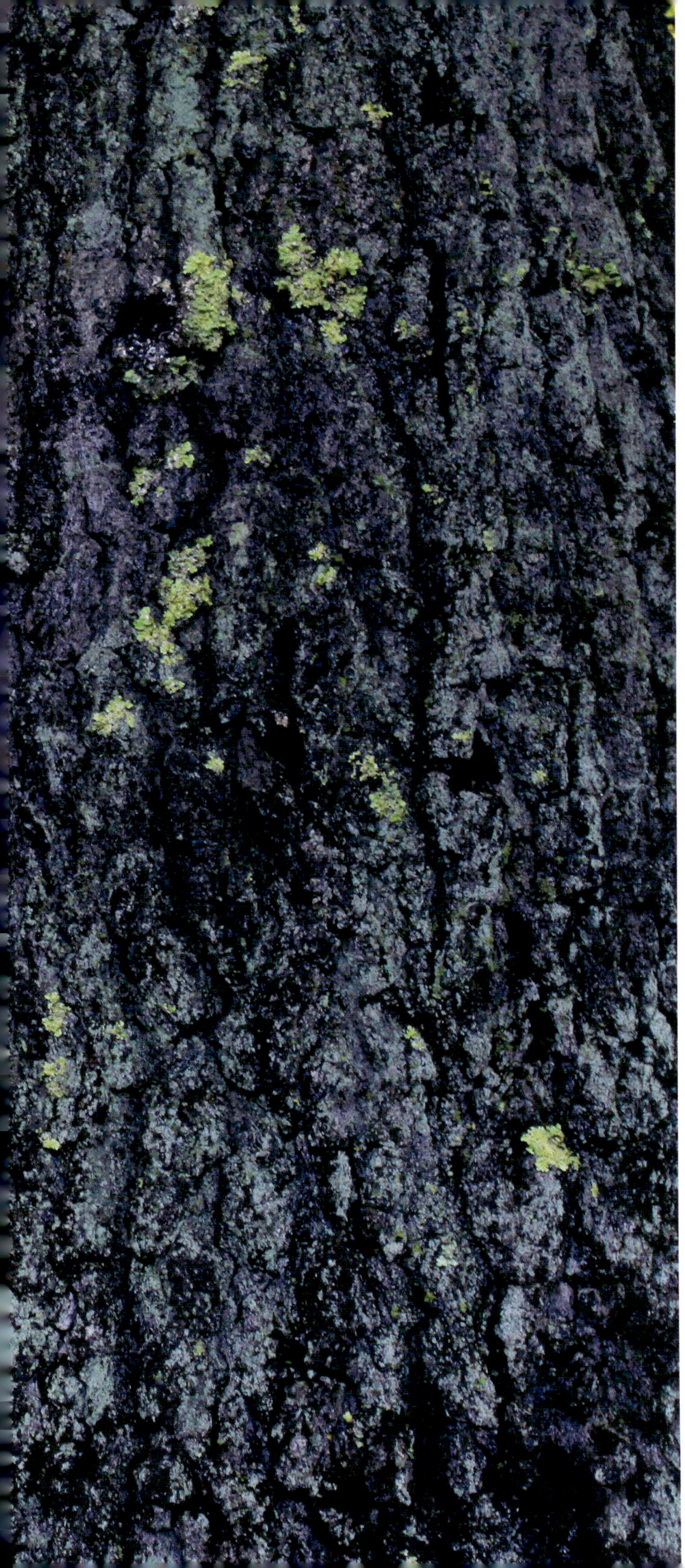

erosion. Overhead, hemlock and white pine needles cast shade that cools the soil and its communities of life beneath the canopy. That chill will become more urgently needed in the distressing age of global warming, which is only getting worse. Consider, in that grim regard, record-breaking temperatures worldwide in summer 2023, to be outdone, no doubt, in the years that follow.

The forest, over time, succeeds toward larger, shadier trees, but not on a straight-line progression. Older forests are subject to winds as strong as hurricanes and to fires where lightning or carelessness strikes. In these ways, openings in the canopy—usually small—become open to daylight, which allows for a mix of other plants to sprout and augment habitat diversity. But overall, hiking through the deepening groves reveals a forest that's becoming more similar to the fabled woods of fairy tales than to the previous tangle of briers or the homogenously adolescent thicket of trees from the recent past.

Owing to all these factors, Ohiopyle State Park might be considered a front line in the efforts to conserve nature, to reinstate ancient forests, and to restore the remarkable plant communities that make the entire Appalachian Range botanically irreplaceable worldwide. Furthermore, the importance of forests extends far beyond recreation, scenery, and wildlife, and onward to the protection of the river, to the health of the entire watershed that drains into the river, and to the fate of the planet as it faces a biodiversity crisis in the global-heating future.

We've seen that trees stand as the basis and backbone of entire ecosystems. They support habitat

An impressive red oak dwarfs other trees in this forest along the Middle Youghiogheny below Confluence.

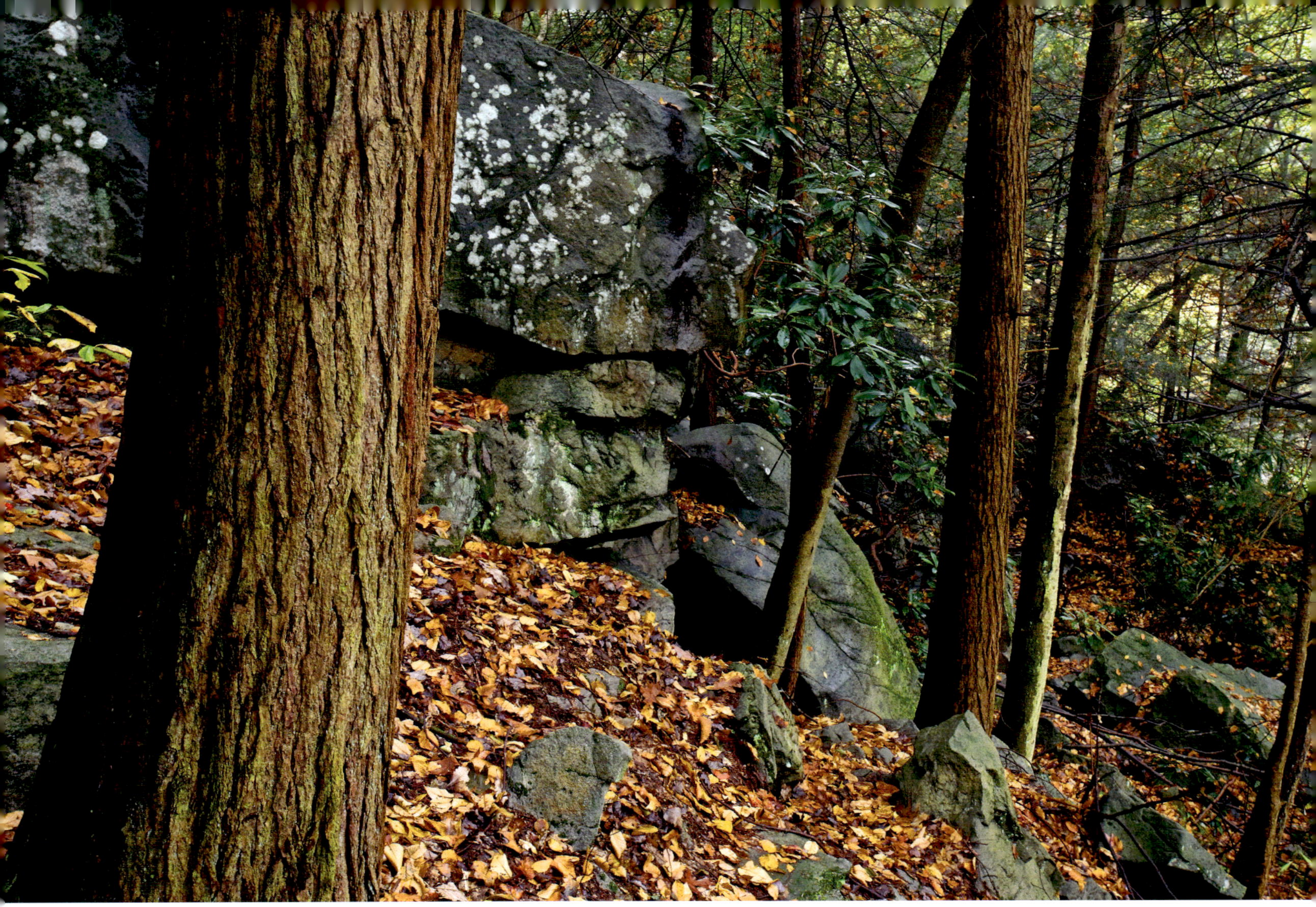

Hemlocks can live for eight hundred years or more, excelling in shade and cool topographic pockets within the Youghiogheny's Great Gorge of Ohiopyle State Park, Swallow Falls State Park in Maryland, and Cathedral State Park in West Virginia.

for hundreds of birds, mammals, reptiles, amphibians, and fish. But they're even more important than that. Through stomata (microscopic holes in the undersides of leaves and needles), trees inhale the carbon dioxide gas that we humans along with other animals exhale. Then the trees exhale the oxygen that we and all other animals need to inhale. Let me describe this inter-species contract another way: trees inhale what we exhale, and vice versa, and it's hard to imagine a more intimate relationship between us and other life-forms than that.

And there's even more to the increasing debt we owe to our forests. Through the complex, mind-boggling biochemical process of photosynthesis (for

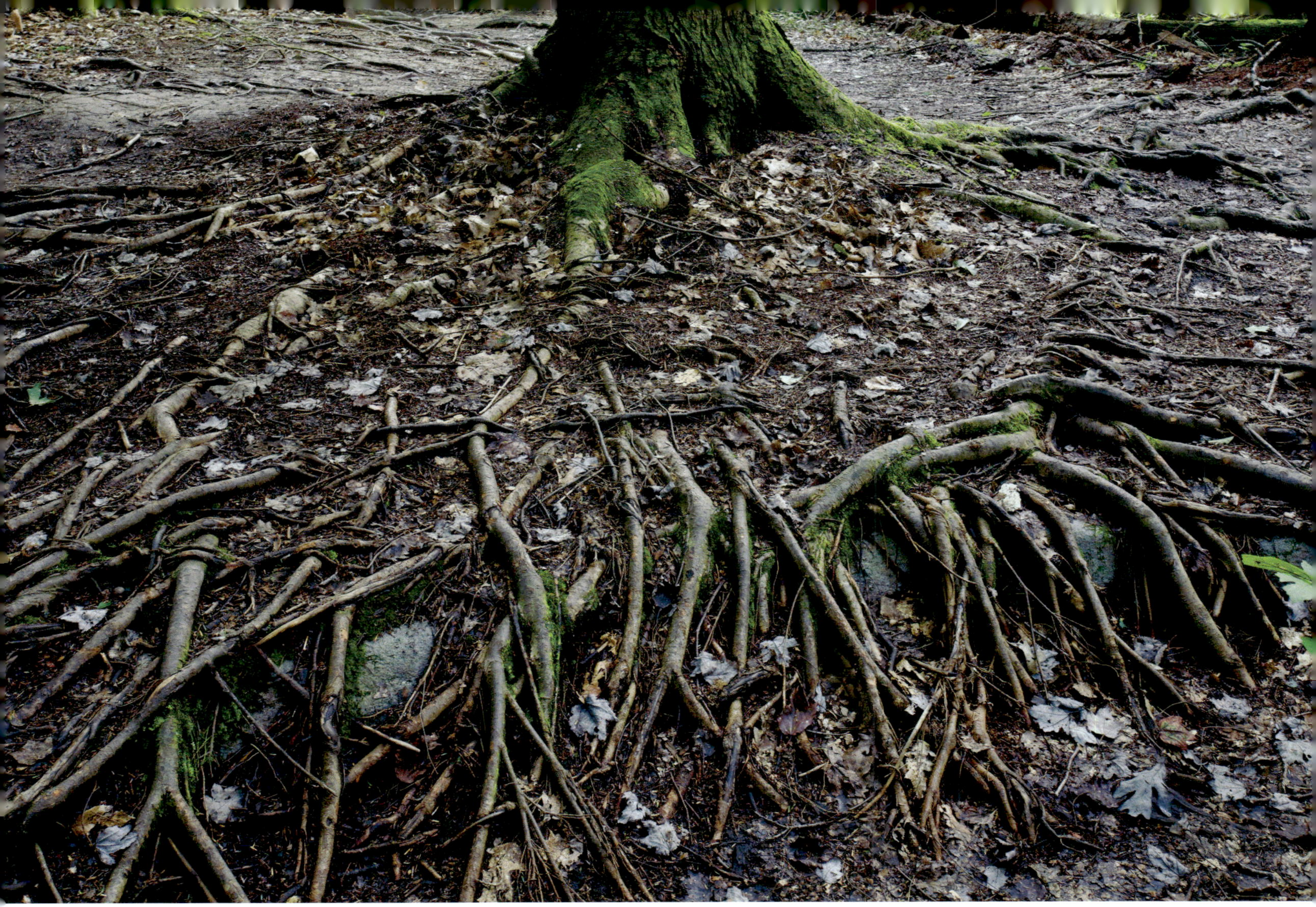

shorthand here let's just call it a miracle), trees convert carbon dioxide gas of the atmosphere into solid carbon when forming trunks, limbs, and roots. In this way, the trees come the closest one might imagine to creating "something" (meaning hard, grainy wood) out of "nothing" (meaning air). This essential process of sucking carbon dioxide out of the air has become a critical component and cornerstone in our efforts to combat the scourge of global warming.

The excessive heating of the climate is caused principally by our burning of coal, oil, and methane gas. That combustion converts both solid and liquid forms of carbon into gaseous carbon dioxide, which, planet-wide, acts like the unwelcome roof of a

Roots of hemlocks protect the soil of steep slopes and riverbanks from erosion here on the Ferncliff Peninsula.

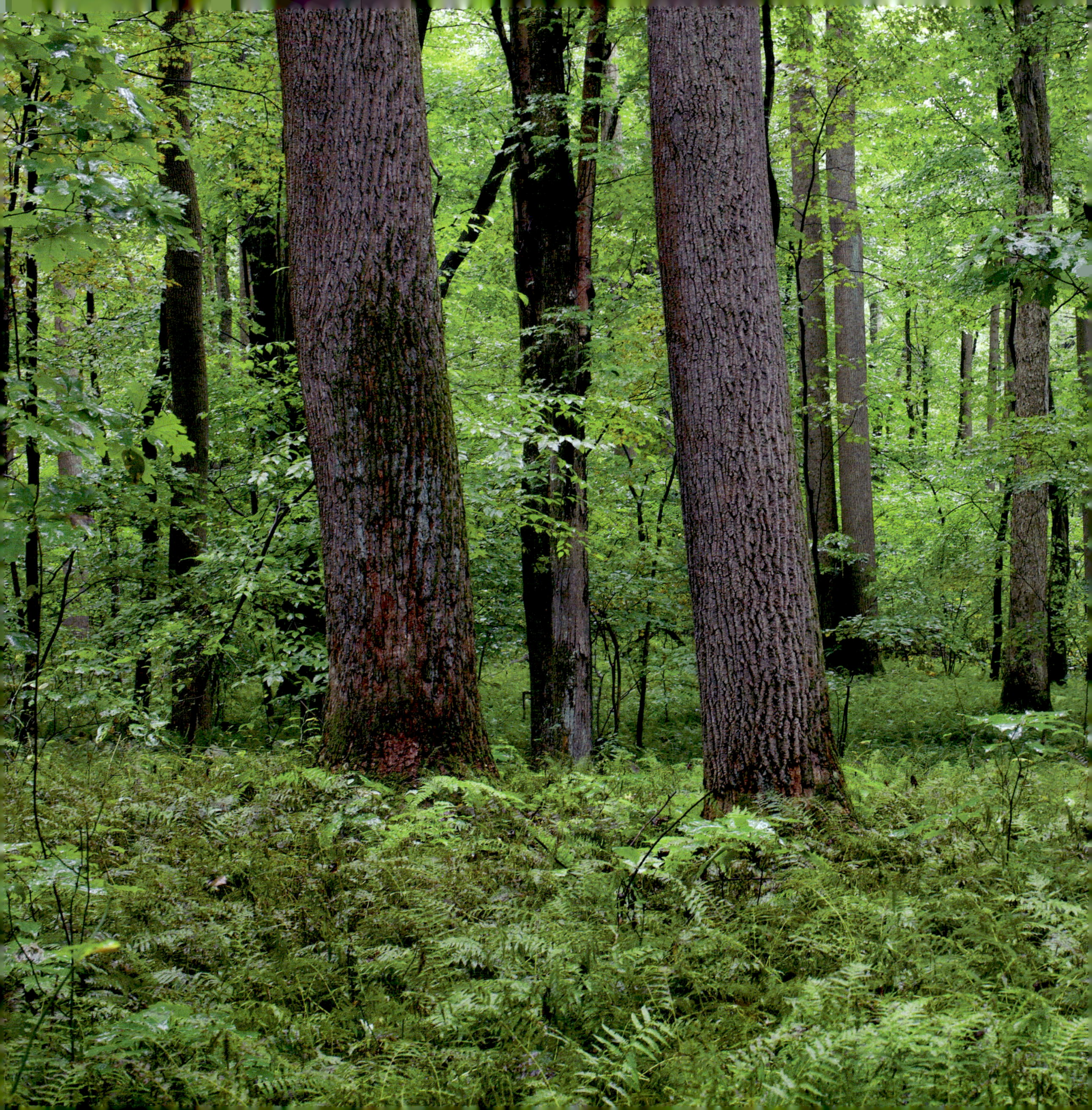

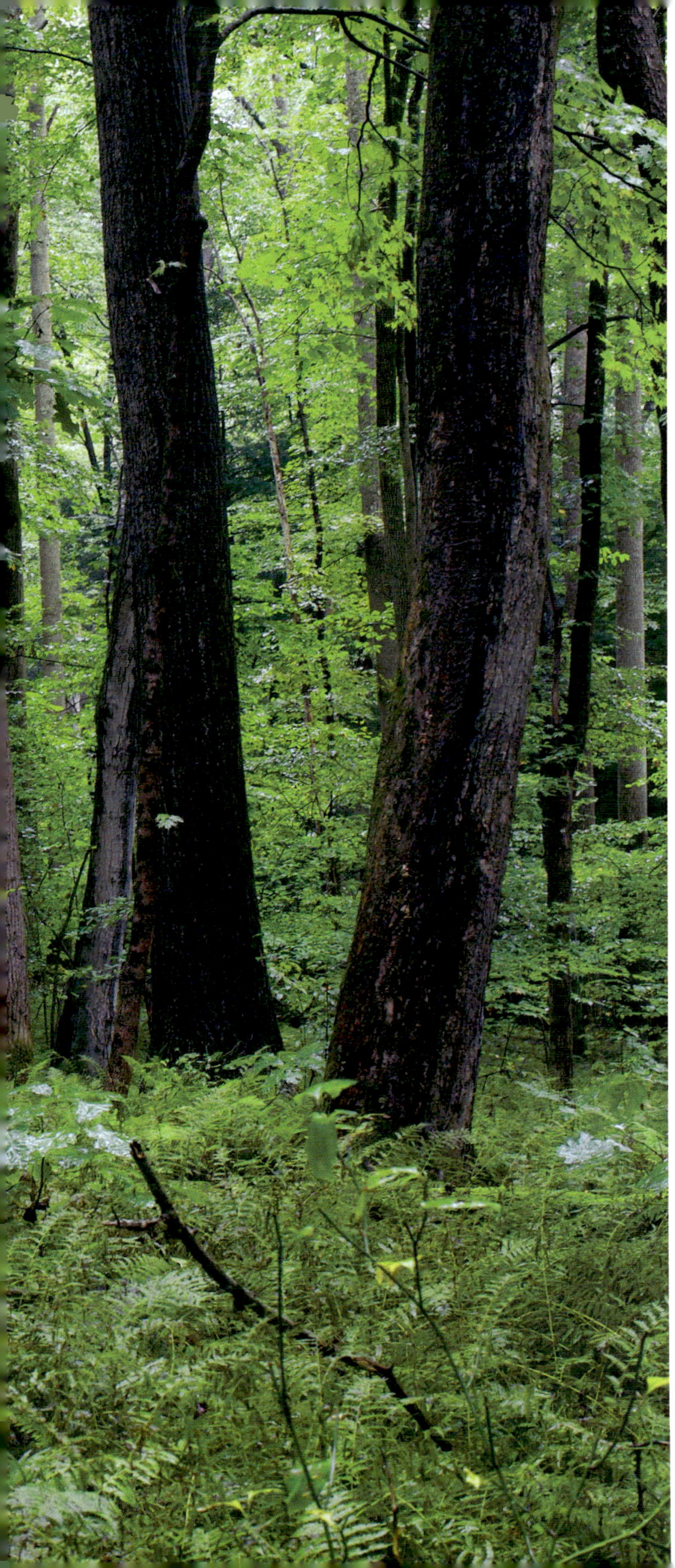

greenhouse. To mix my metaphors a bit here, think of the carbon-saturated atmosphere as a bubble completely enclosing the earth, benevolently allowing the sun's heat to penetrate but perversely preventing its escape. By converting that carbon gas into a solid form, trees are the earth's primary means of reversing the troublesome—one might ultimately say suicidal—increase in atmospheric carbon dioxide that we are causing by burning so much fossil fuel. However, only large trees generate the adequate mass of wood needed to sequester significant amounts of carbon. Each annual ring on an oak, maple, and most effectively a tuliptree because it is so fast-growing, means that increasing amounts of carbon are being converted from a planet-heating gas and changed into solid carbon, which lingers for centuries in the benign, stable, planet-protecting shape of a tree. Which, besides all that, is beautiful to behold.

As I said, old forests with their large trees remove considerably more carbon from the atmosphere than do young forests, and the numbers here are amazing. As reported by Northeast Wilderness Trust in 2021, because of its increasing volume, a girthy eighty-year-old tree sequesters as much additional carbon in one year of growth as an entire medium-size tree contains in its trunk and branches totaled together. Researchers found that it takes from twenty to thirty years for a young forest to begin to deliver net gains of carbon storage in woody tissues as well as in the soil. While

Maturing tuliptrees, oaks, and maples convert large amounts of atmospheric carbon into solid carbon that's beneficially locked up in the heartwood and cellulose of trunks, limbs, and roots, all providing an essential and efficient means to combat the perils of global warming.

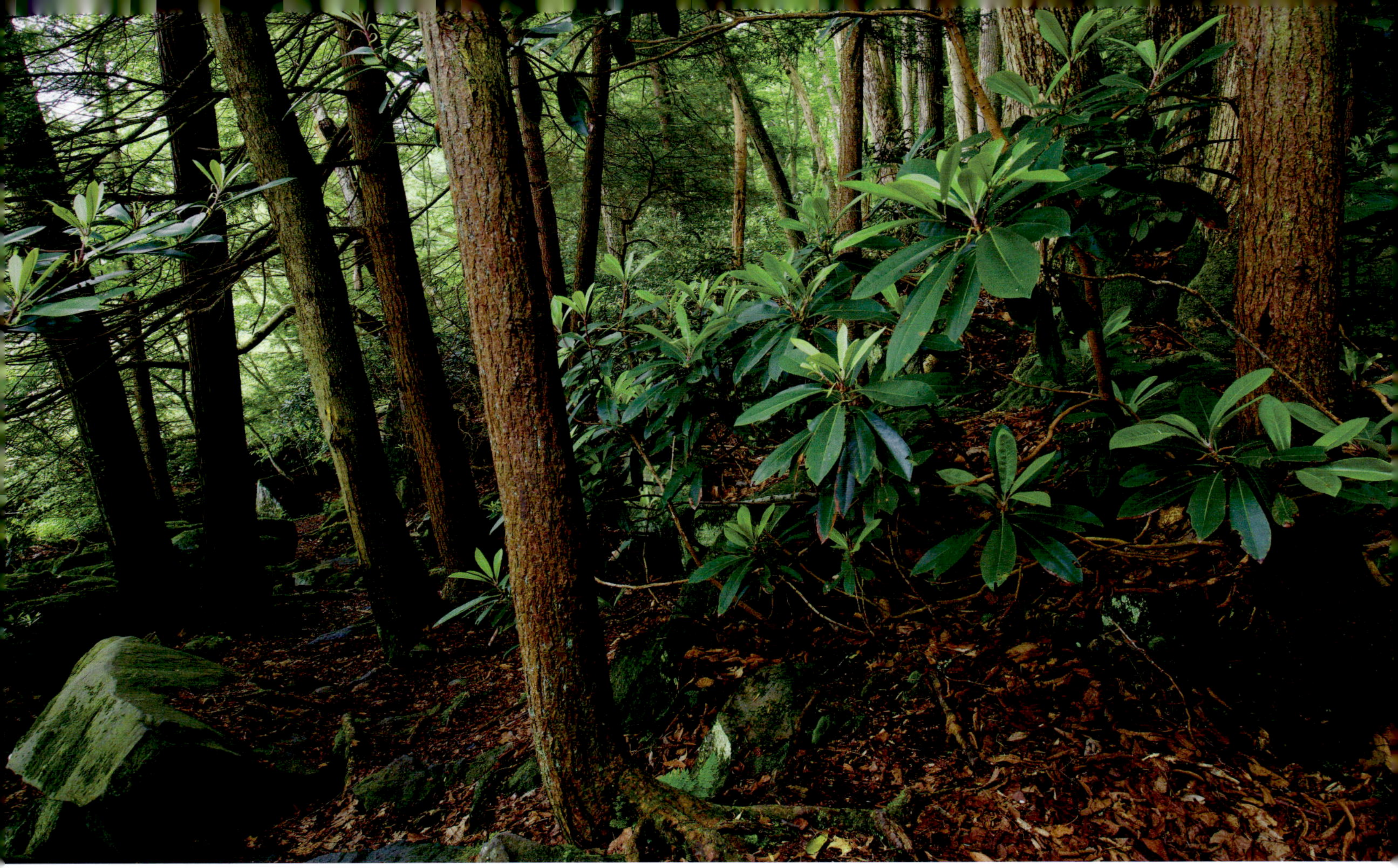

Invasion of the exotic woolly adelgid—a minute insect that arrived in the United States via infected nursery stock—kills hemlock trees throughout the East. State park officials carefully apply insecticide to hundreds of hemlocks with hopes for benign biological solutions such as the cautious introduction of beetles that feed on the adelgids.

it's most essential that we reduce the burning of fossil fuels, the other mandatory need in combating global heating is to keep intact the reservoirs of solid carbon that large old trees create and sequester, and to expand the acreage of those old-growth reservoirs wherever possible.

While 60 percent of the forest in Ohiopyle State Park is now crowned by maturing trees (from sixty to a hundred years old as I write this in 2024), the impressive large boles remain uncommon in forests virtually anywhere outside the state park system. So, these parklands are important. And they are scarce, accounting for only 1 percent of Pennsylvania's landscape. Even in state forests, under the jurisdiction of the State Bureau of Forestry rather than the Bureau of State Parks, management typically includes commercial timber harvest, with the result that large uncut groves of trees are rare, even on those public lands. Estimates of old-growth forest acreage statewide total about one one-hundreth of 1 percent of what they originally were when European settlers first stepped off the boat.

Reflecting further on what we see versus what the observant George Washington saw as an intrepid twenty-one-year-old explorer crossing Youghiogheny headwaters, no one has to look very hard to realize that exotic plants of many kinds have been introduced to the river's shorelines and, for that matter, to the entire state and nation. However, large trees, with their deep shade and undisturbed depth of humus, create an environment that's less vulnerable to hostile takeover by the invasive rogue plants. Ohiopyle State Park's management recommendations recognize that "reduction of invasive species will be most successful if management aims to restore the ecological conditions that favor native species."

But, making this aim hard to achieve, ancient ecological balances that long typified native plant and animal communities have been upset, often at the scale of a large bull in a small china shop. In the ensuing biological mayhem, many of the invading species egregiously outcompete the natives. Japanese knotweed, purple loosestrife, and reed canary grass infest many Youghiogheny floodplains. In this real-life biological nightmare, exotics displace natives, shade them out, reproduce faster, and thrive in conditions most typically found at logged, developed, mined, and scraped-over sites. For key areas of Ohiopyle State Park, the official agency plan wisely directs that "management of future old growth conditions should be the all-encompassing natural resource goal for these Critical Resource Areas."

The fate of hemlocks provides a particularly poignant example of what's at stake. These magnificent trees, otherwise living for centuries if not cut down, are being decimated by a minute insect, the hemlock woolly adelgid, imported to North America on infected nursery stock from Asia and first noticed in Virginia. State park staff carefully treat hundreds of local hemlocks with insecticide in hopes of keeping them alive until more sustainable means of control, such as introduced beetles that eat the alien insects, become effective.

Likewise threatened, white ash trees—with their hard yet resilient wood famously lathed into baseball bats since the time of Abner Doubleday—have been eradicated by the emerald ash borer, yet another exotic insect carelessly exported from Asia. A lone mature ash can still be seen on the lawn of the historic superintendent's home in Ohiopyle State Park, just above the mouth of Meadow Run. Park staff treat this grand specimen with insecticide in a last-ditch effort to keep it alive.

Ohiopyle State Park's spectacular groves of beech trees—elegant with smooth grey bark—are similarly vulnerable to sinister forces of an exotic insect and fungal pathogen teaming up to kill nearly all mature beeches in an Appalachian-wide swath across the East. As of this writing, beech bark disease has not yet arrived to magnificent beech groves such as one on the Great Gorge Trail heading up Kentuck Mountain, but the disease fatally infects trees in northern Pennsylvania and will likely reach the Youghiogheny any time now. Appreciate these trees while you can.

It's often said, as the wise and talented singer Joni Mitchell proclaimed, that we don't know the value of what we've got until it's gone. That sad axiom is surely true in terms of trees, forests, and the waters that flow from them. Knowing that much of nature is endangered and knowing that its health requires greater care can together become a motivator to see more of this exceptional place and thereby to know it

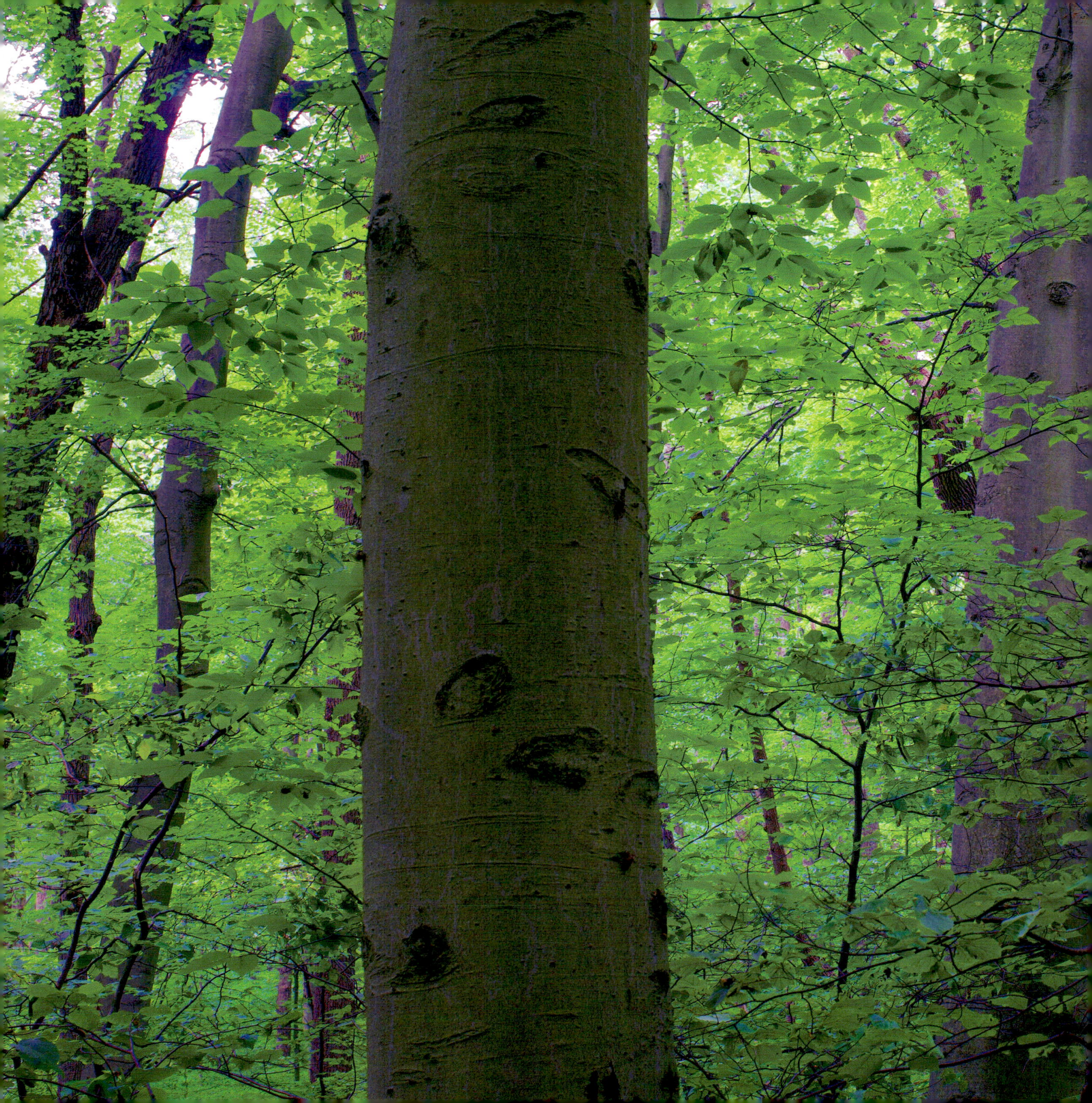

better. That's important because, as others have said, we only protect what we love, and we only love what we know. Let me add—here in the spirit of this book's title—that we only protect, love, and know what we stop to see and to watch.

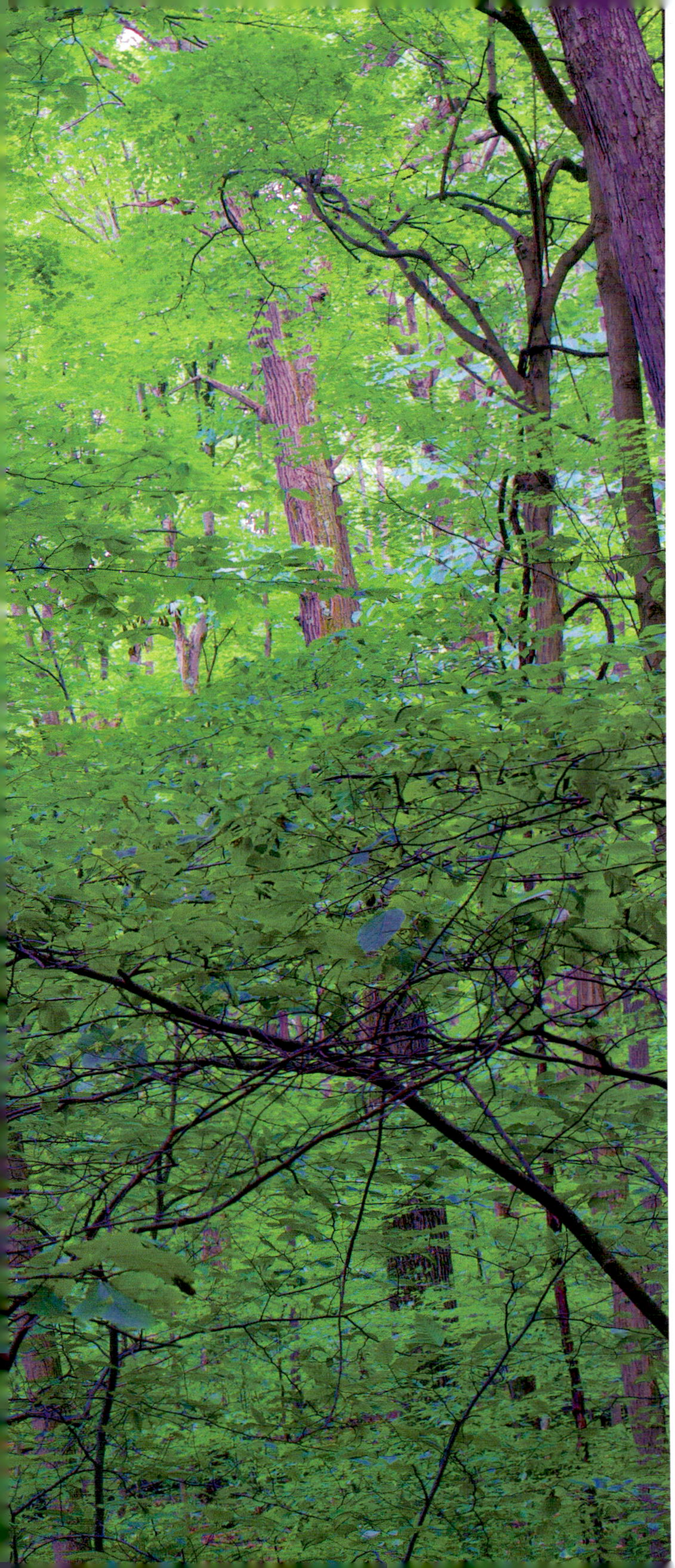

Beech trees form a green canopy over many Youghiogheny basin forests, such as here along the Ohiopyle State Park trail ascending Kentuck Mountain. These trees will be threatened by beech bark disease, now spreading through the Appalachians from north to south.

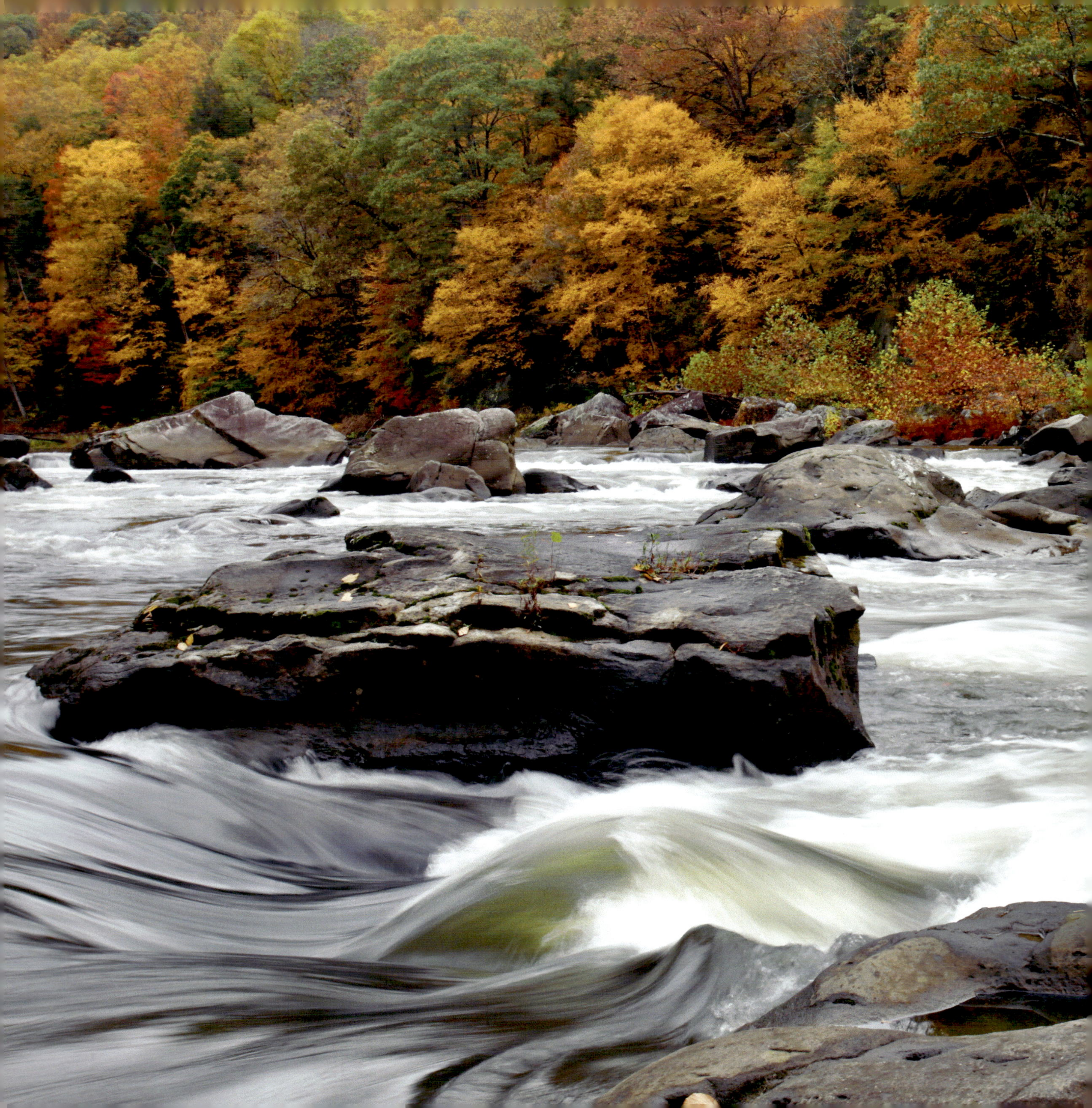

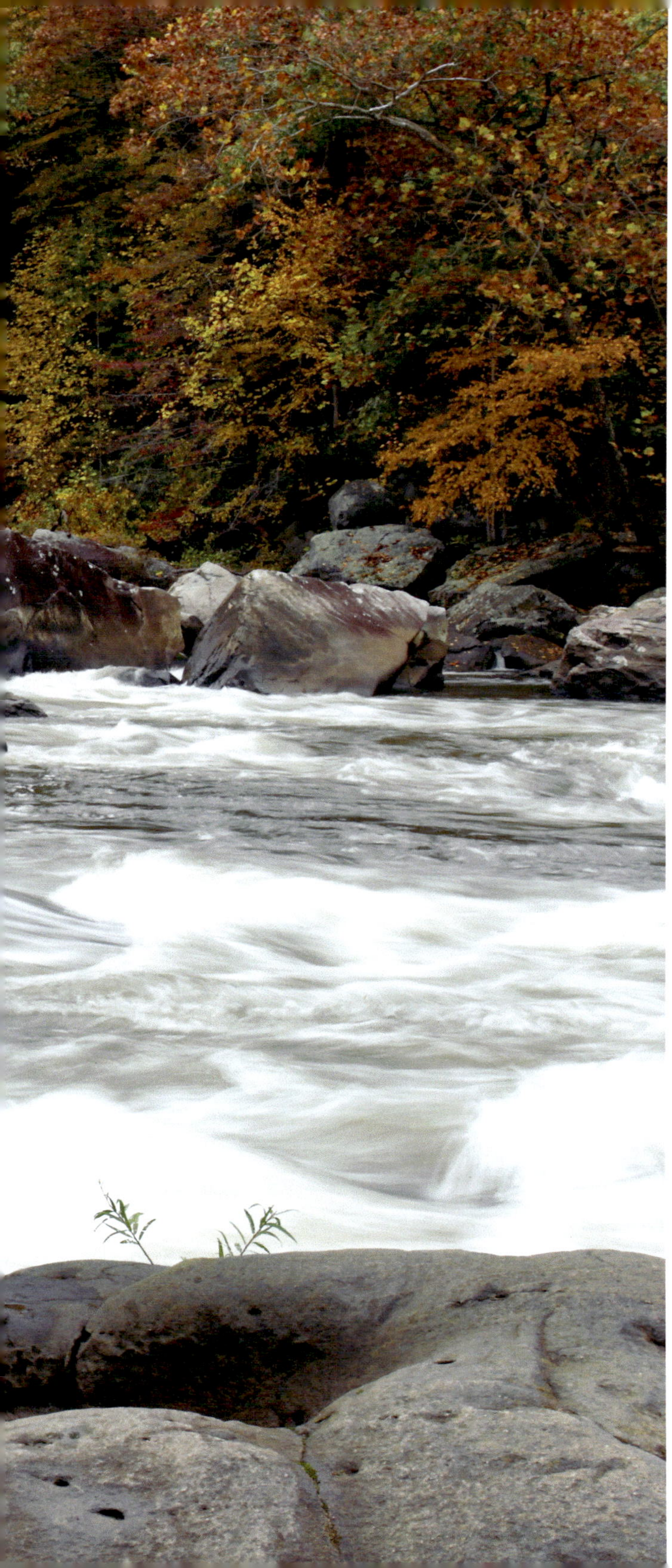

Chapter Four

The Whole River

With eyes on the river, and with attention to the tributaries, the mountains, and the forests, I've photographed the Youghiogheny from top to bottom. This chapter features scenes that impressed me as striking, emblematic, and accessible among unlimited views we can each discover by exploring this river with our eyes wide open and with curiosity about what lies upstream, what awaits downstream, and what will change with the seasons, with the time of day, and with the passage of years and decades to come.

Autumn colors of red oaks, tuliptrees, basswoods, and red maples brighten Railroad Rapid's intricate passage at the lower end of the Youghiogheny's 1.7-mile-long Ohiopyle Loop.

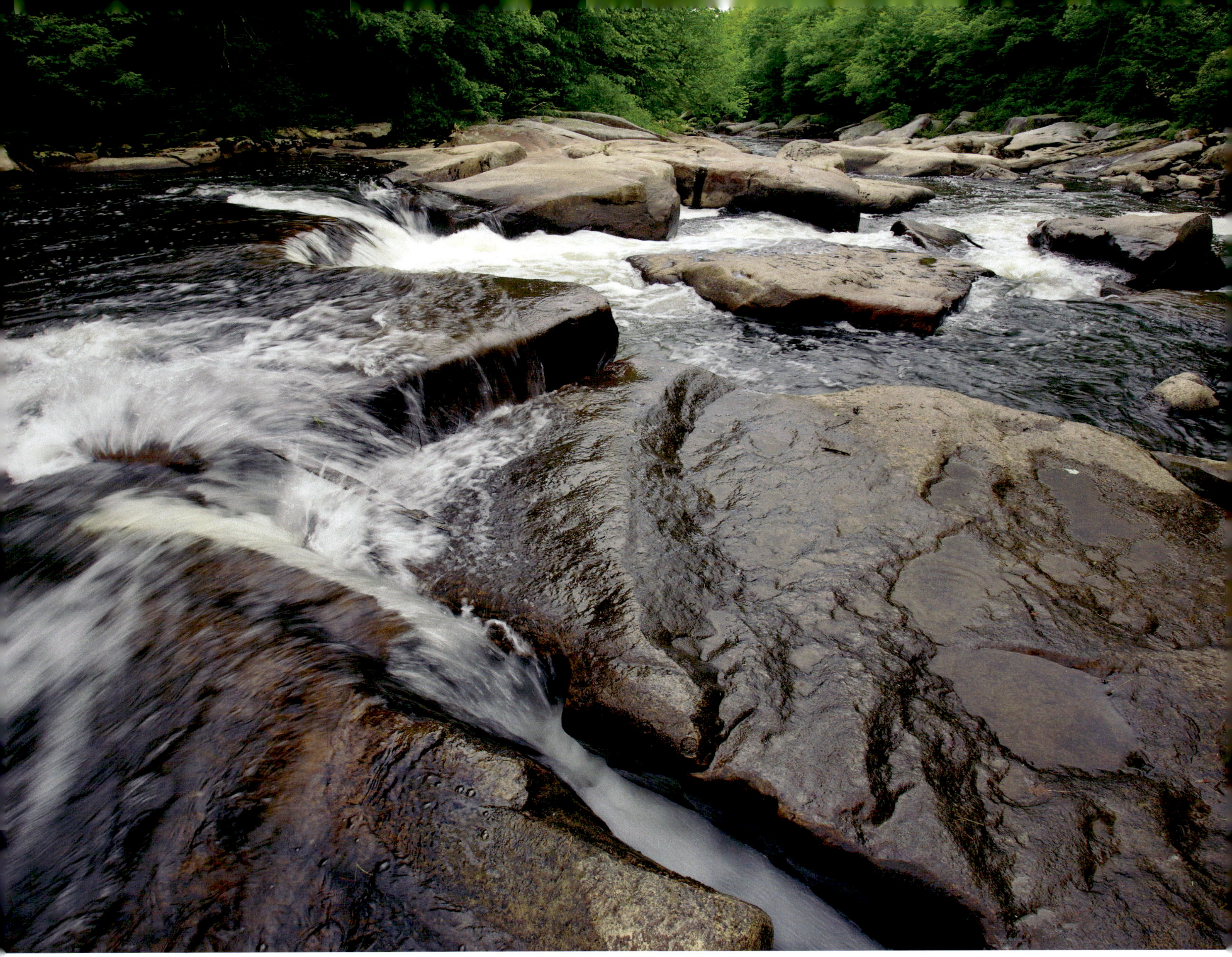

Steep, rock-riddled rapids north of Oakland descend toward their climax in Swallow Falls State Park.

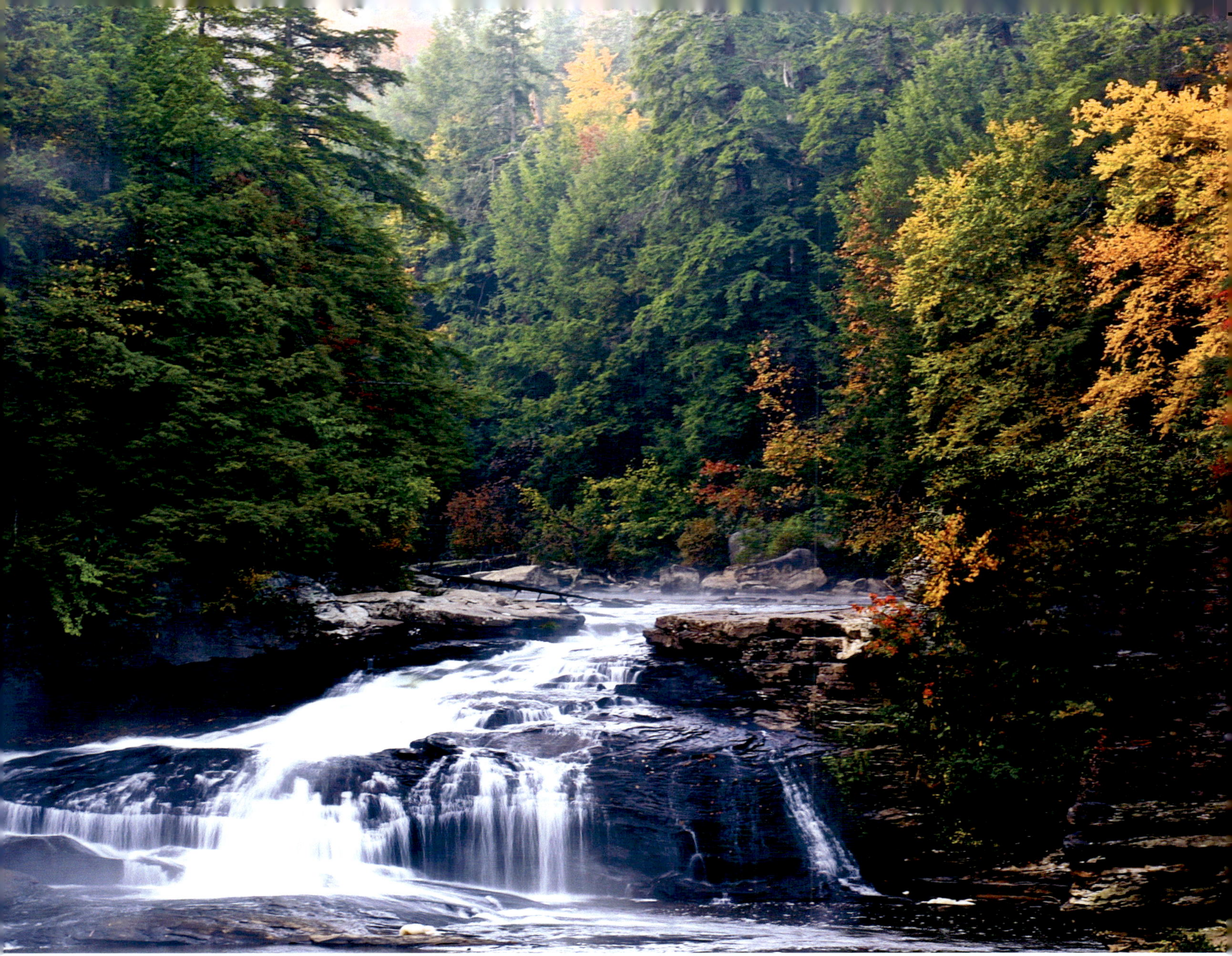

The Youghiogheny's second-steepest drop in gradient forms the centerpiece of Swallow Falls State Park in Maryland.

Below Swallow Falls the Youghiogheny sweeps against undercut boulders.

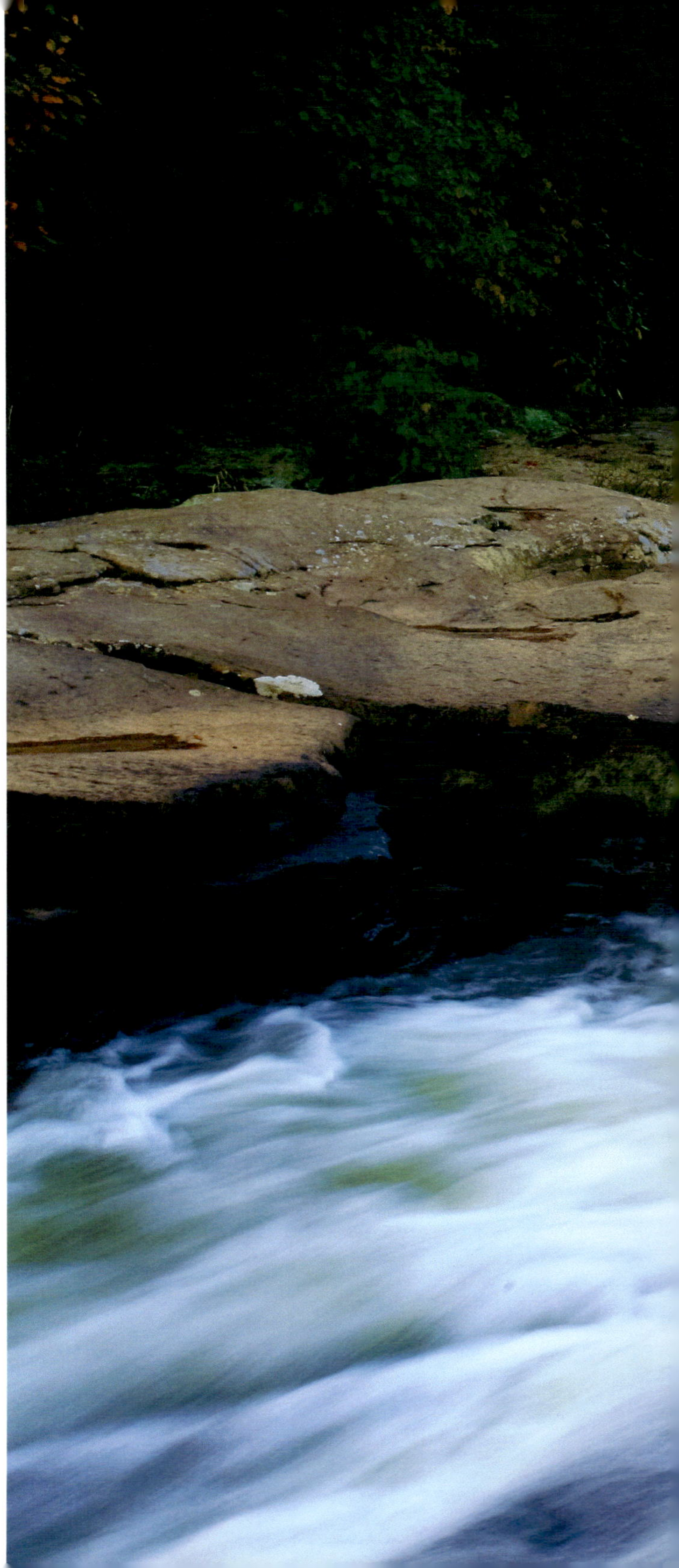

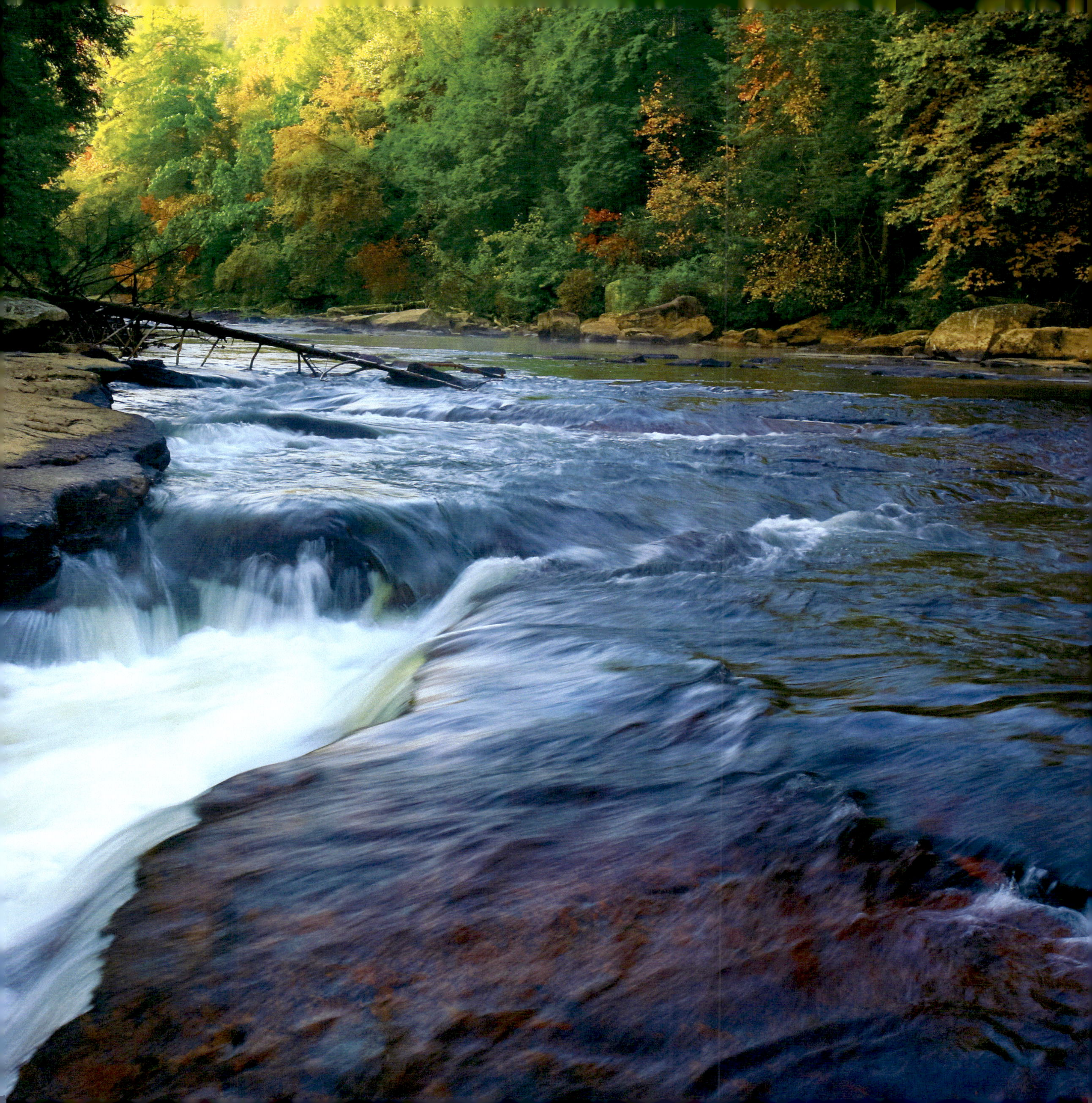

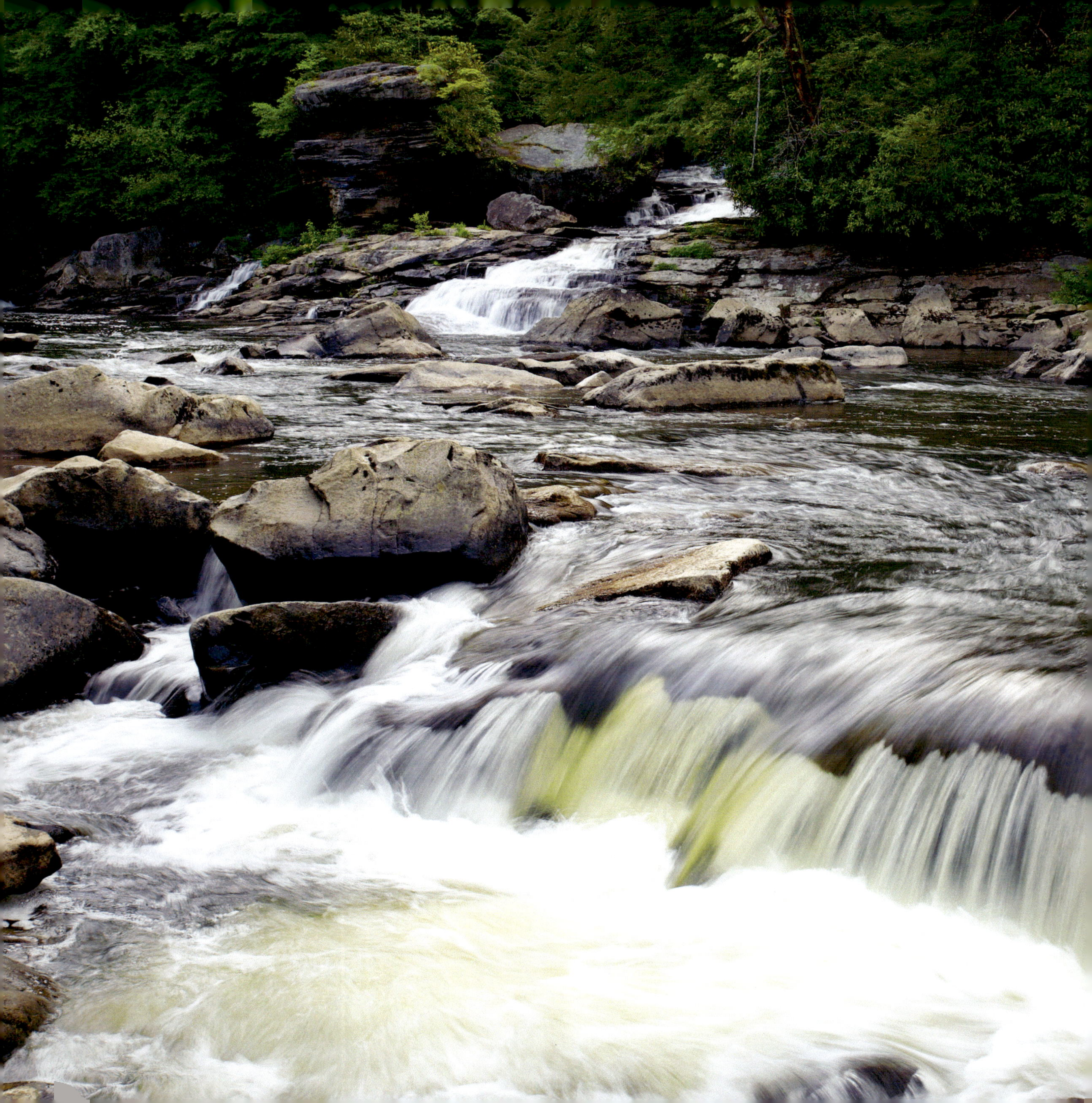

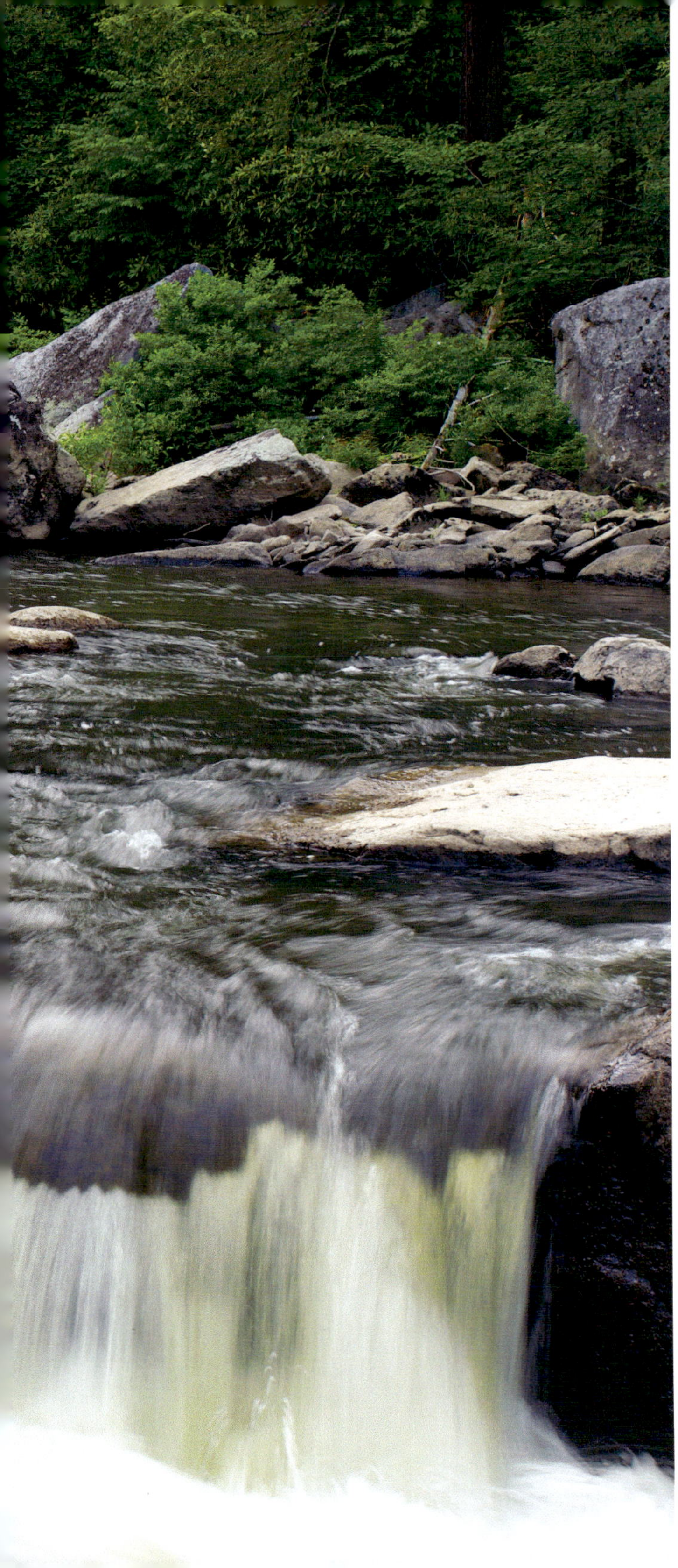

Muddy Creek enters the Youghiogheny in the background, downstream from Swallow Falls, while the river charges through rocky sieves and over ledges of ancient sandstone.

From headwaters on Meadow Mountain, the Casselman River riffles north and westward beneath a historic stone bridge near Grantsville.

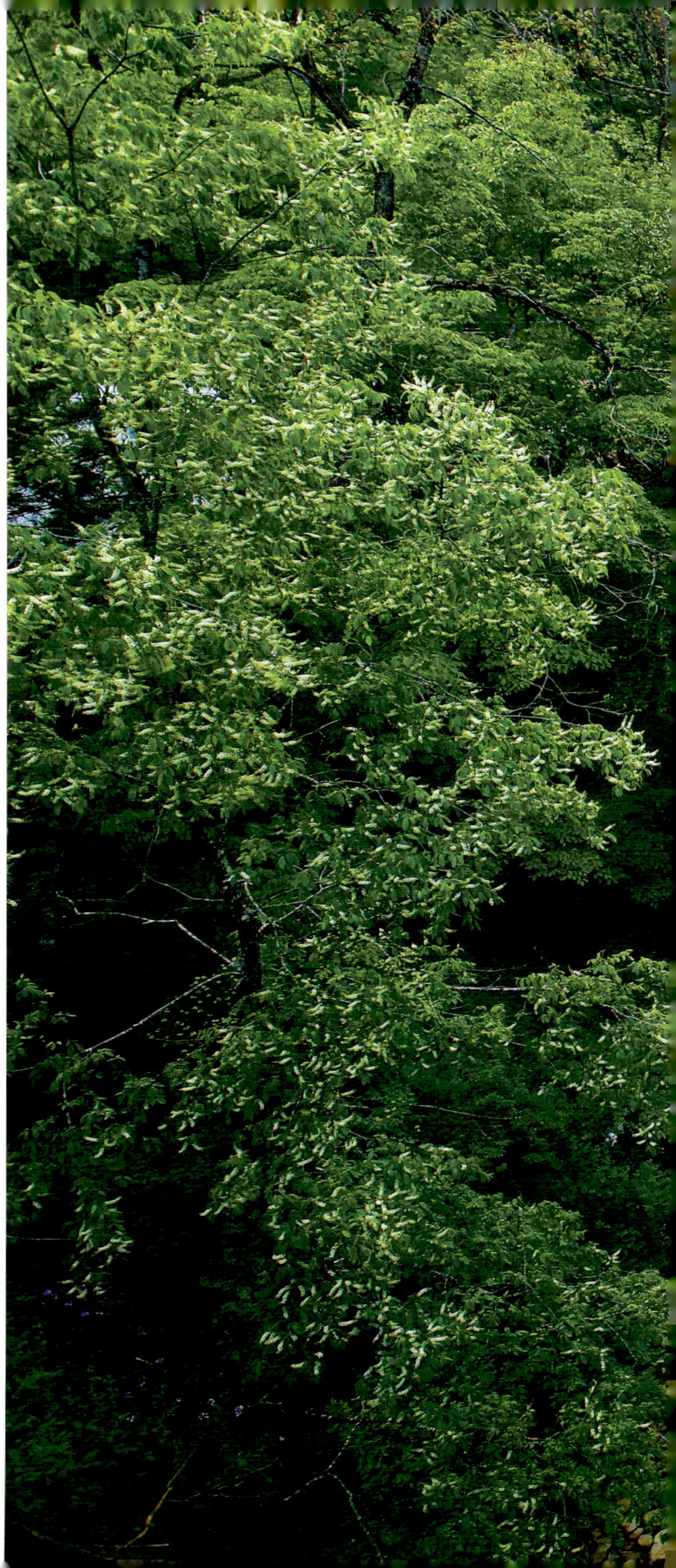

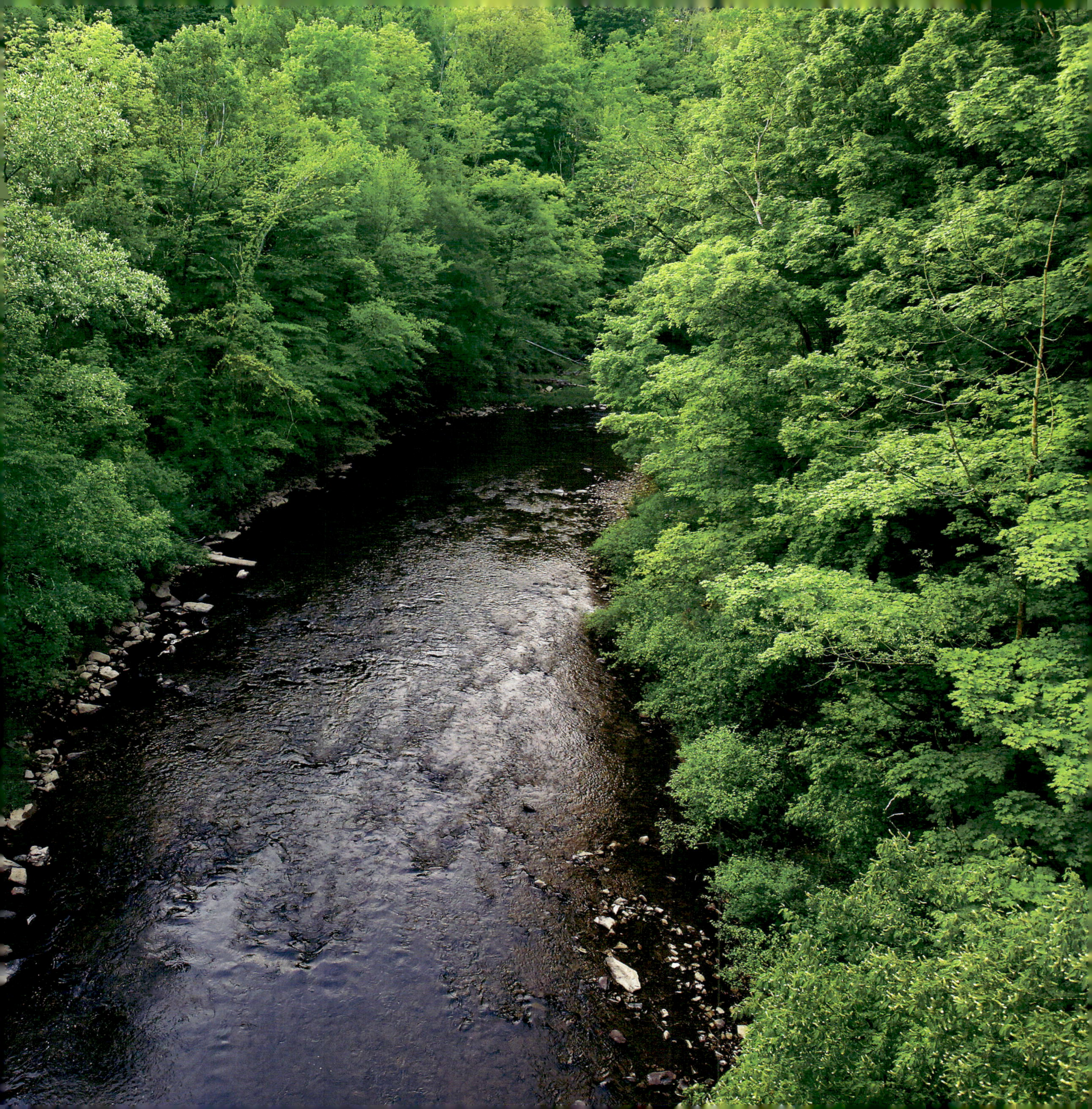

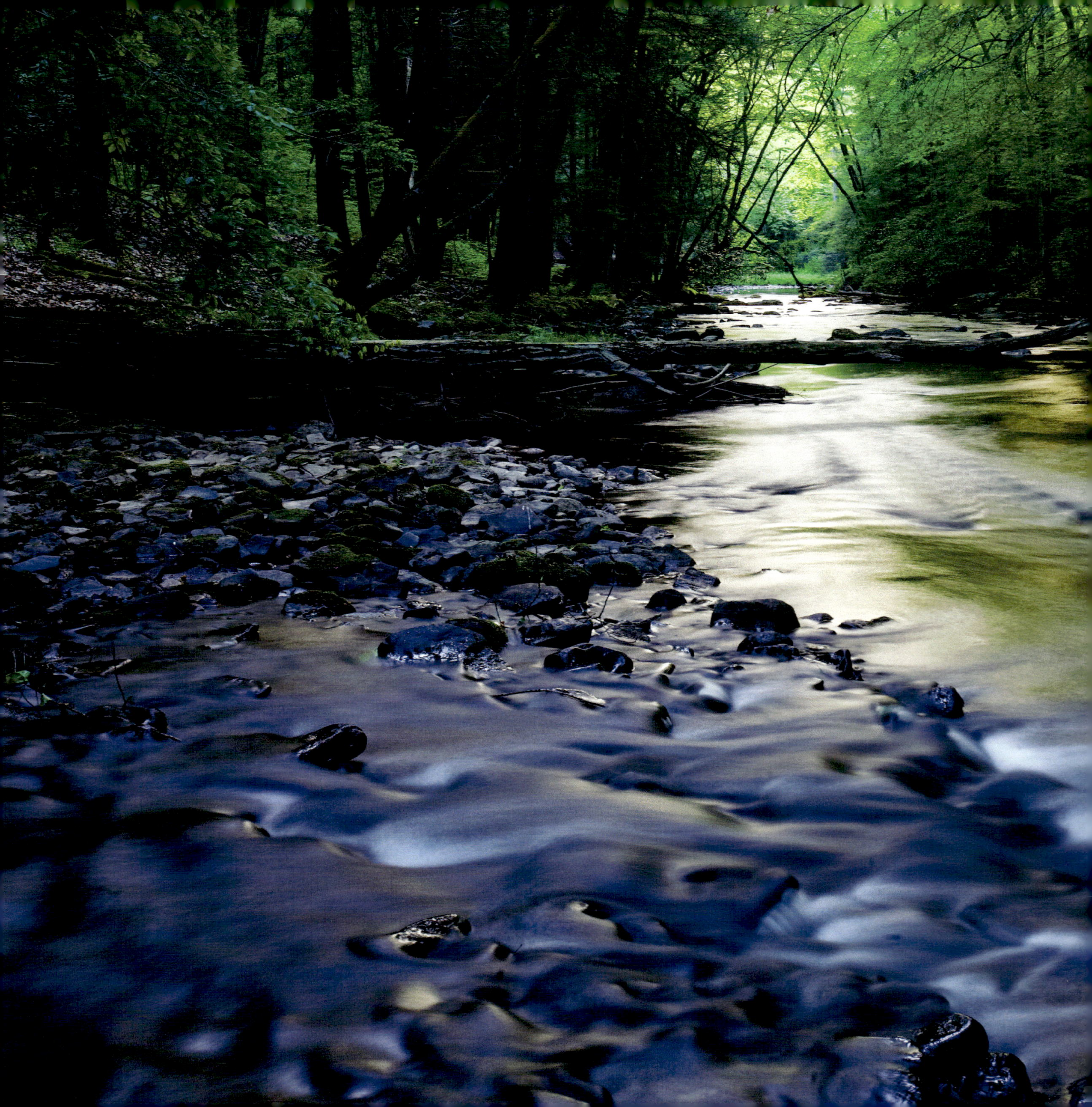

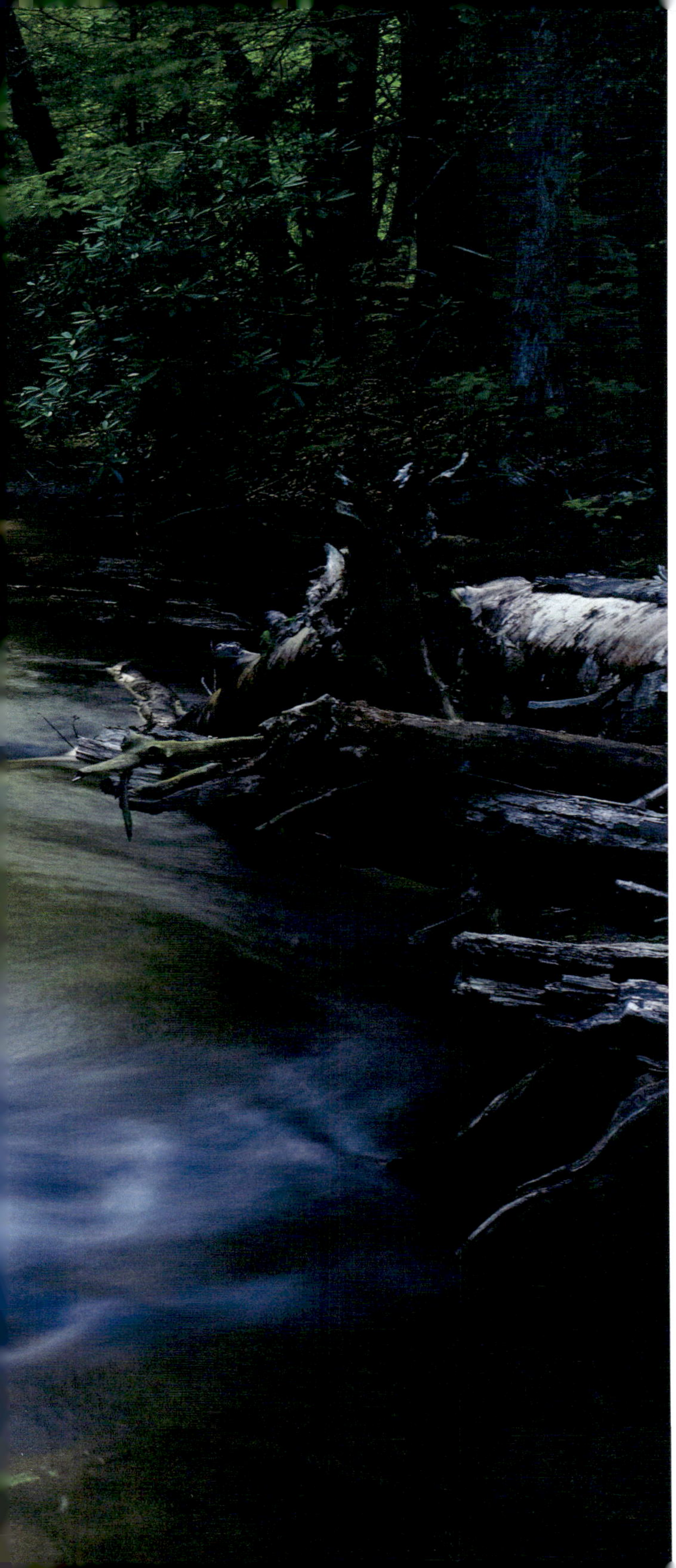

Refreshingly cool waters of the Casselman tributary Whites Creek support a trout fishery in the shaded woodland stream.

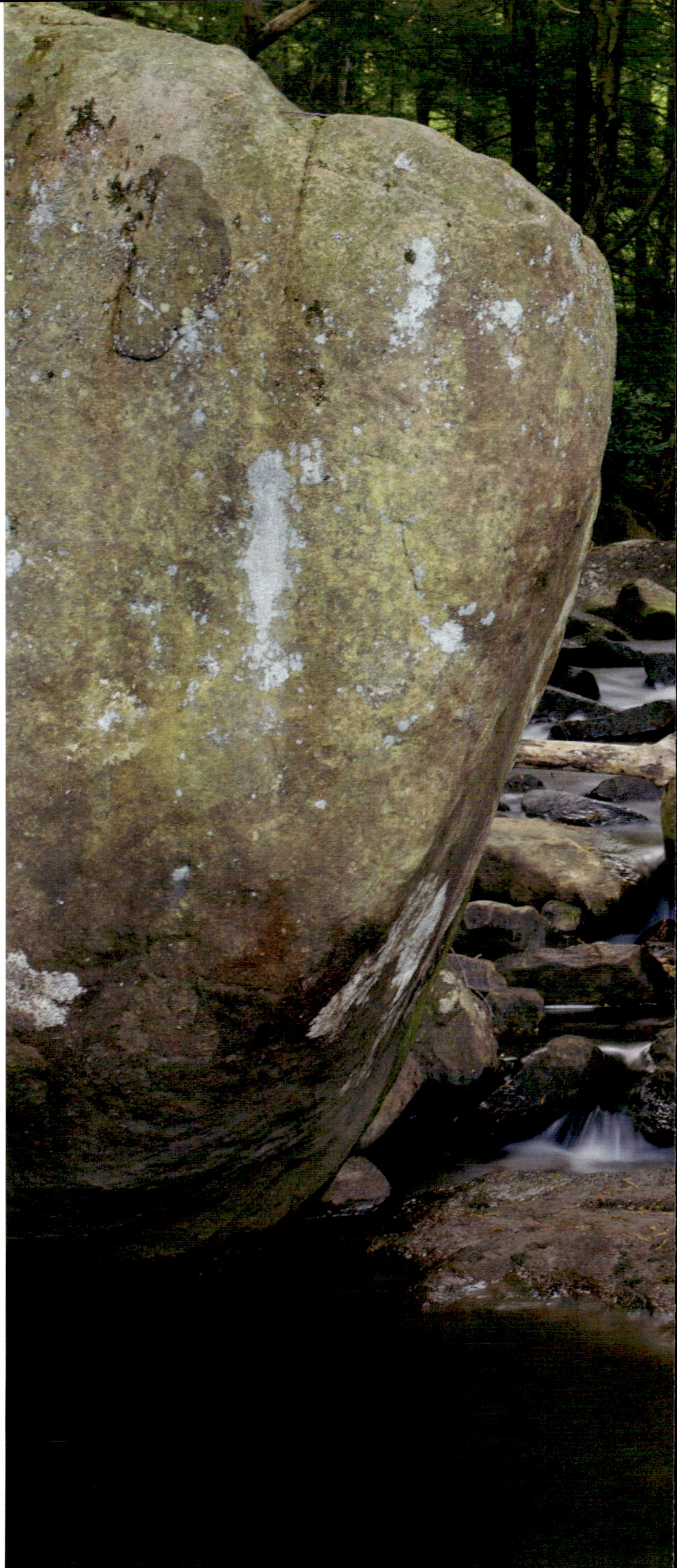

Town Line Creek emerges among boulders from its forested basin before entering the Casselman River at the settlement of Markleton.

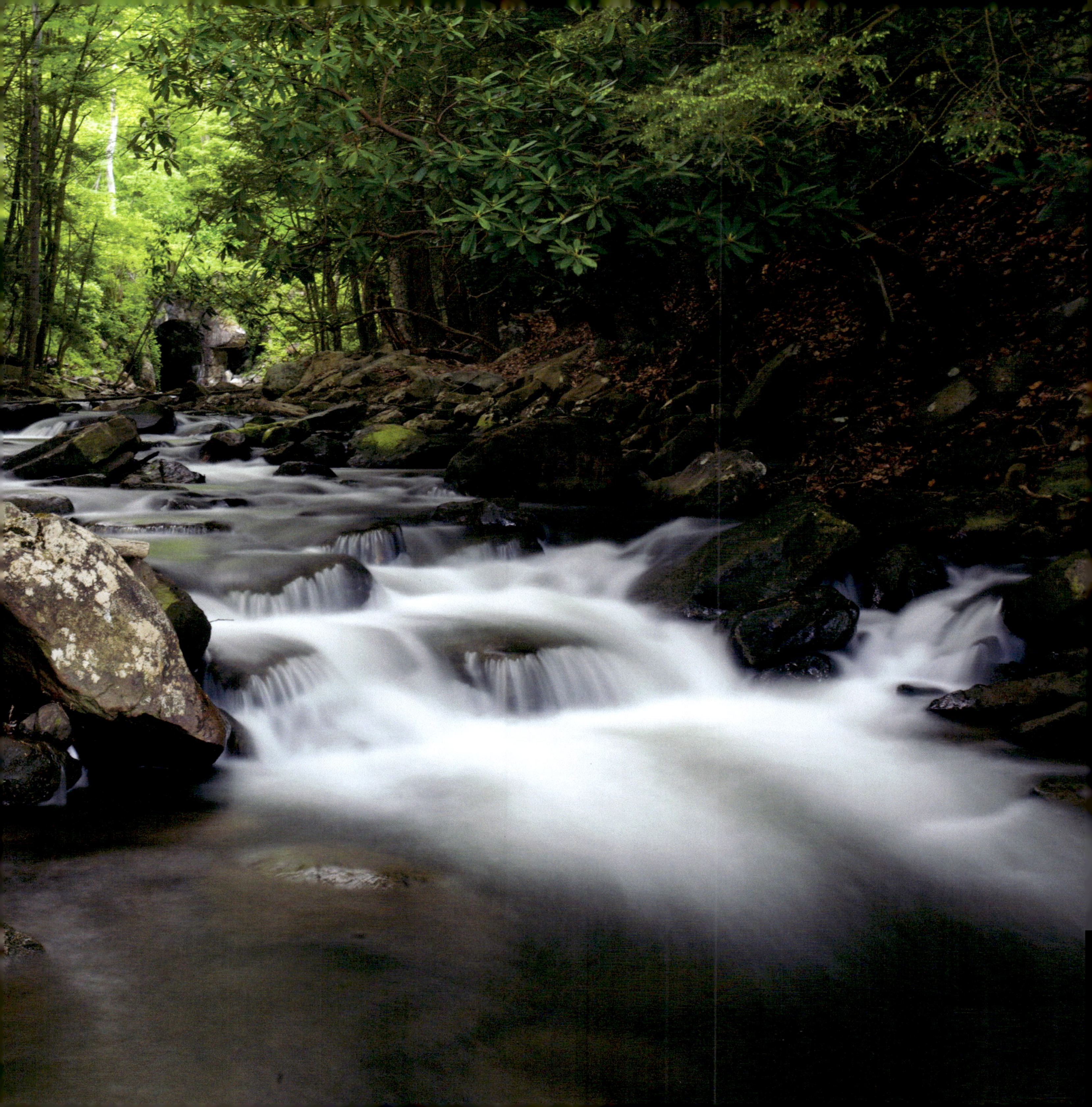

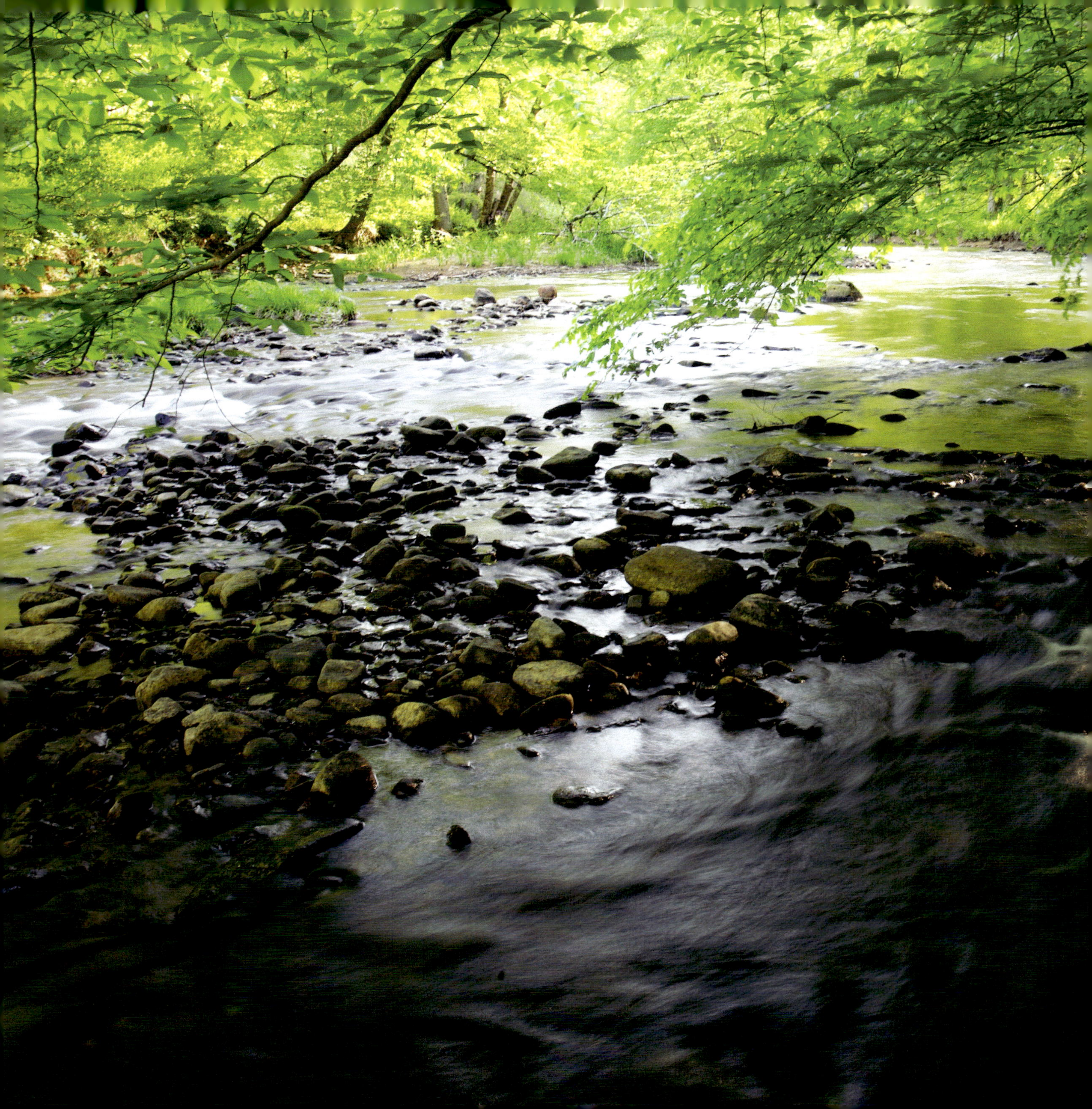

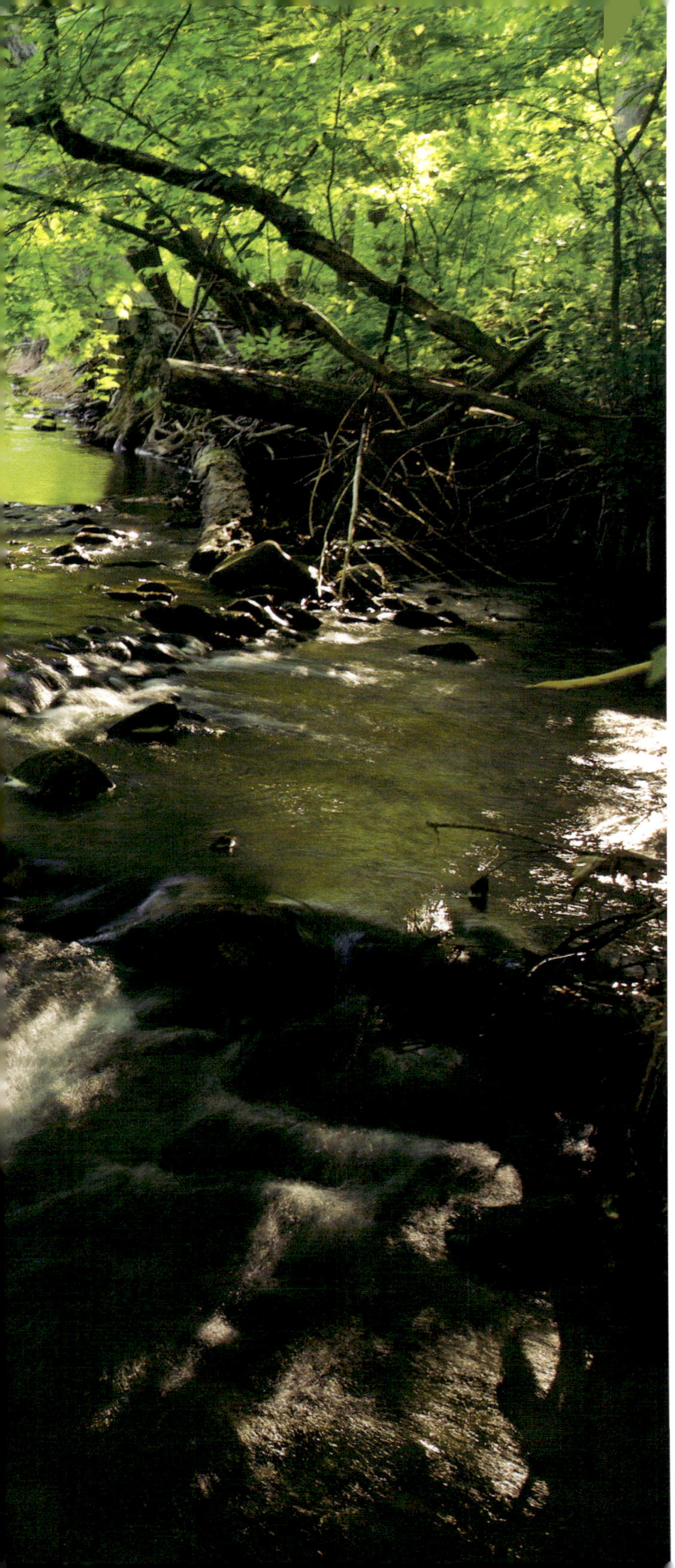

Laurel Hill Creek collects runoff from a broad basin on the east side of Laurel Hill's prominent northeast–southwest-aligned mountaintop. This shining shoal of gravel has settled upstream of the town of Confluence at State Game Lands, which complement the Youghiogheny basin's state parks and state forests as public property where all people are welcome to walk, appreciate nature, and fish.

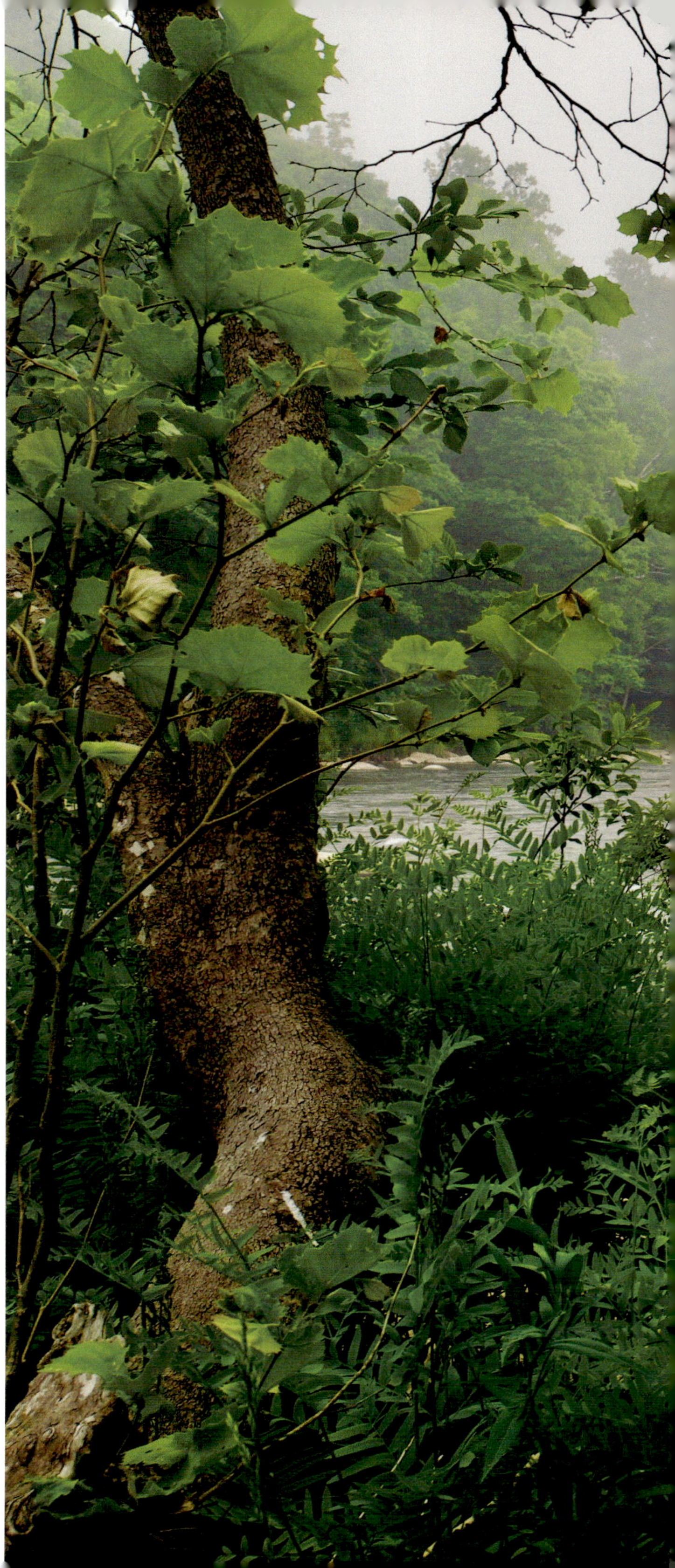

Mist clings to the Youghiogheny valley on a cool summer morning at Paddler's Lane campground and river resort downstream from Confluence.

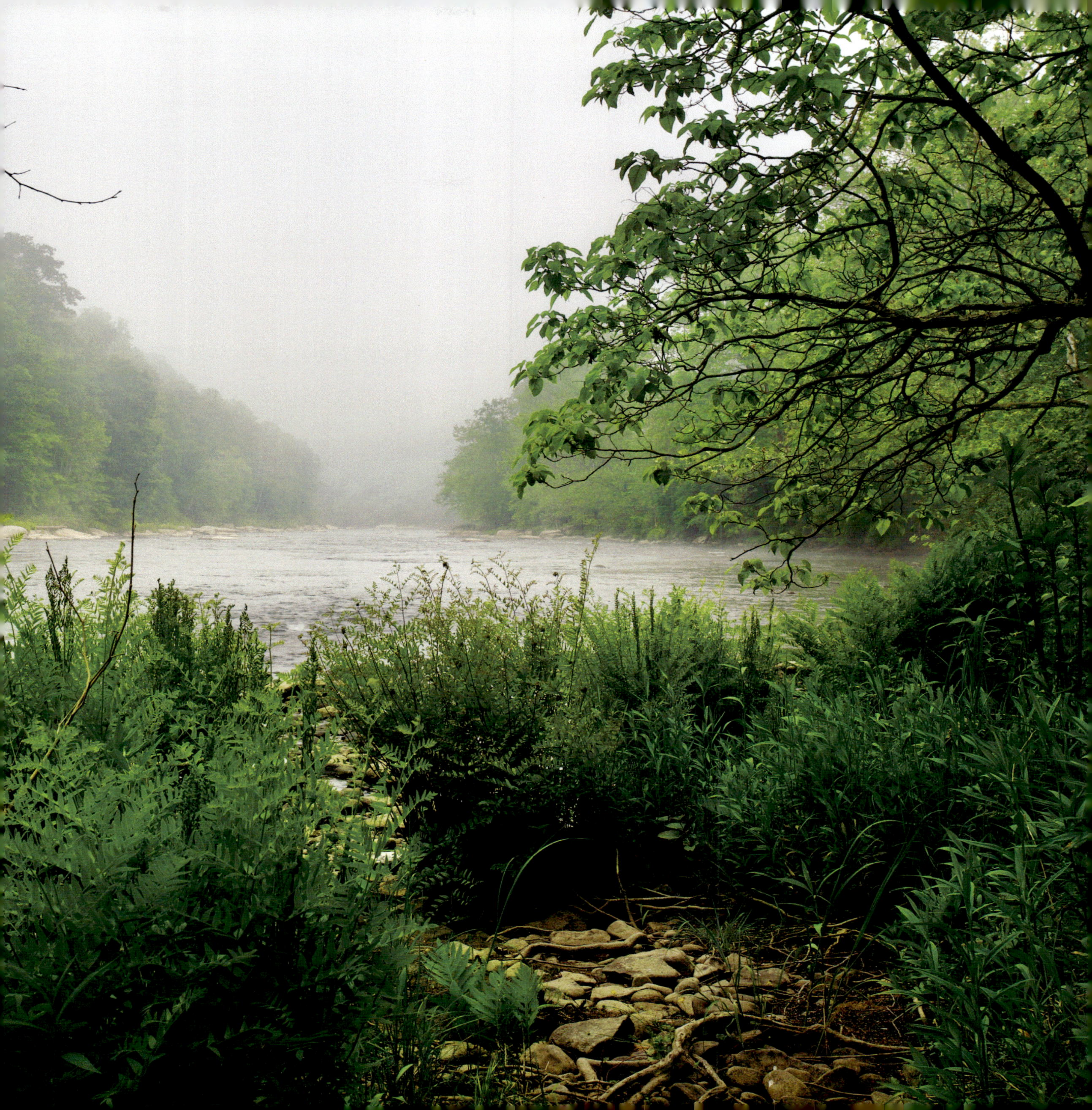

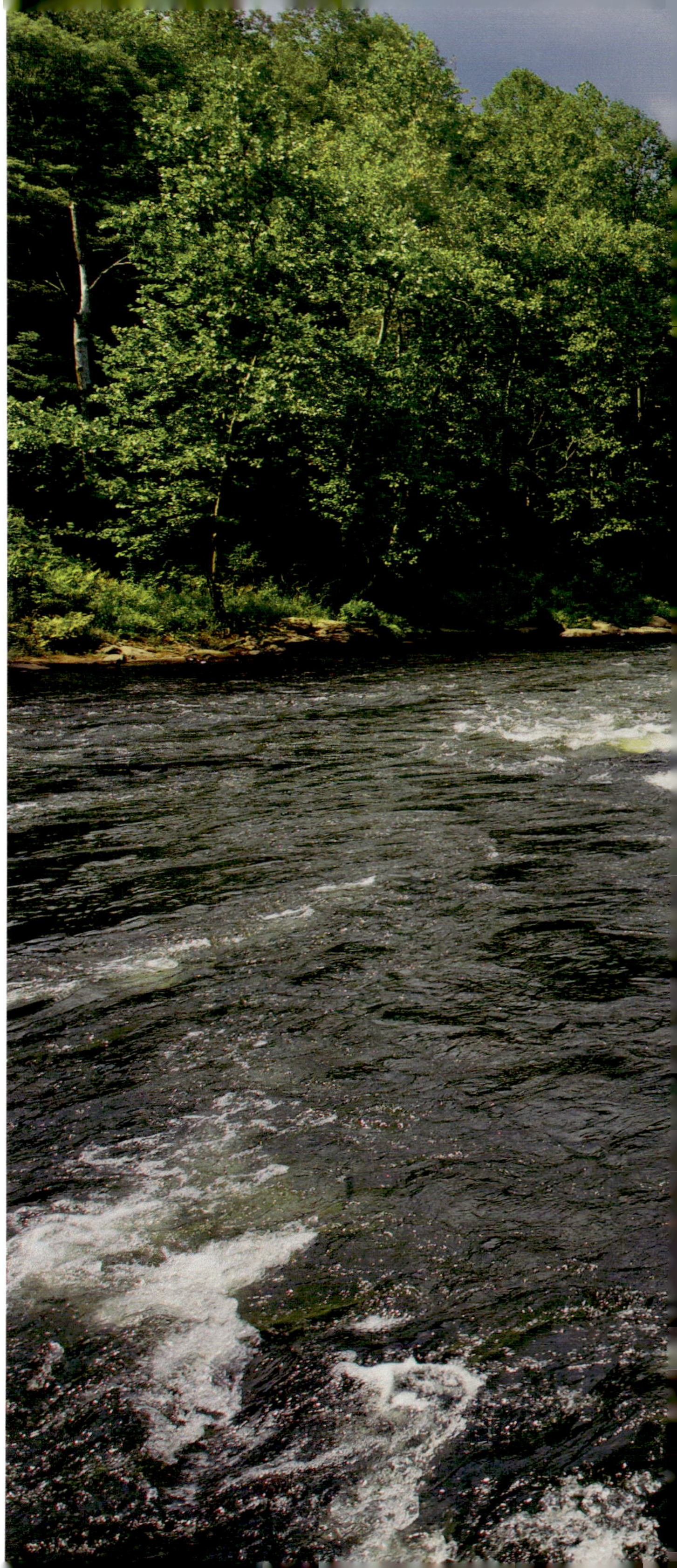

The Middle Youghiogheny's 12 miles of Class 1 and 2 riffles and rapids appeal as excellent paddling water for novice rafters, beginning kayakers, and capable canoeists.

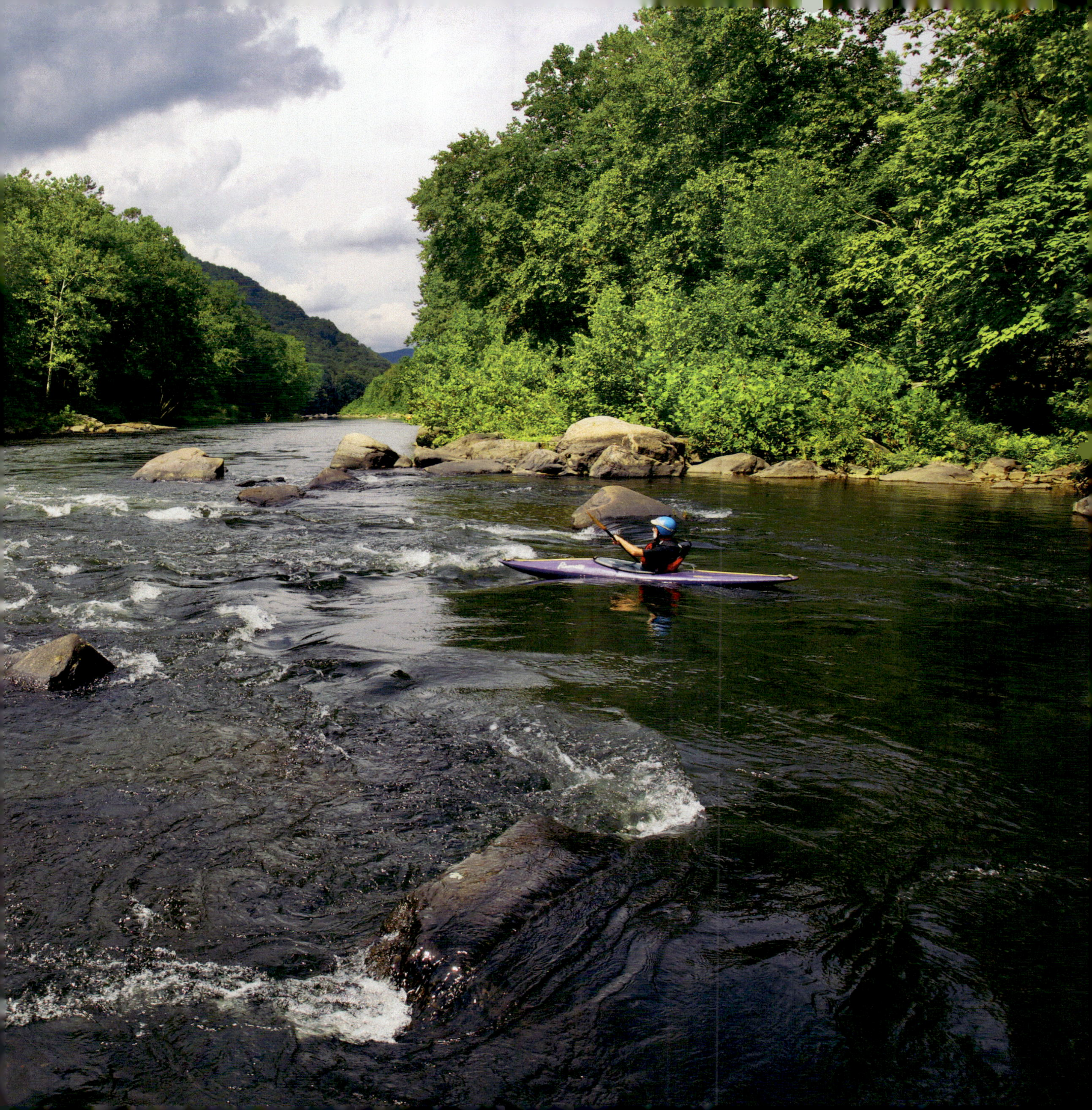

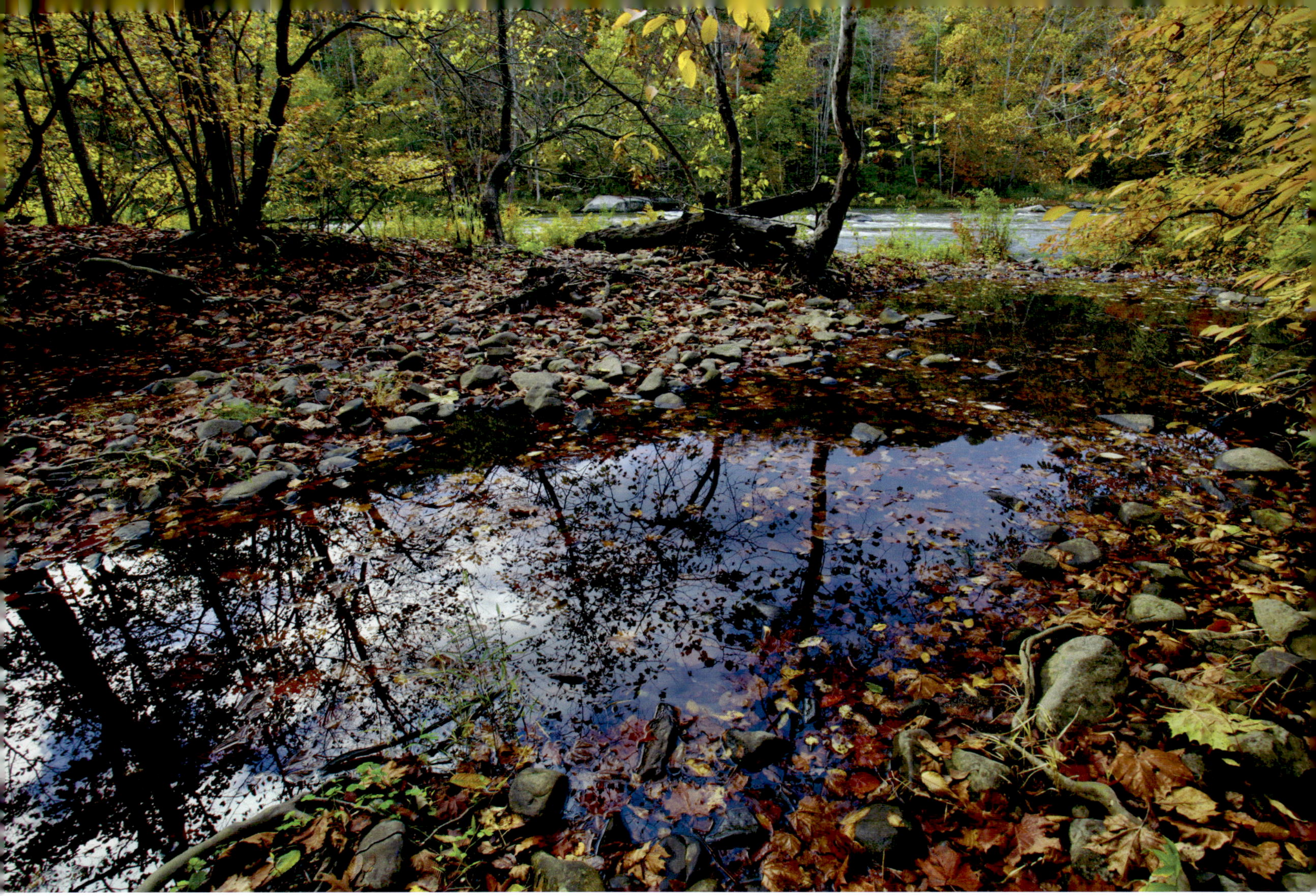

ABOVE

Backwater sloughs mirror autumn's sweet birch and sycamore trees rising above the Middle Yough at Paddler's Lane campground and resort.

OPPOSITE PAGE

The Middle Youghiogheny cuts its deep gap through the Appalachian landmark of Laurel Hill while a forest of red oak, red maple, beech, and sweet birch grows greener when the weather warms in May.

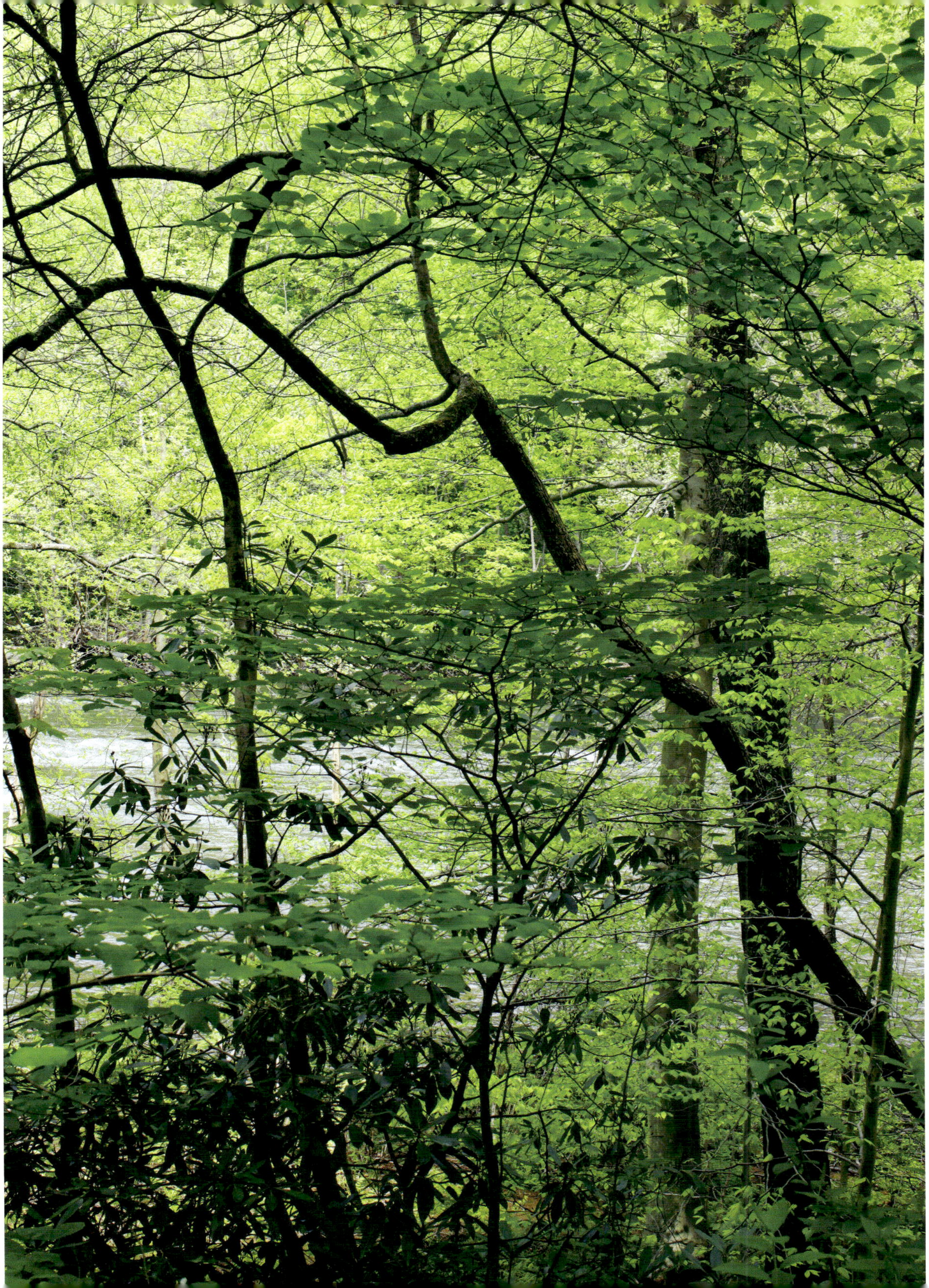

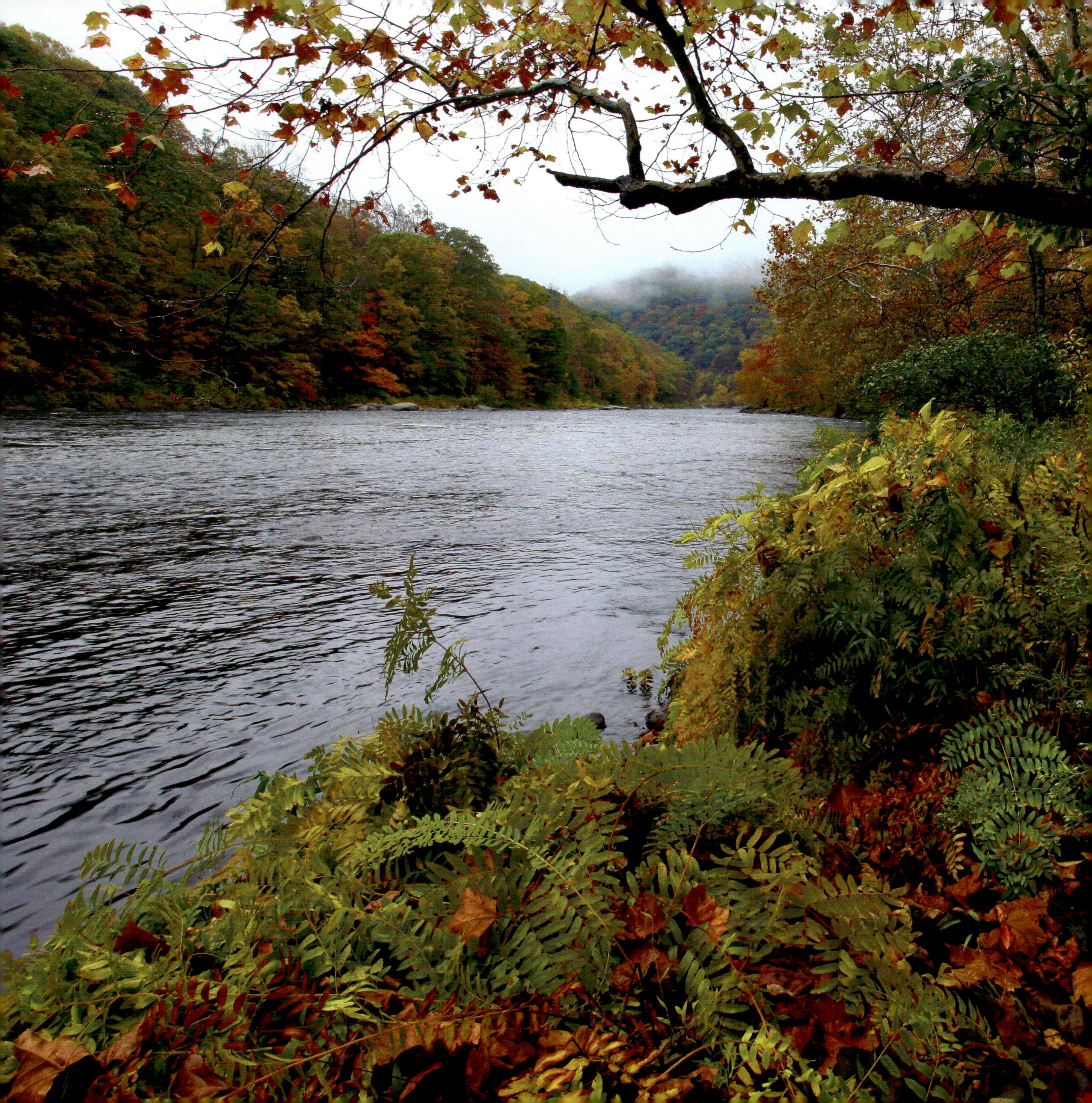

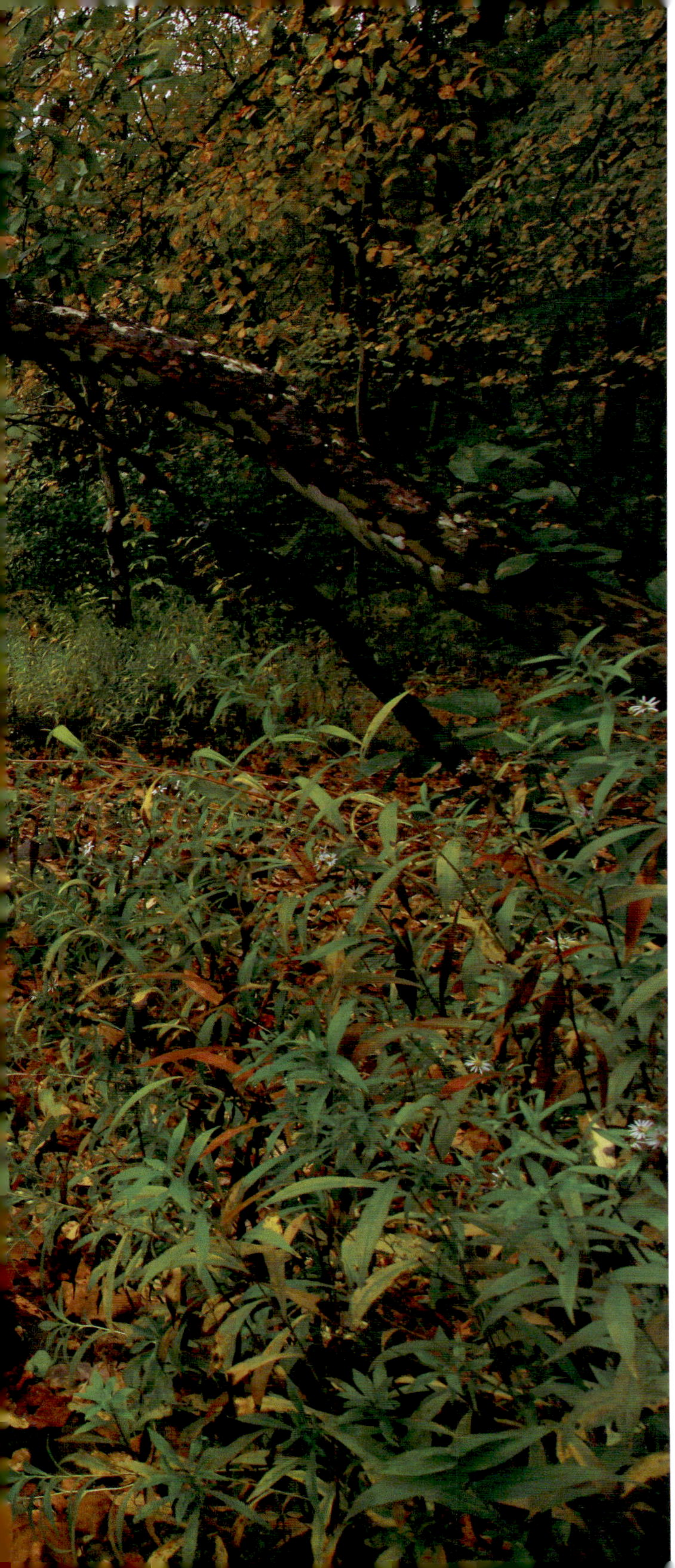

Smooth-moving water alternates with moderate rapids through the 8-mile paddling run between Ramcat access and the boaters' takeout at the upstream end of Ohiopyle.

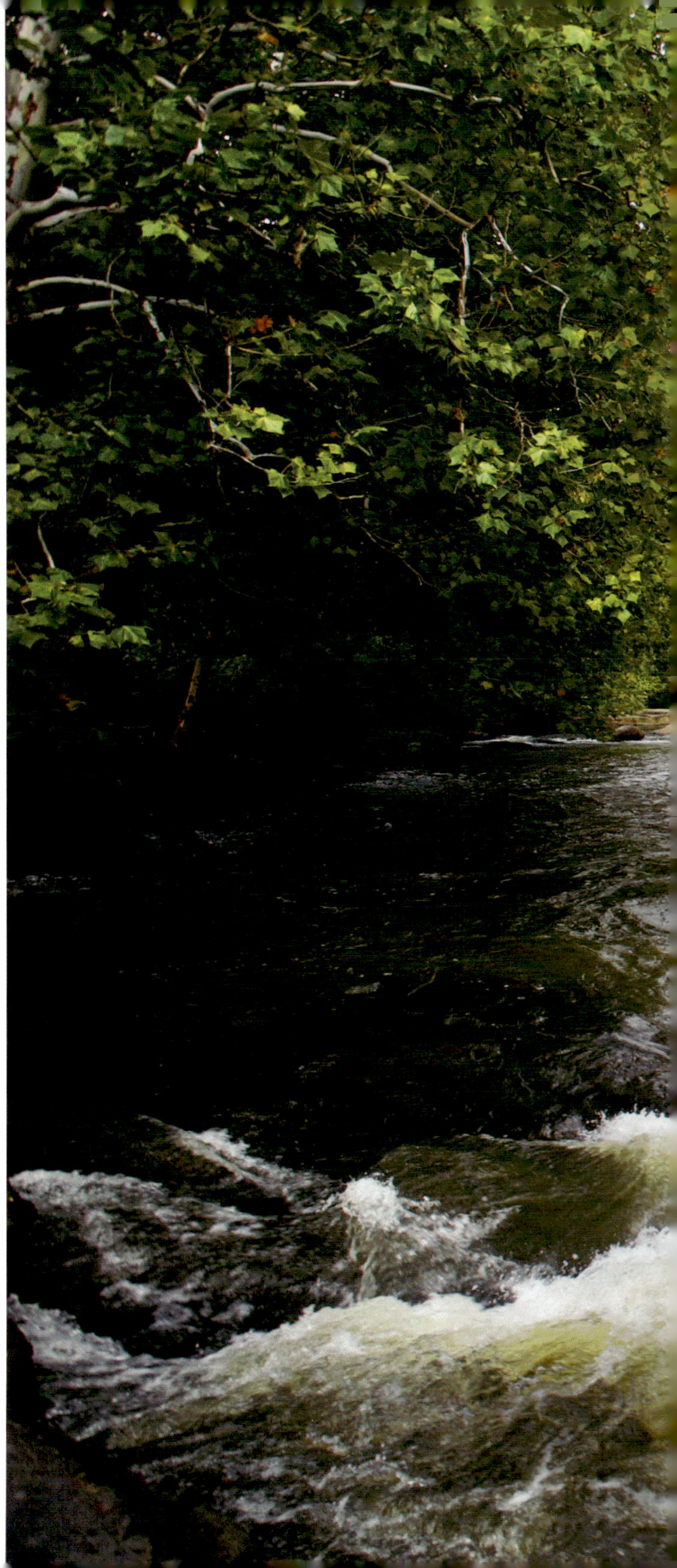

An intricate rapid, descriptively called Z, lies below the Ohiopyle State Park takeout for the Middle Yough and can be seen just upstream from the Highway 381 bridge entering Ohiopyle.

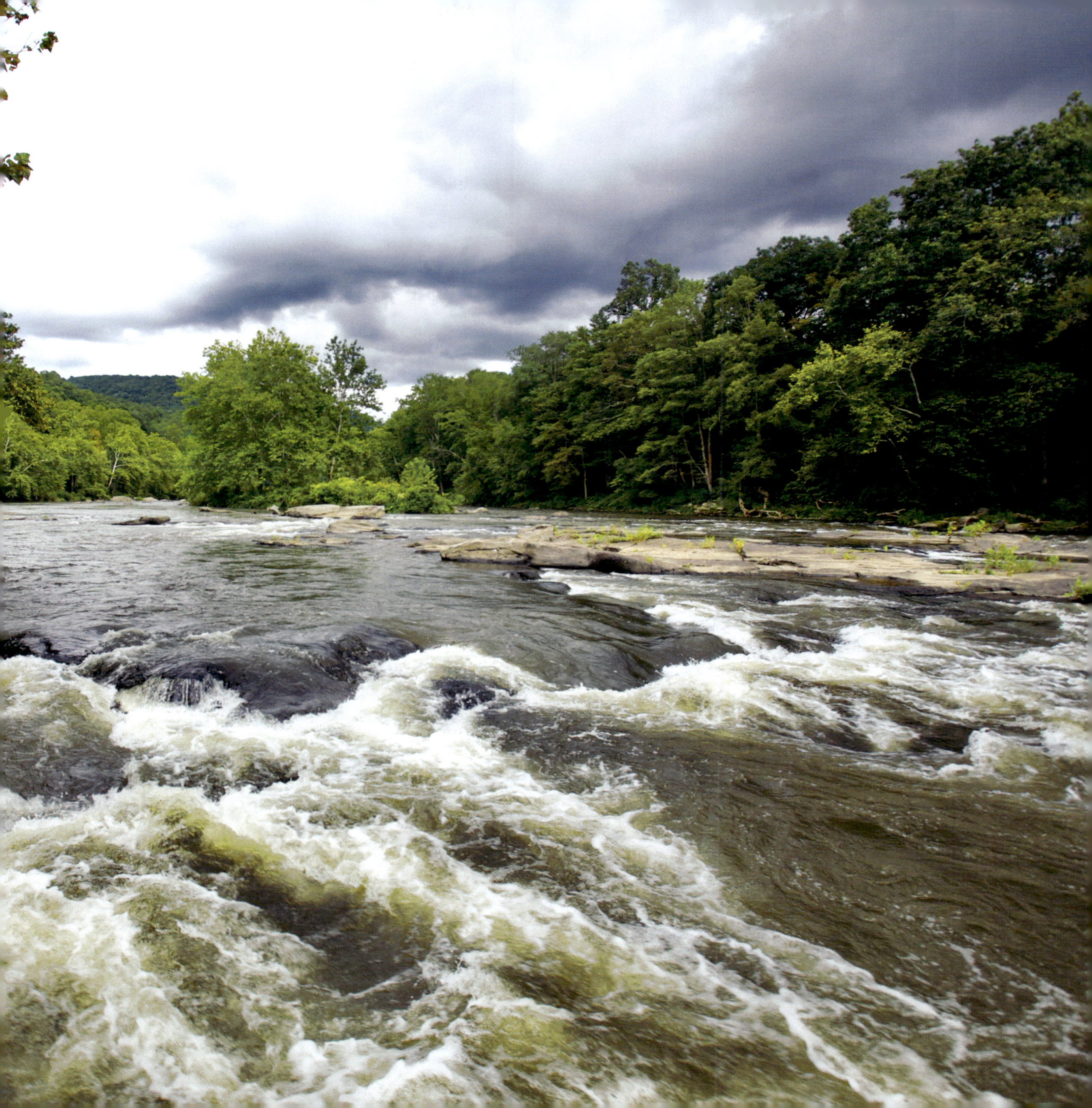

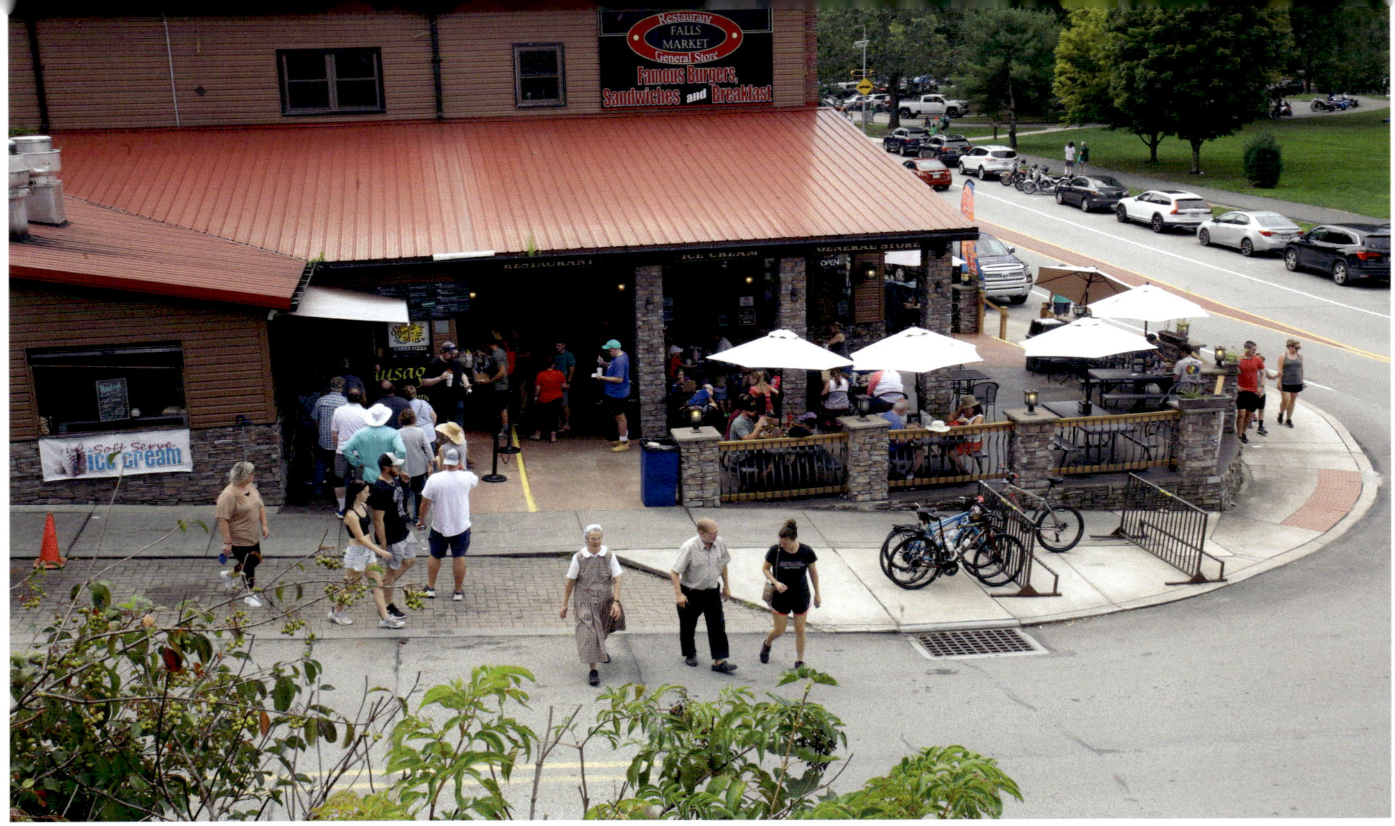
Restaurant
FALLS
MARKET
General Store
Famous Burgers,
Sandwiches and Breakfast
Soft Serve
Ice Cream

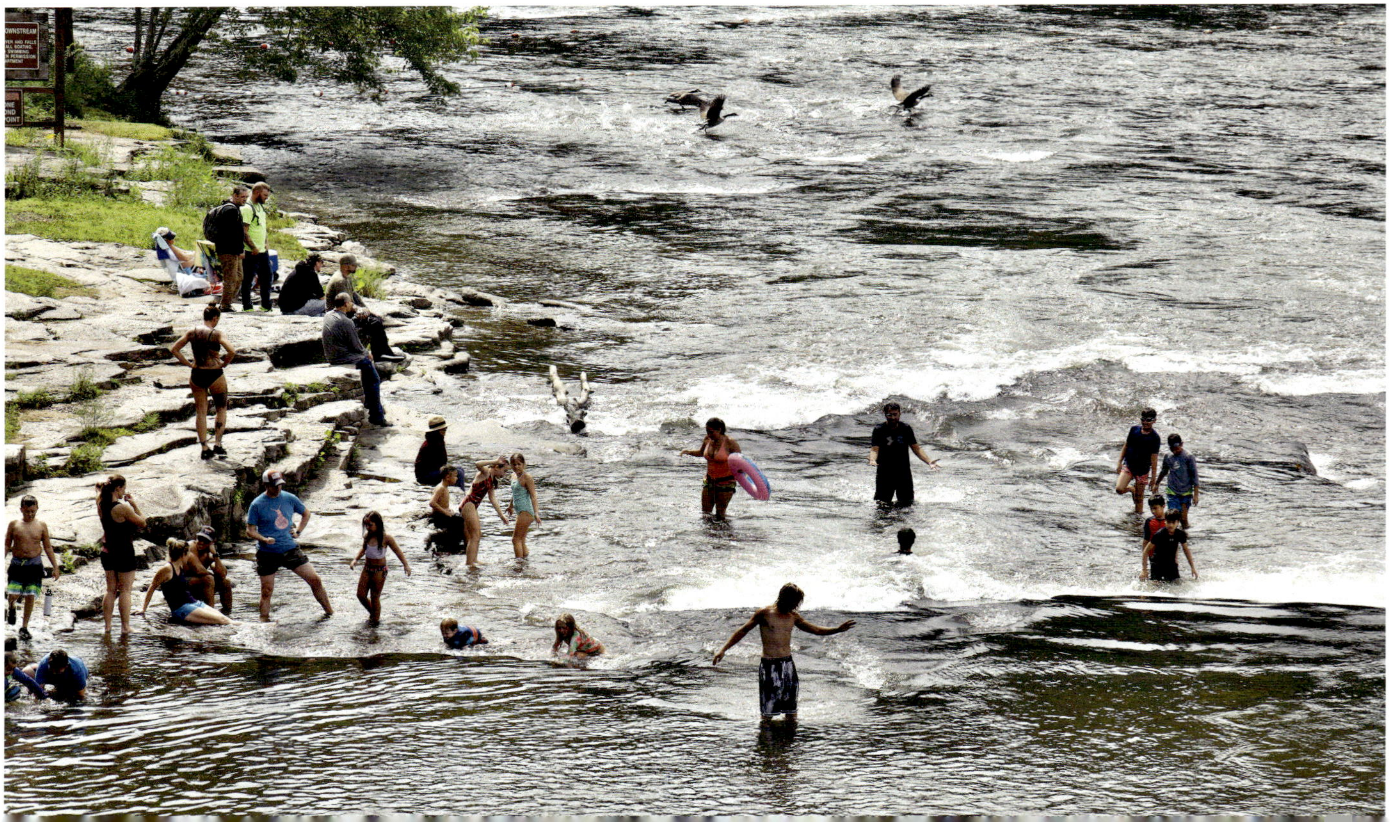

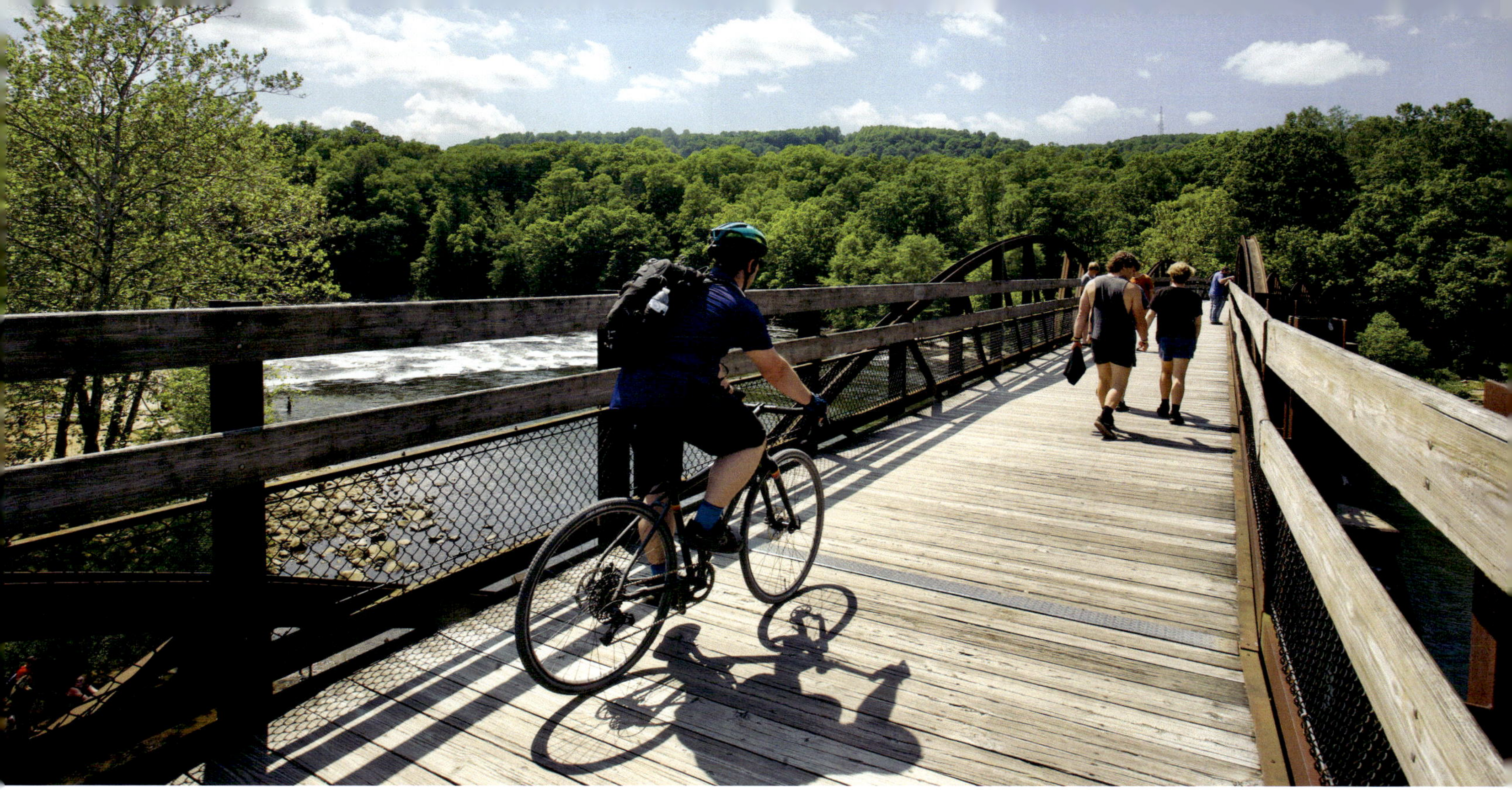

OPPOSITE PAGE, TOP

The quiet Appalachian town of Ohiopyle has been transformed into one of Pennsylvania's most popular state parks. Holt's Department Store, at this site, was converted to Falls Market and later facelifted with outdoor dining, serving crowds of visitors who come to see the Youghiogheny.

OPPOSITE PAGE, BOTTOM

Ohiopyle State Park evolved from the Western Pennsylvania Conservancy's initial goal of protecting the Ferncliff Peninsula to become Pennsylvania's largest park in land area. On hot summer days, visitors come to paddle, to bicycle, and to wade in the river's irresistible pools and riffles.

ABOVE

Conversion of Western Maryland Railroad's right-of-way to a trail has led to 126,000 bicyclists peddling through Ohiopyle State Park in 2022. Many more enjoy additional sections of the 150-mile-long Great Allegheny Passage trail between Pittsburgh and Cumberland.

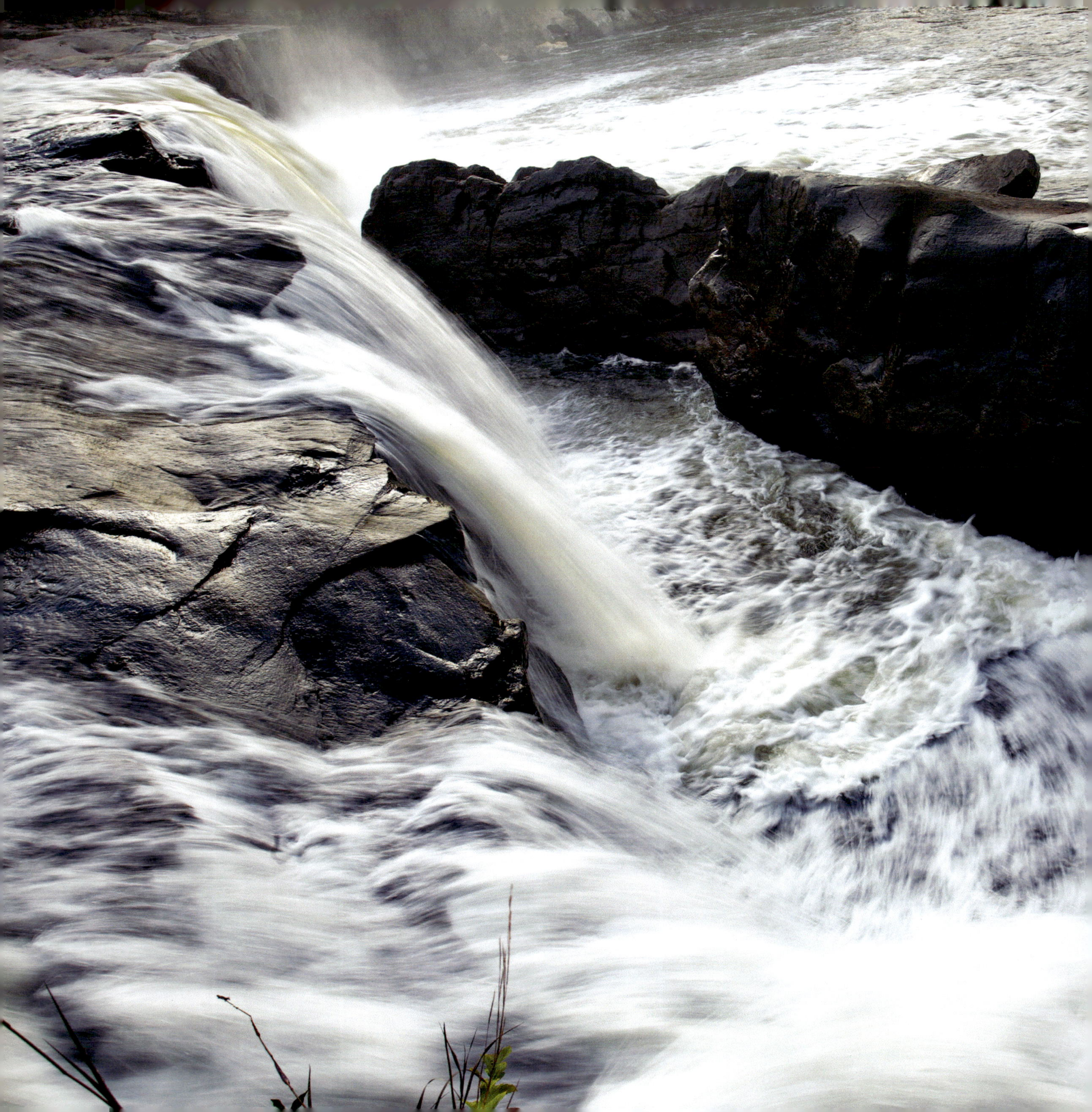

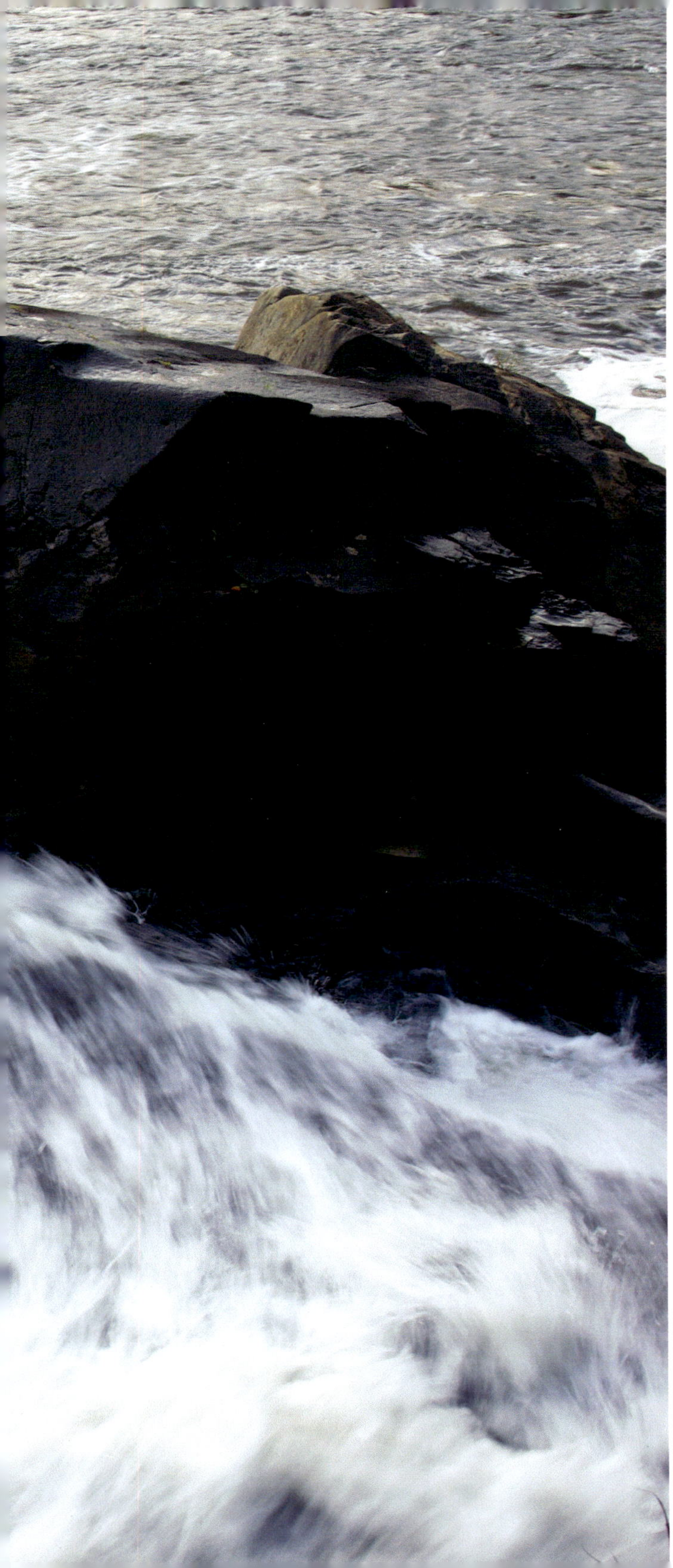

The Youghiogheny's landmark waterfall has been the main attraction at Ohiopyle ever since the town's earliest days of recreation in the 1800s. On the river-right side (opposite the State Park Visitor Center), large blocks of sandstone have broken off from the eighteen-foot precipice.

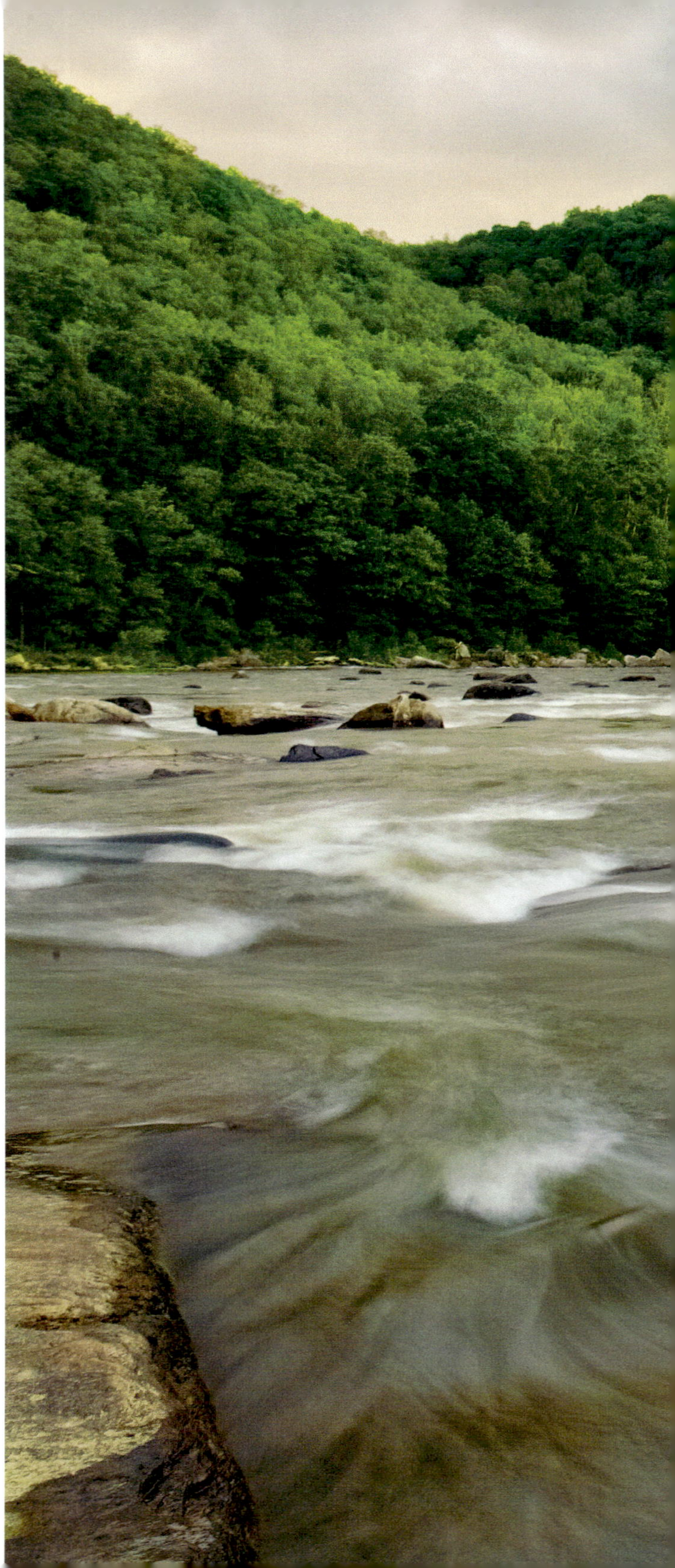

Below the Ohiopyle Falls, boulders and ledges of sandstone mark the top of Entrance Rapid.

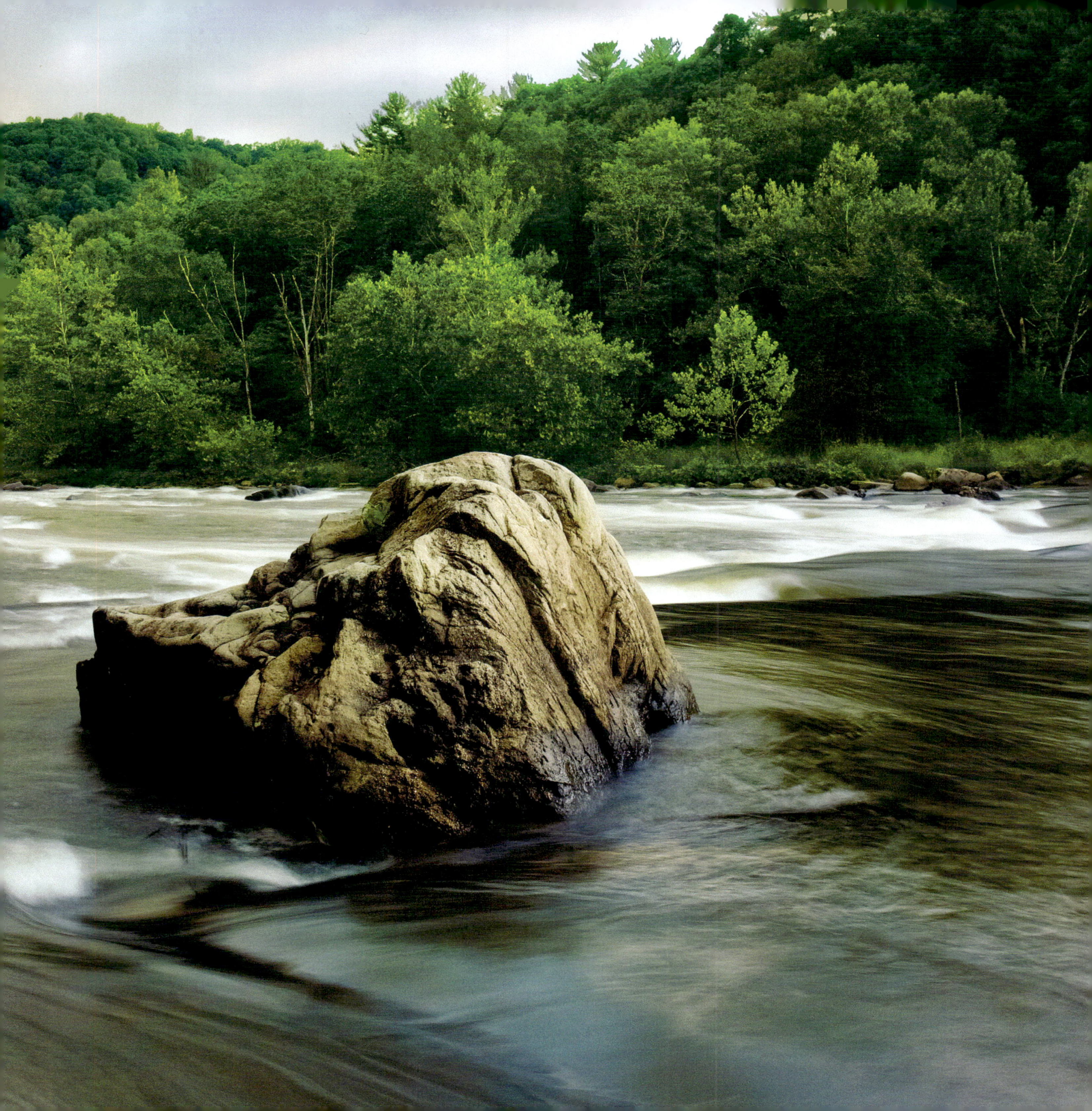

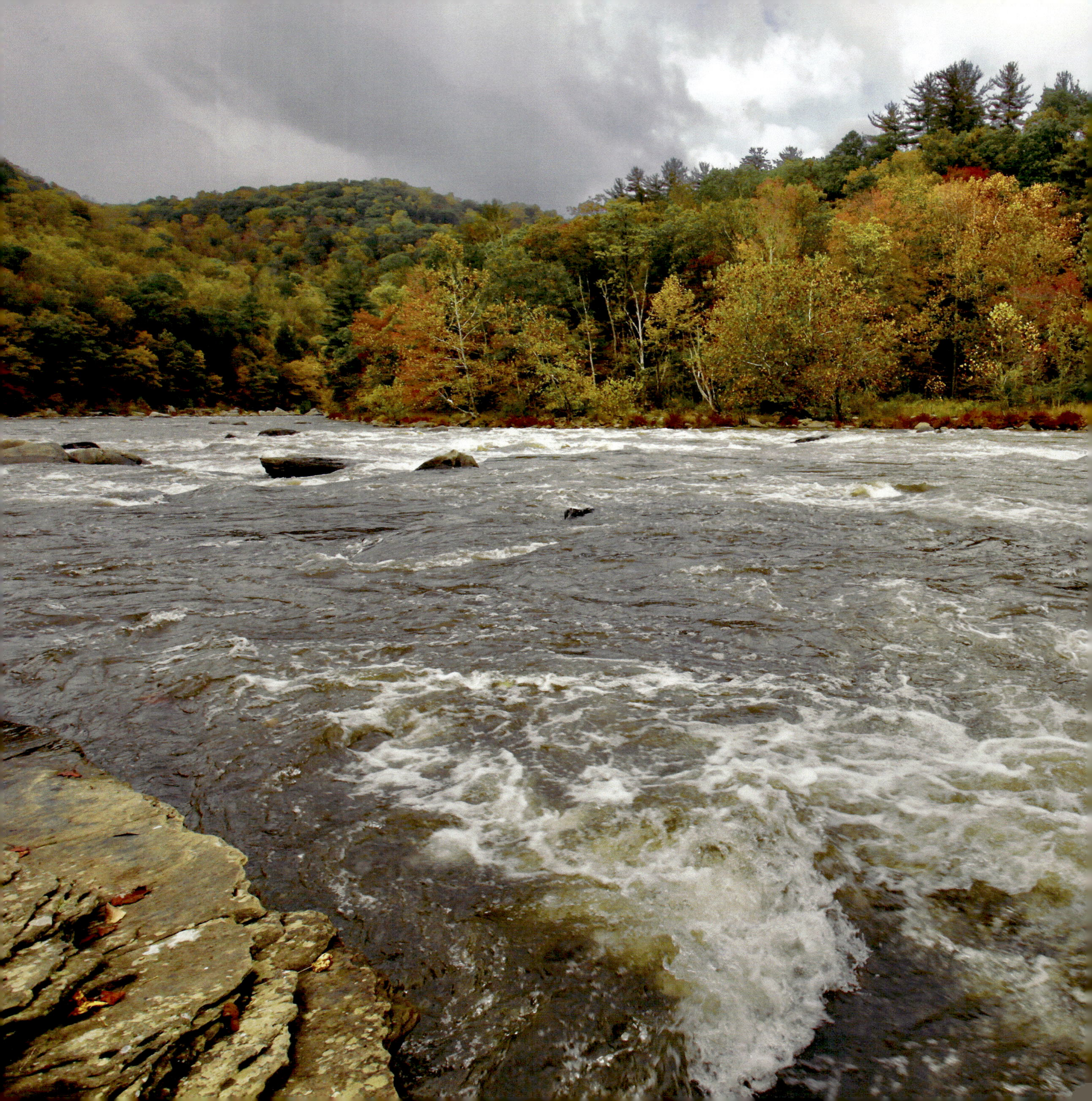

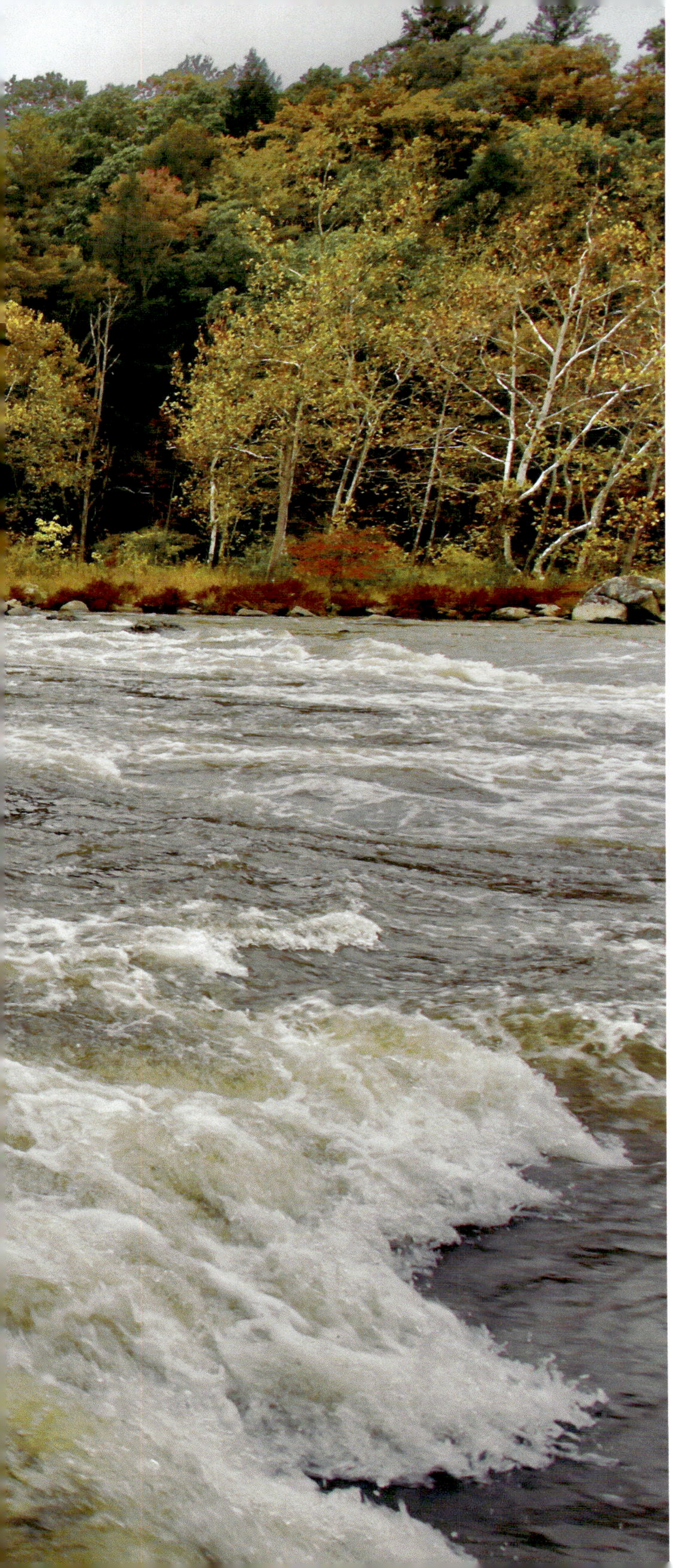

Entrance Rapid's complex route looms first in a sequence of drops spanning 7.5 miles that in the 1970s became one of the most paddled sections of whitewater in America.

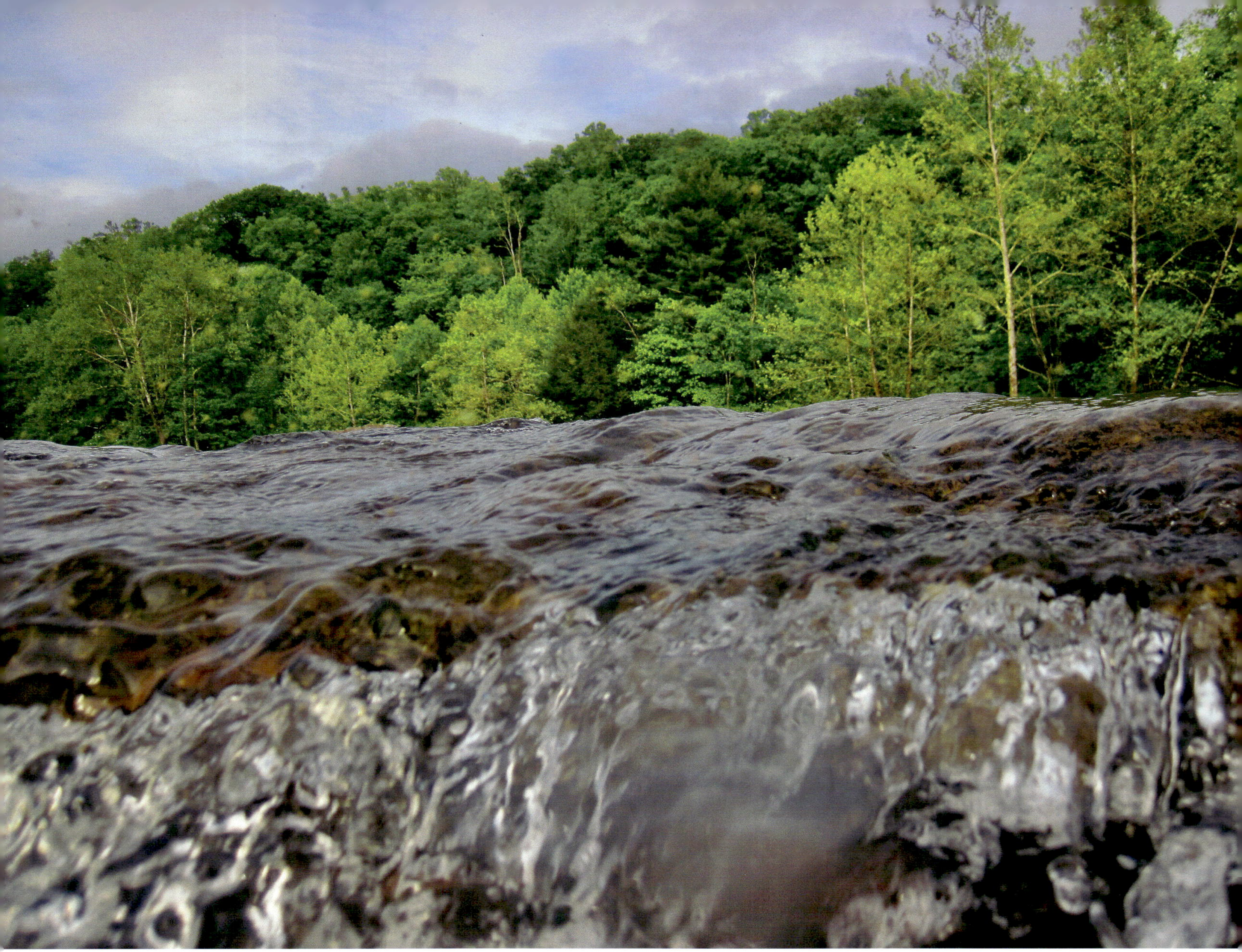

Close-up views of Youghiogheny rapids reveal details of ledges with summertime green of the Appalachian forest rising in the background.

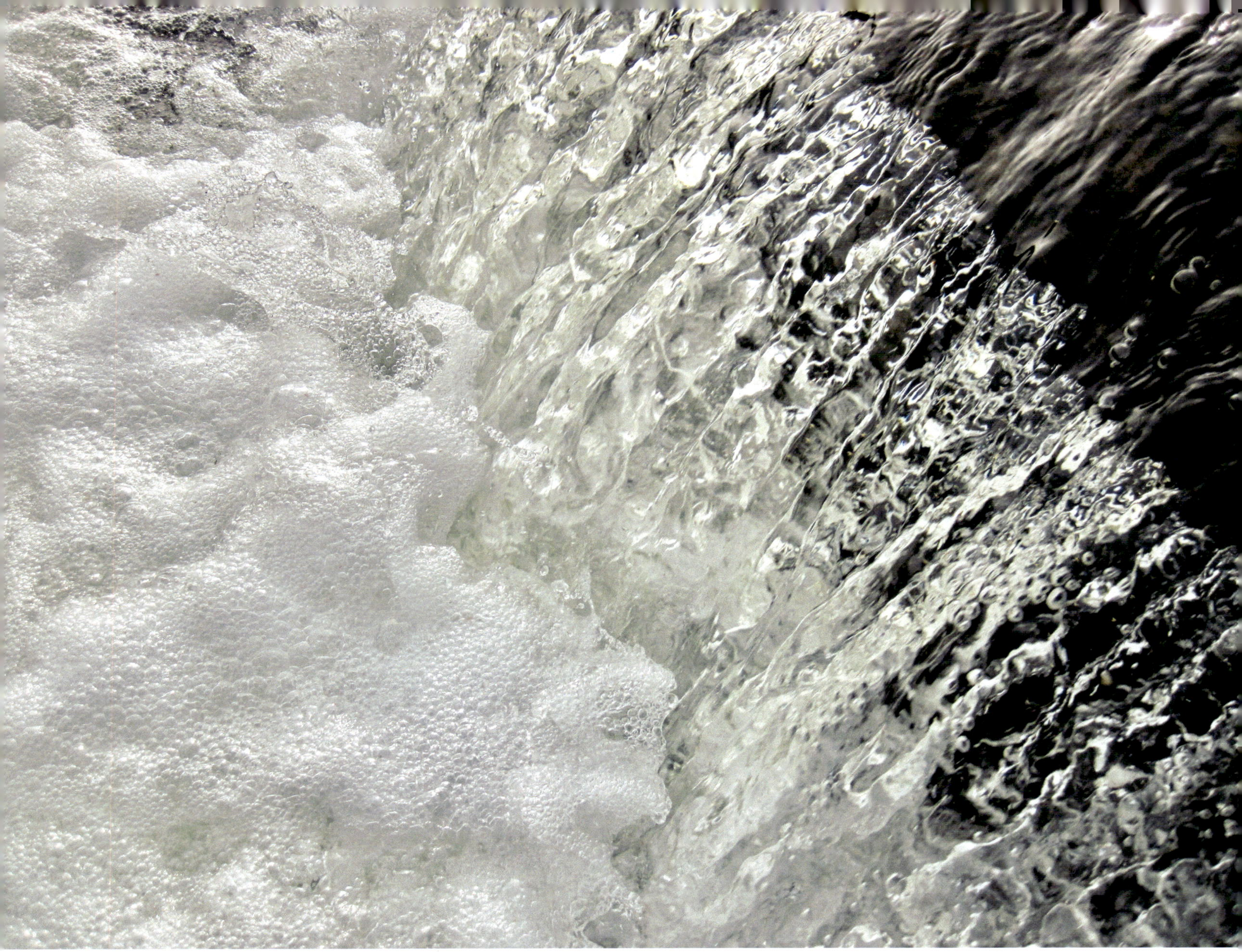

Meadow Run bubbles over ledges and around boulders to join the Youghiogheny just downstream from Ohiopyle.

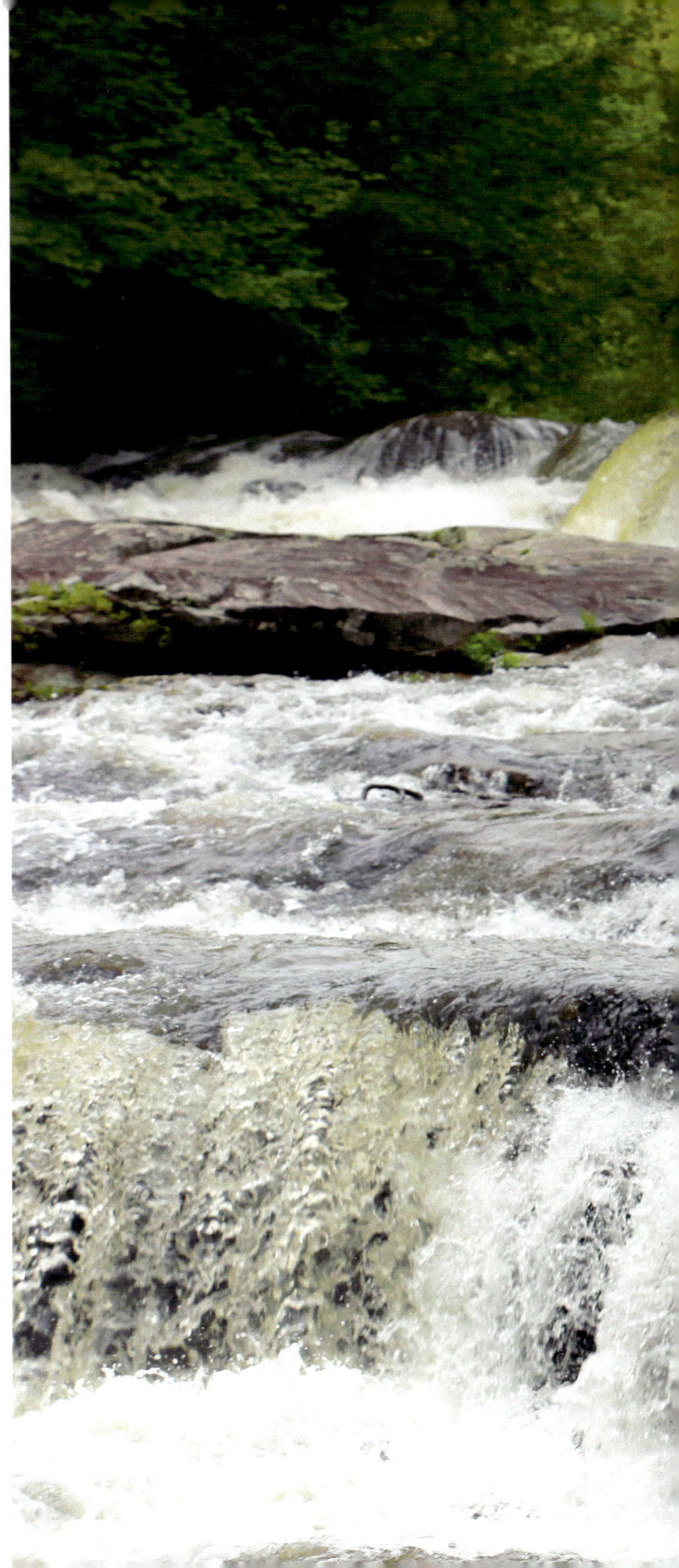

The Meadow Run Cascades are reached by hiking from the Meadow Run trailhead just downstream from Ohiopyle.

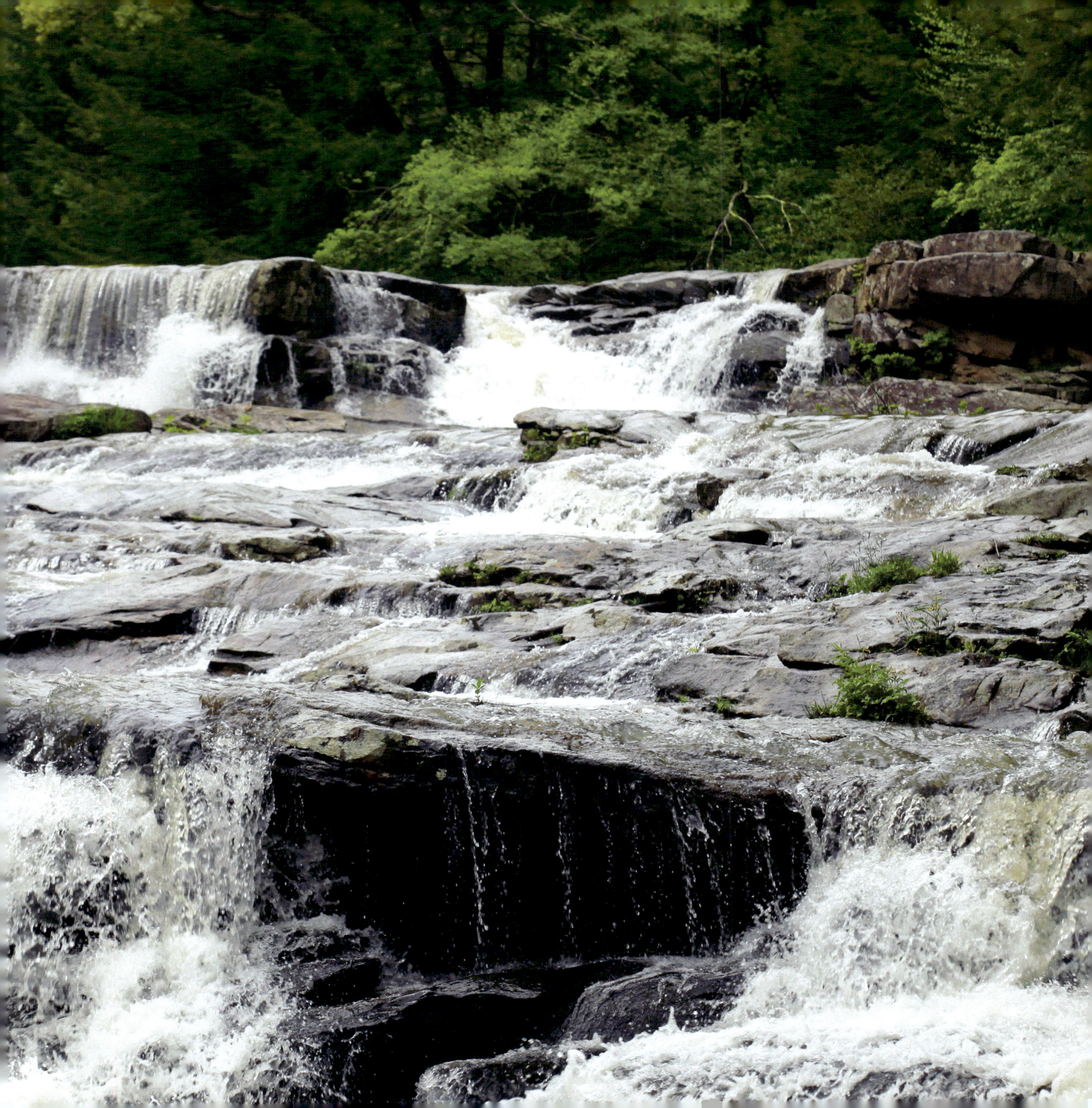

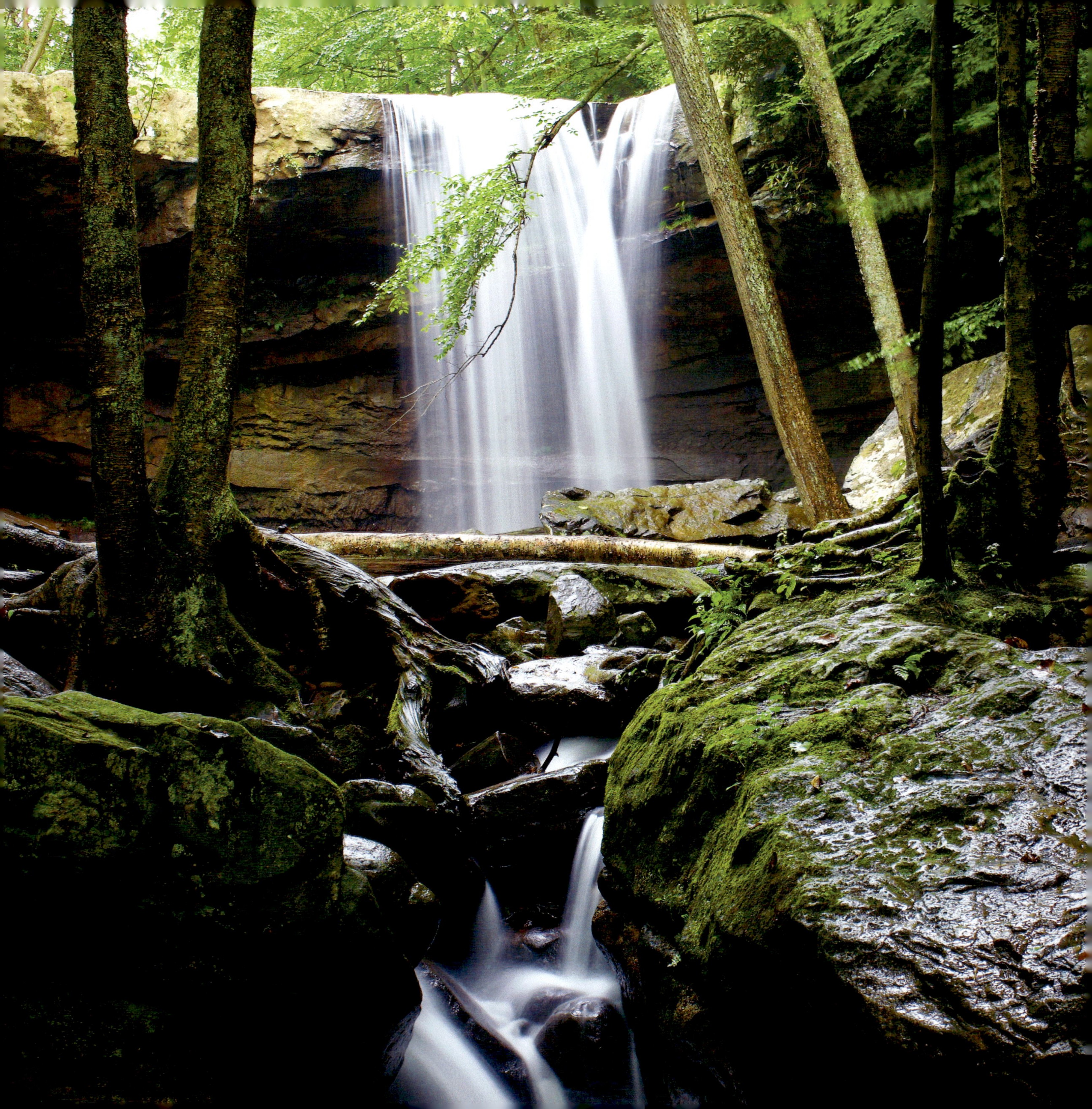

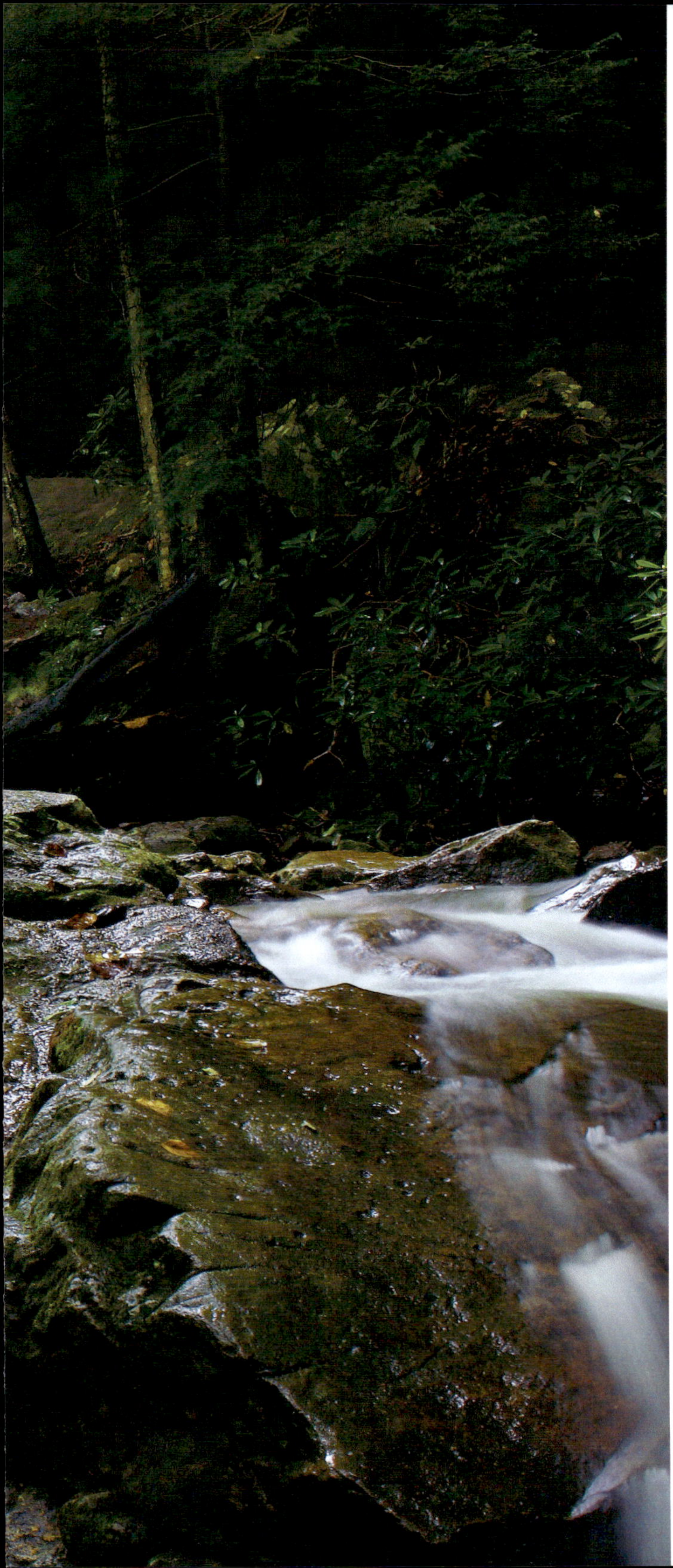

The classic white veil of Cucumber Run has become one of Pennsylvania's most photographed waterfalls, reached by a short walk above the tributary's confluence at the Youghiogheny or by car along the Chalk Hill Road south of Ohiopyle.

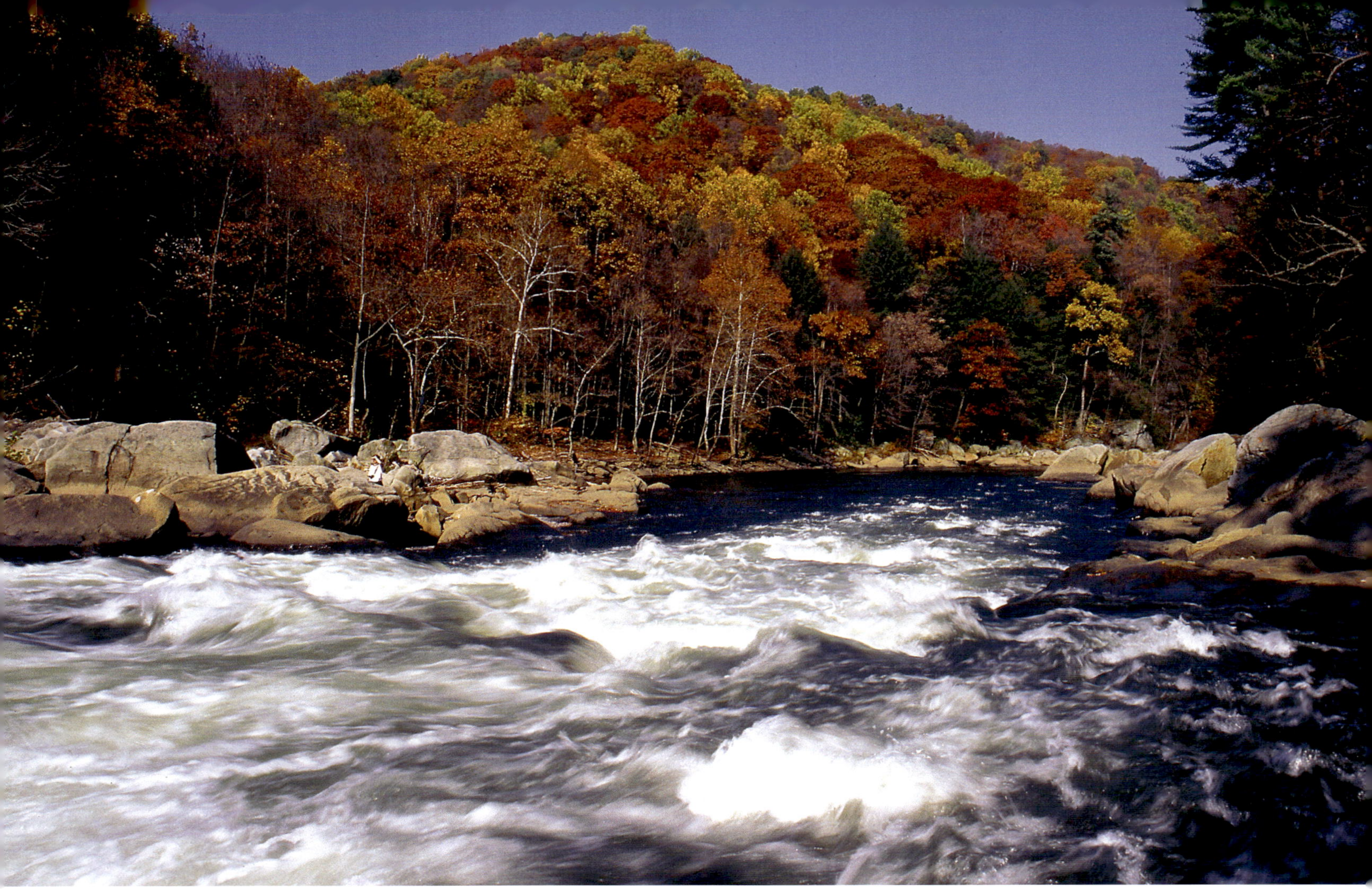

As the second formidable drop of what paddlers call the "Lower" Yough, Cucumber Rapid foams downward on a dazzling October afternoon.

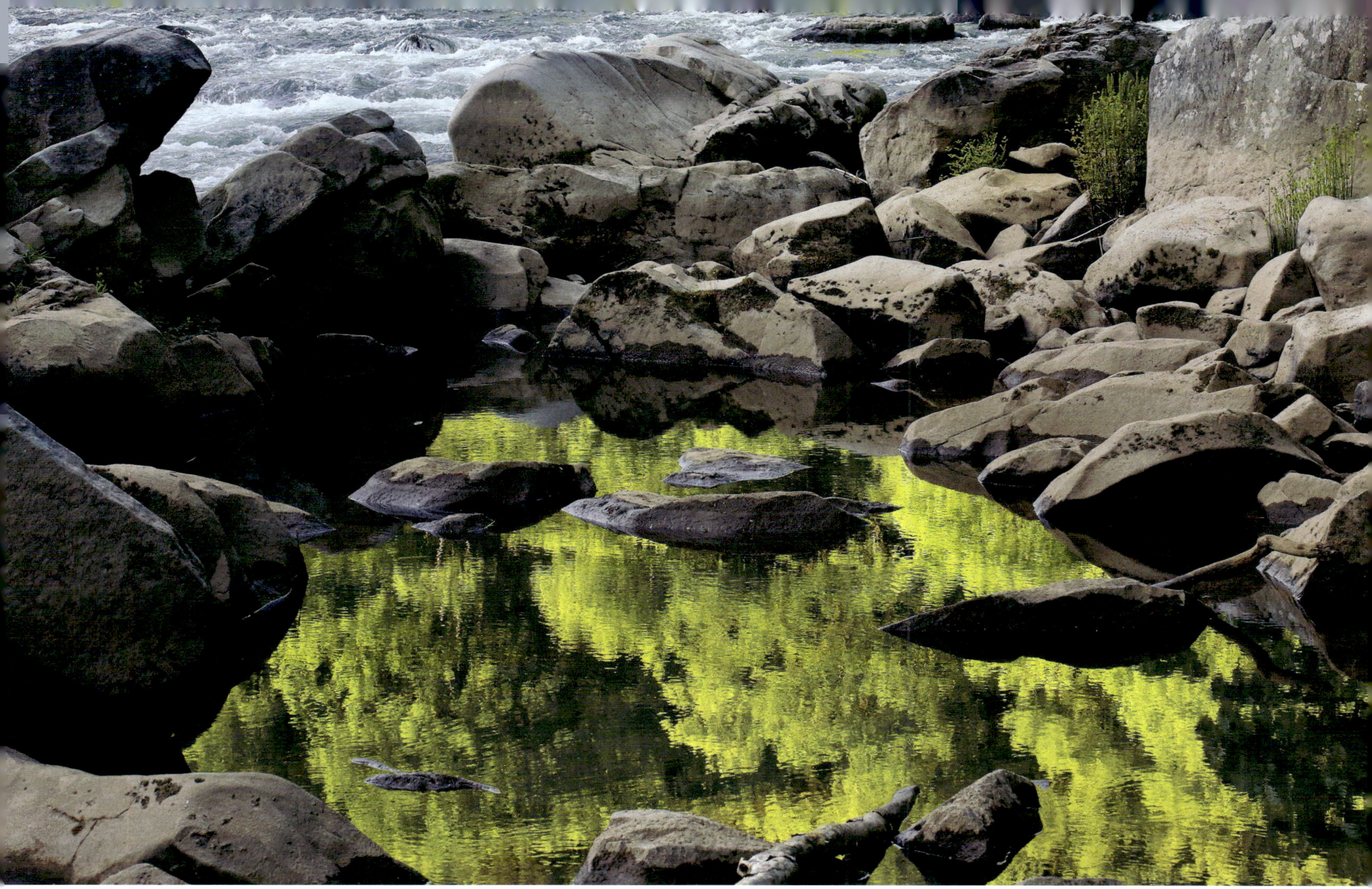

A glassy pool at the edge of Cucumber Rapid reflects the lush growth of trees that rise up from the river.

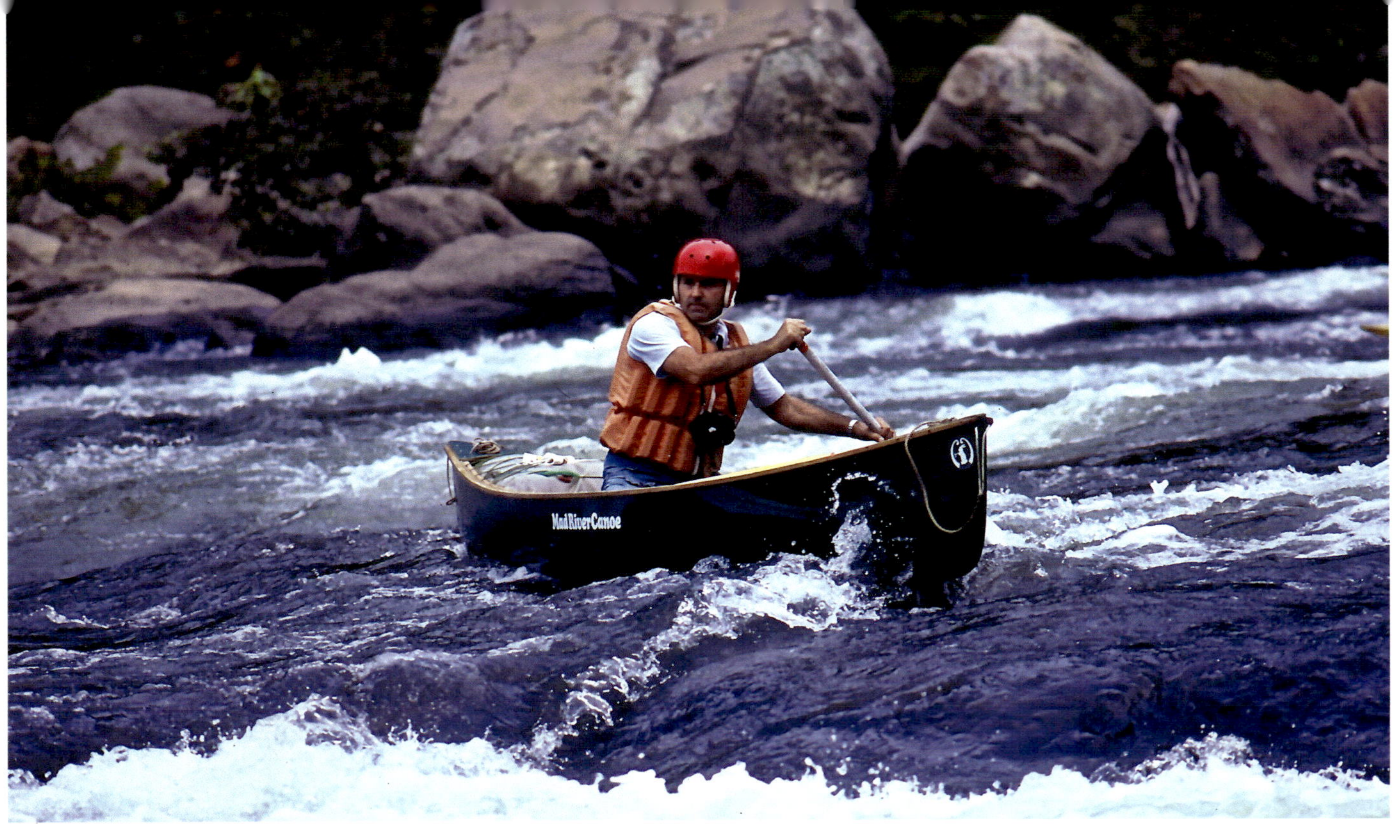
MadRiverCanoe

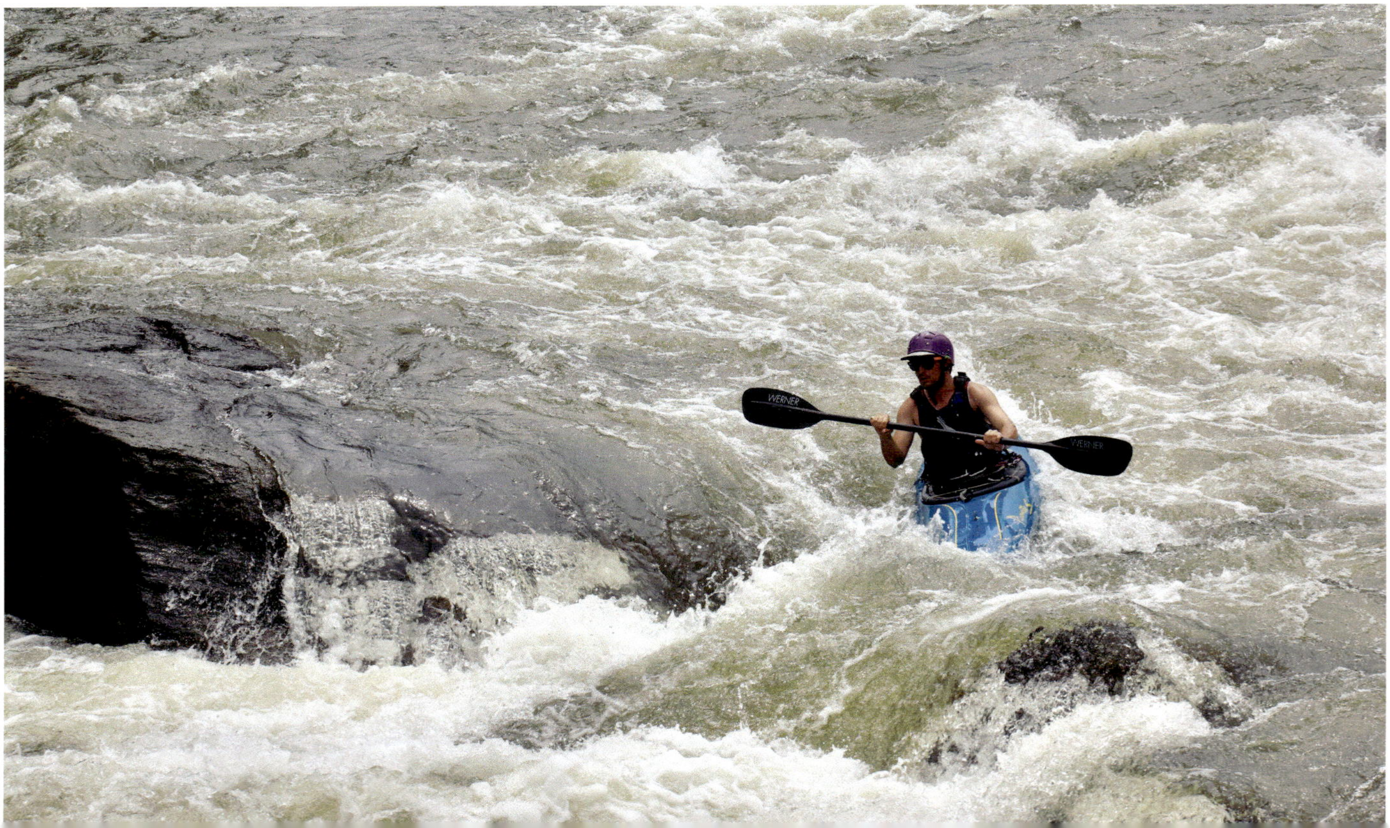
WERNER

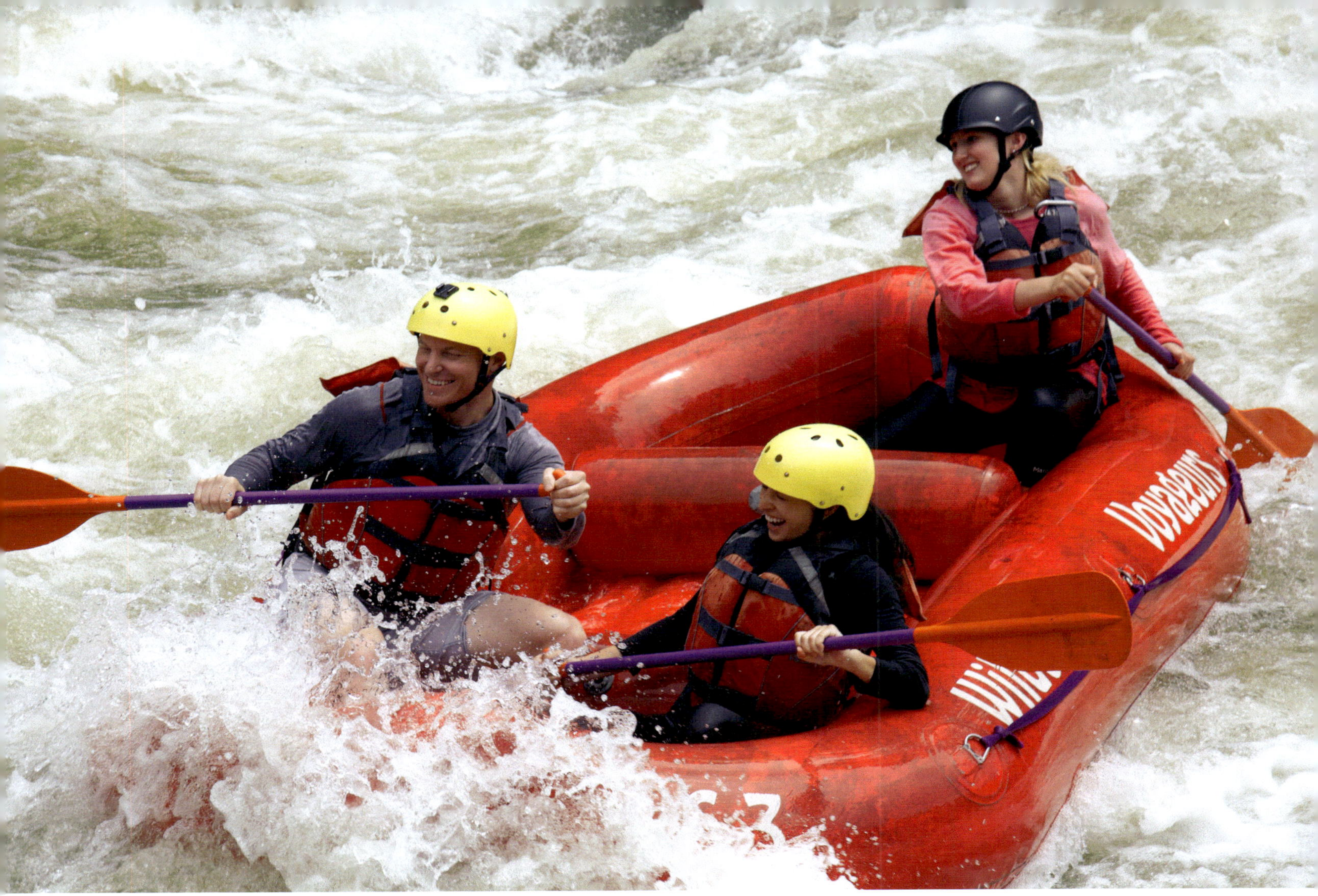

OPPOSITE PAGE, TOP

The Lower Youghiogheny ranks as one of America's premier expert canoe runs, though paddlers of these elegant and traditional craft have become far less common after a 1980s surge in popularity.

OPPOSITE PAGE, BOTTOM

Kayakers relish the Lower Youghiogheny's Class 3 and 4 whitewater. This paddler capably opts for the intricate left channel at Cucumber Rapid.

ABOVE

Expertly guided, this commercial raft and its happy crew go for the gusto in the main channel of Cucumber Rapid.

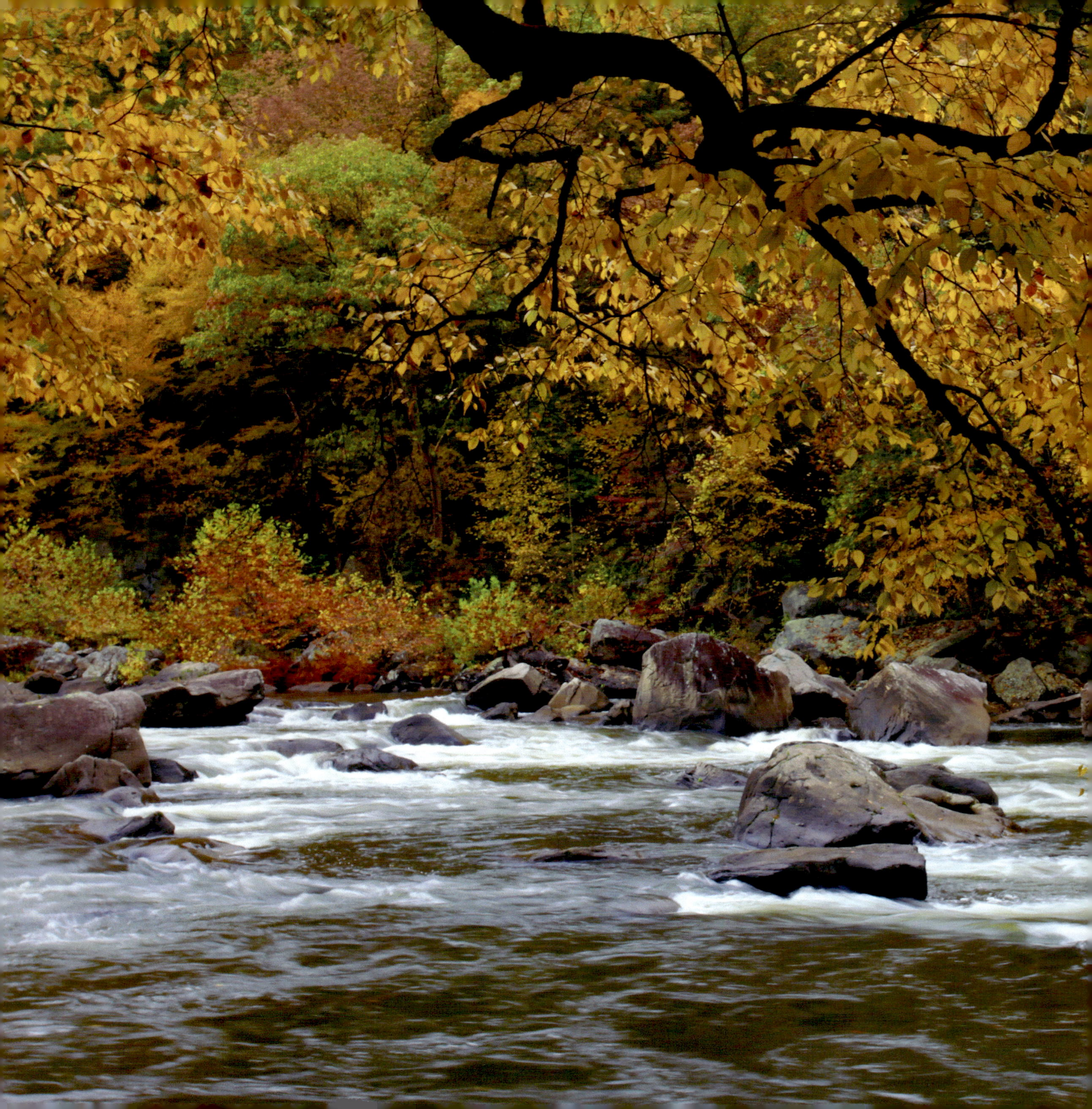

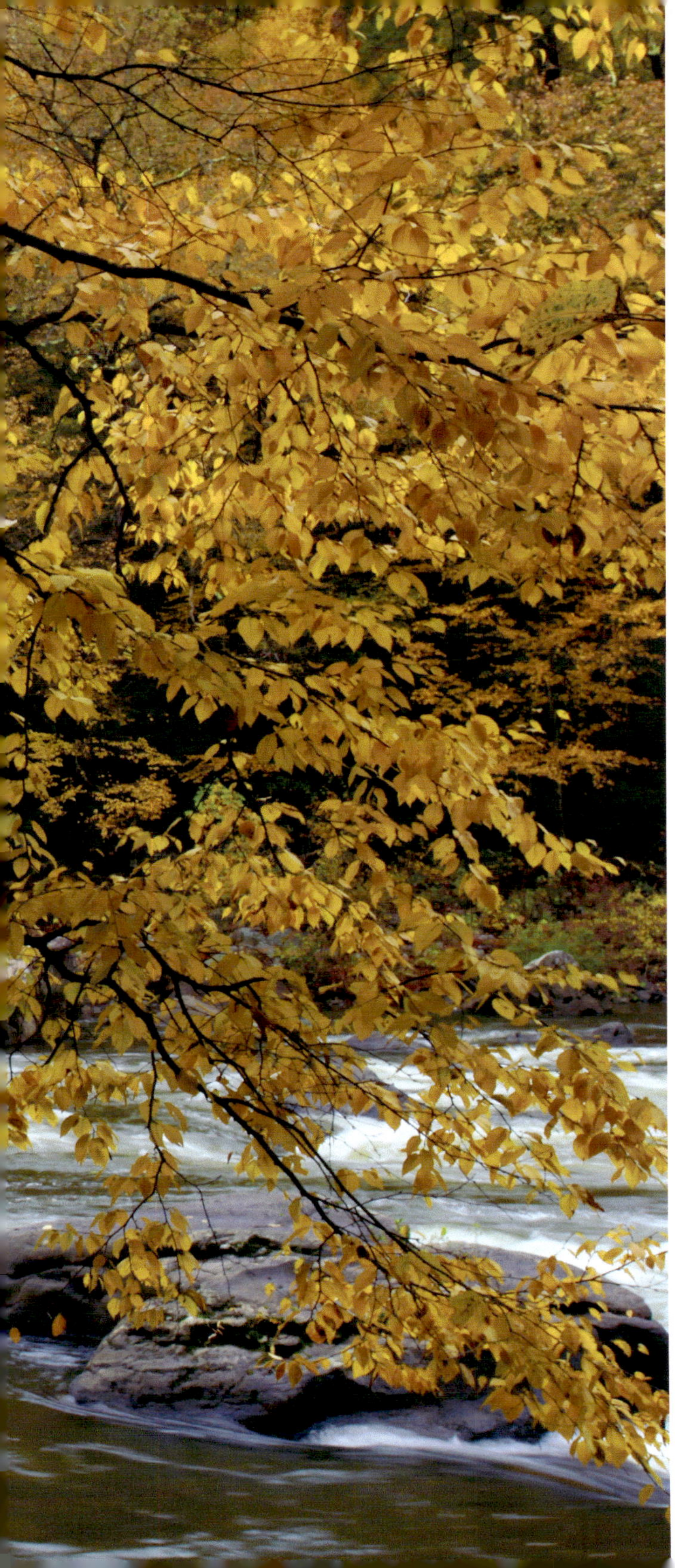

The 1.7-mile-long Ohiopyle Loop, with stairsteps of six distinctive rapids, drops 60 turbulent feet as the Youghiogheny curves around Ferncliff Peninsula.

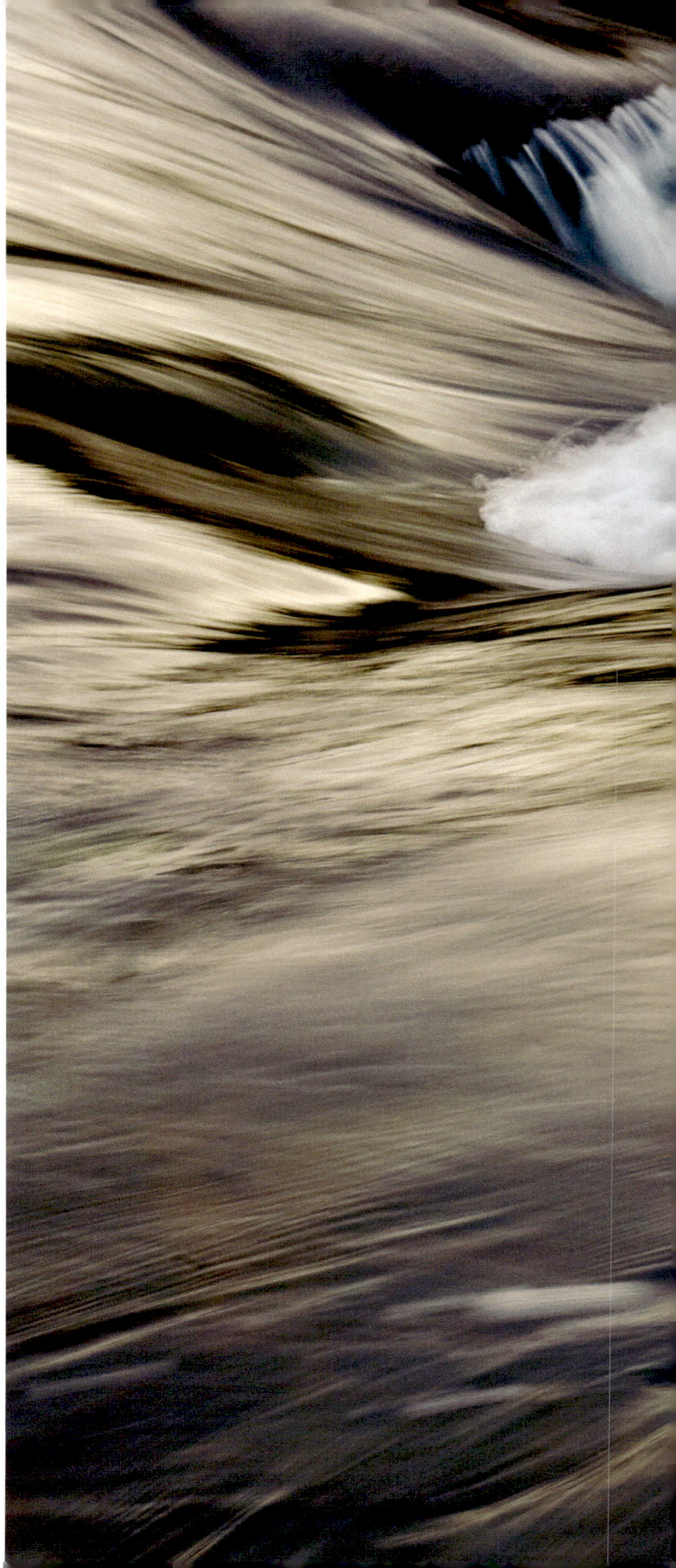

Railroad Rapid's multiple passageways present paddlers with multiple choices on routes to take.

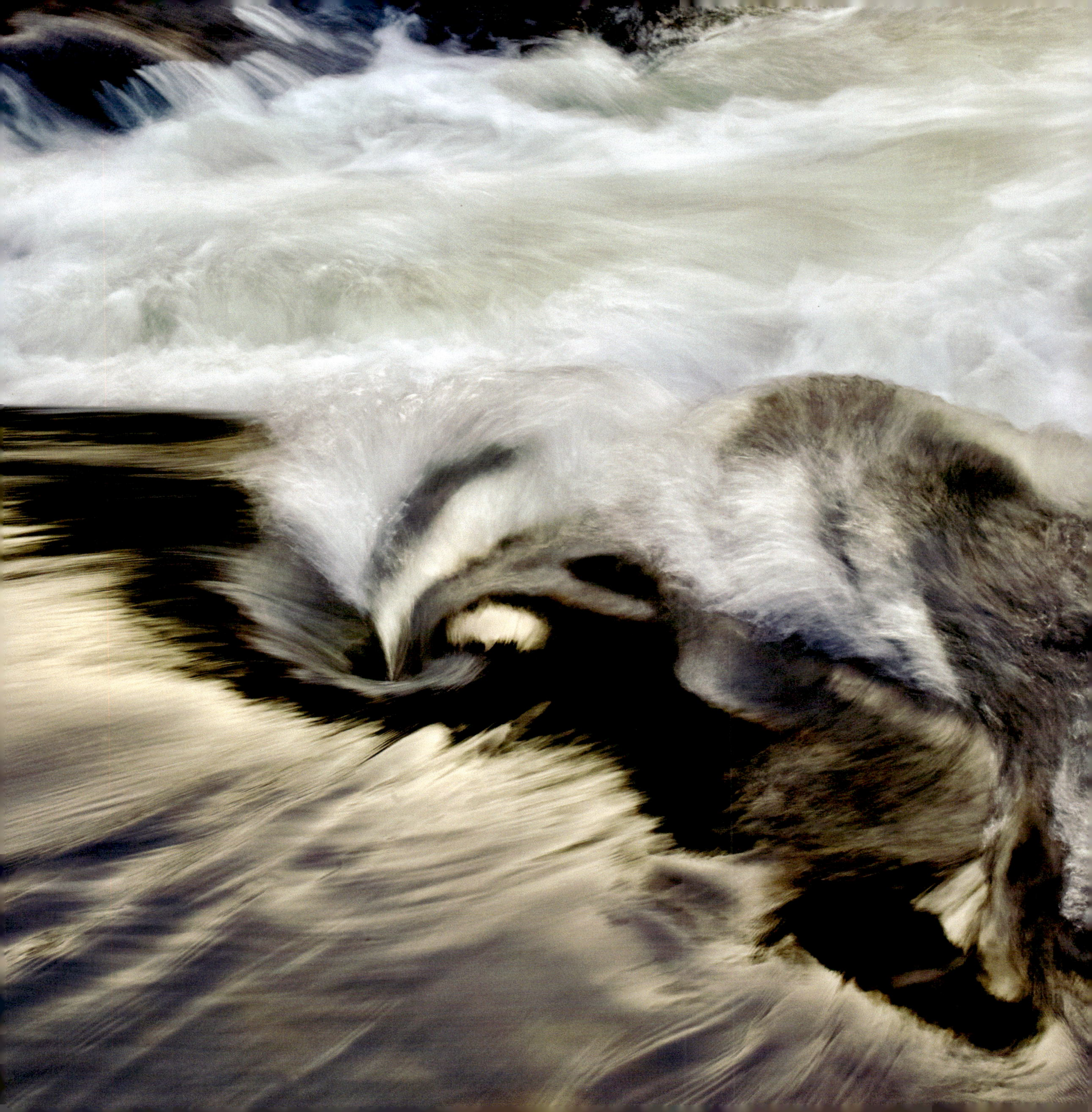

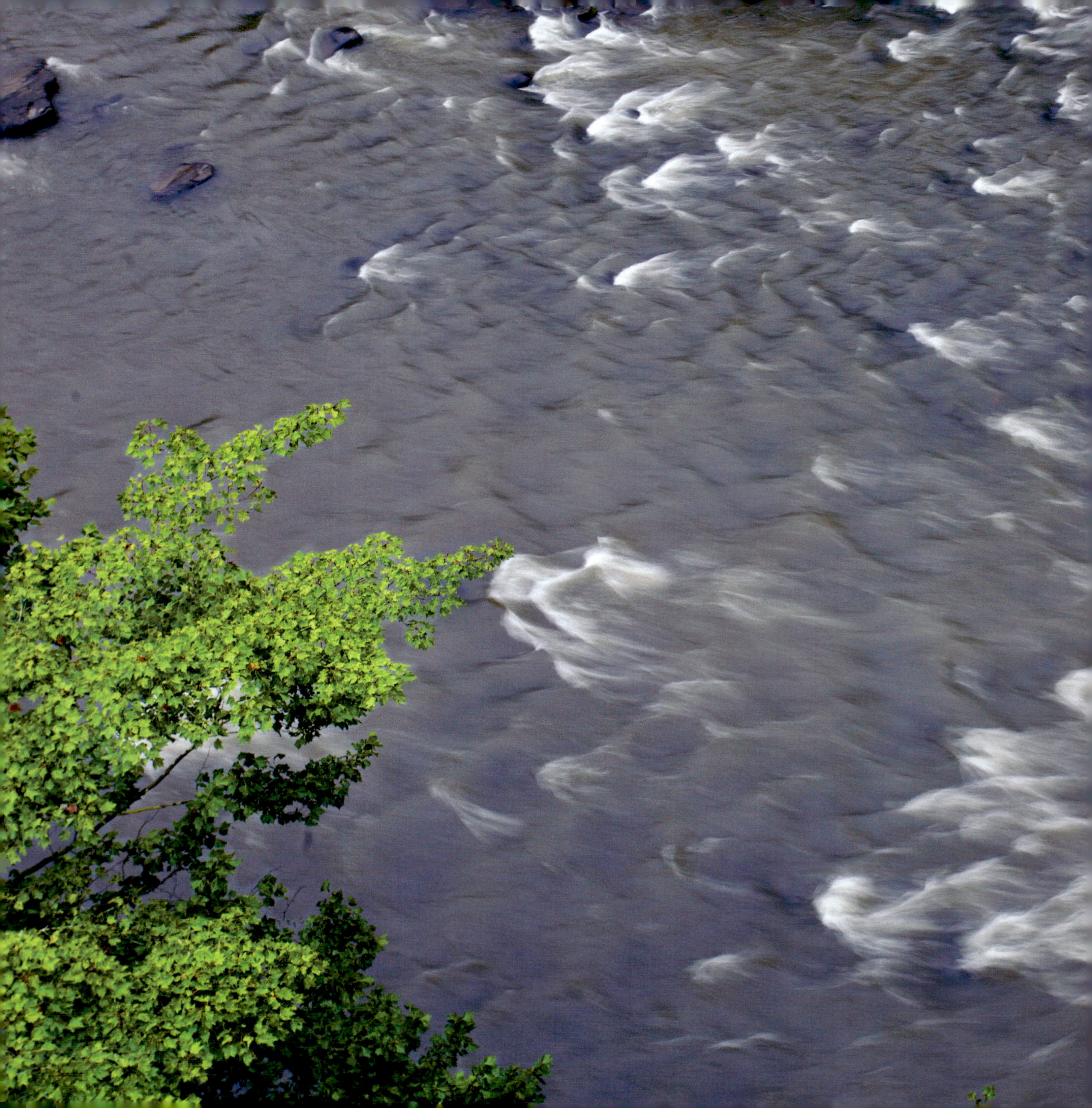

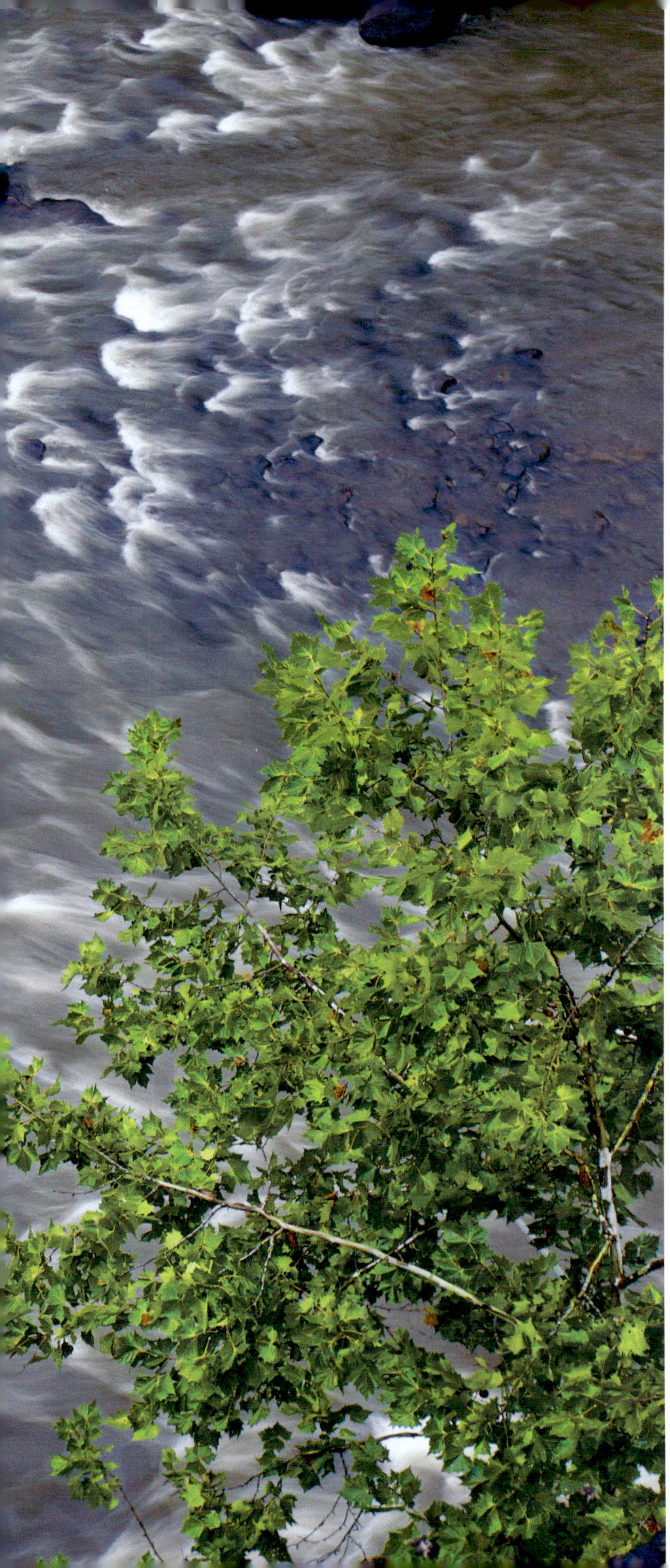

The High Bridge for hikers and bicyclists west from Ohiopyle offers an unequaled view straight down one hundred feet to the river and its rushing whitewater.

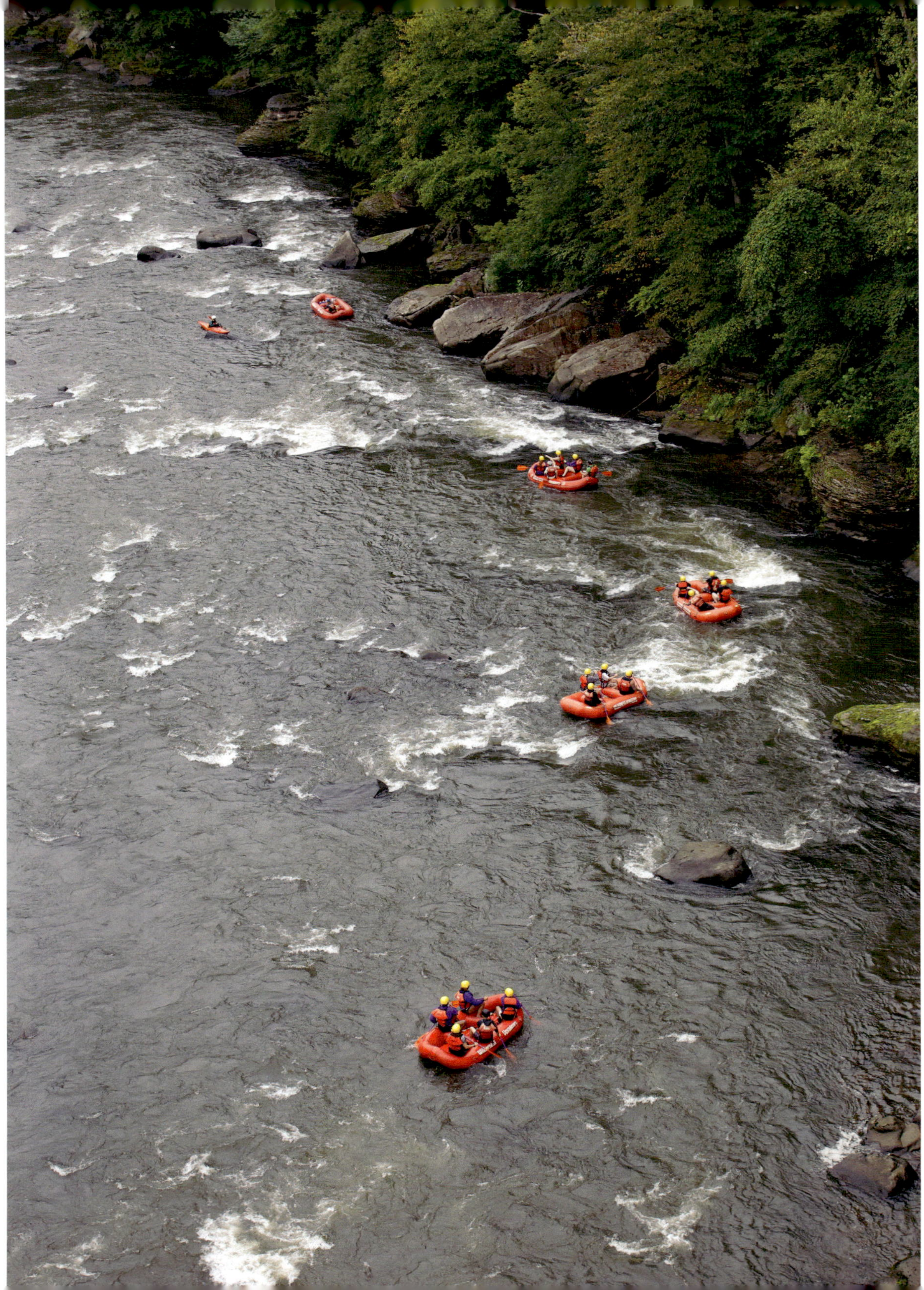

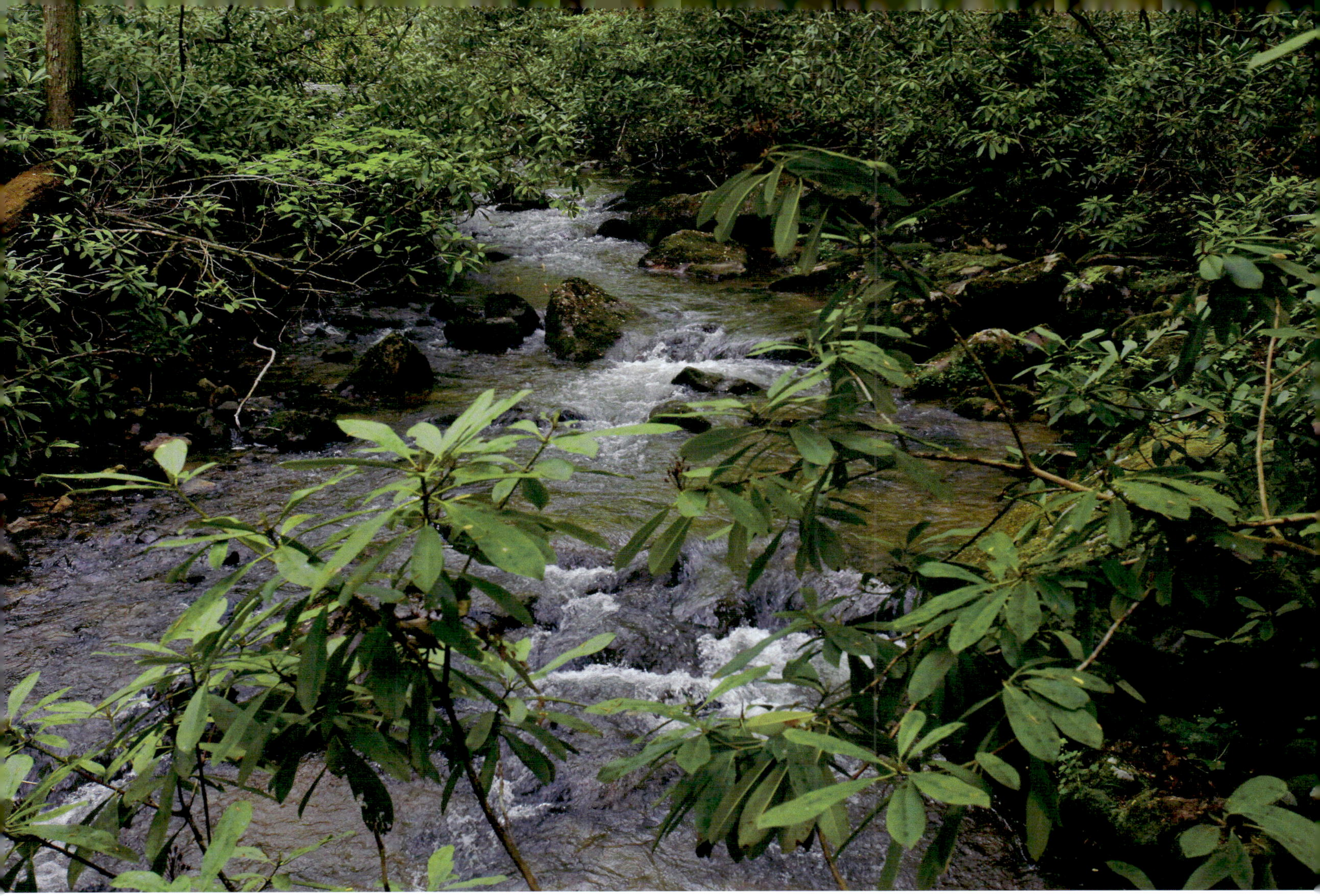

OPPOSITE PAGE
Guides of the Youghiogheny's oldest commercial rafting company—Wilderness Voyageurs—negotiate the maze of Lower Railroad Rapid, seen here from the bike trail's High Bridge west of Ohiopyle.

ABOVE
Bear Run riffles through its deep, recovering forest at the Western Pennsylvania Conservancy's Bear Run Nature Reserve.

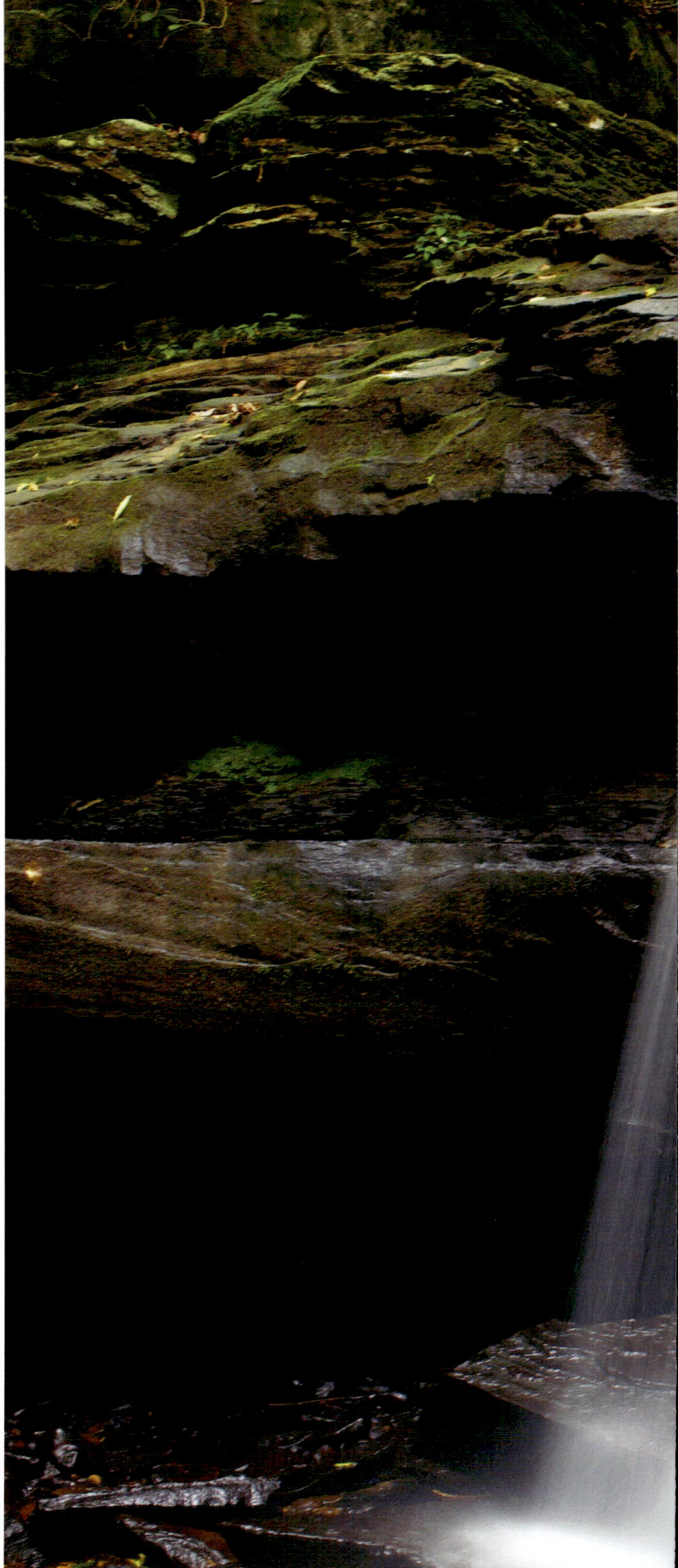

The Jonathan Run waterfall hides in a remote tributary gorge of the Lower Youghiogheny. A fine trail follows much of this stream, but a good view of the falls is reached only with a challenging descent of steep slopes or with a difficult scramble up the stream from below.

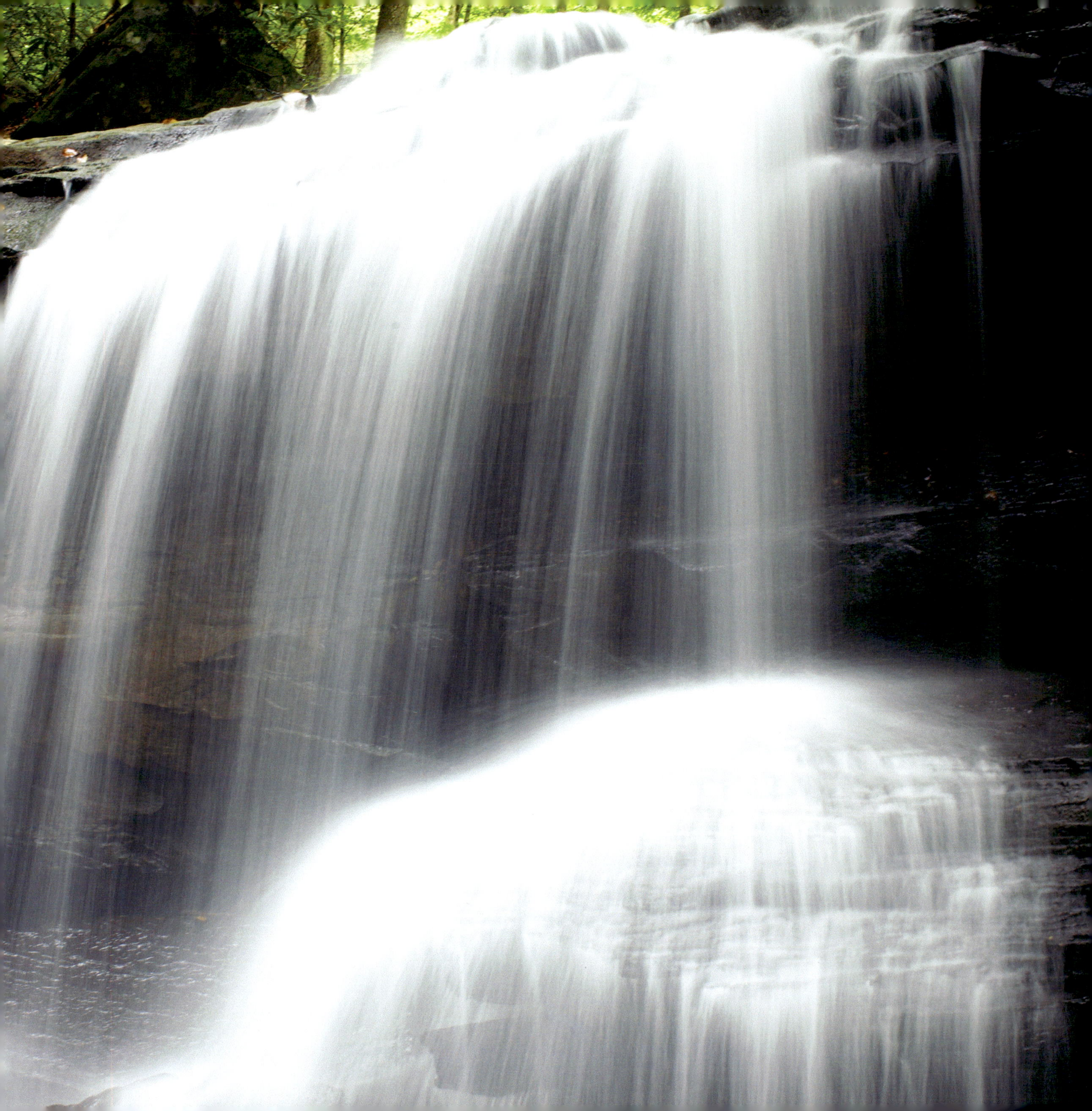

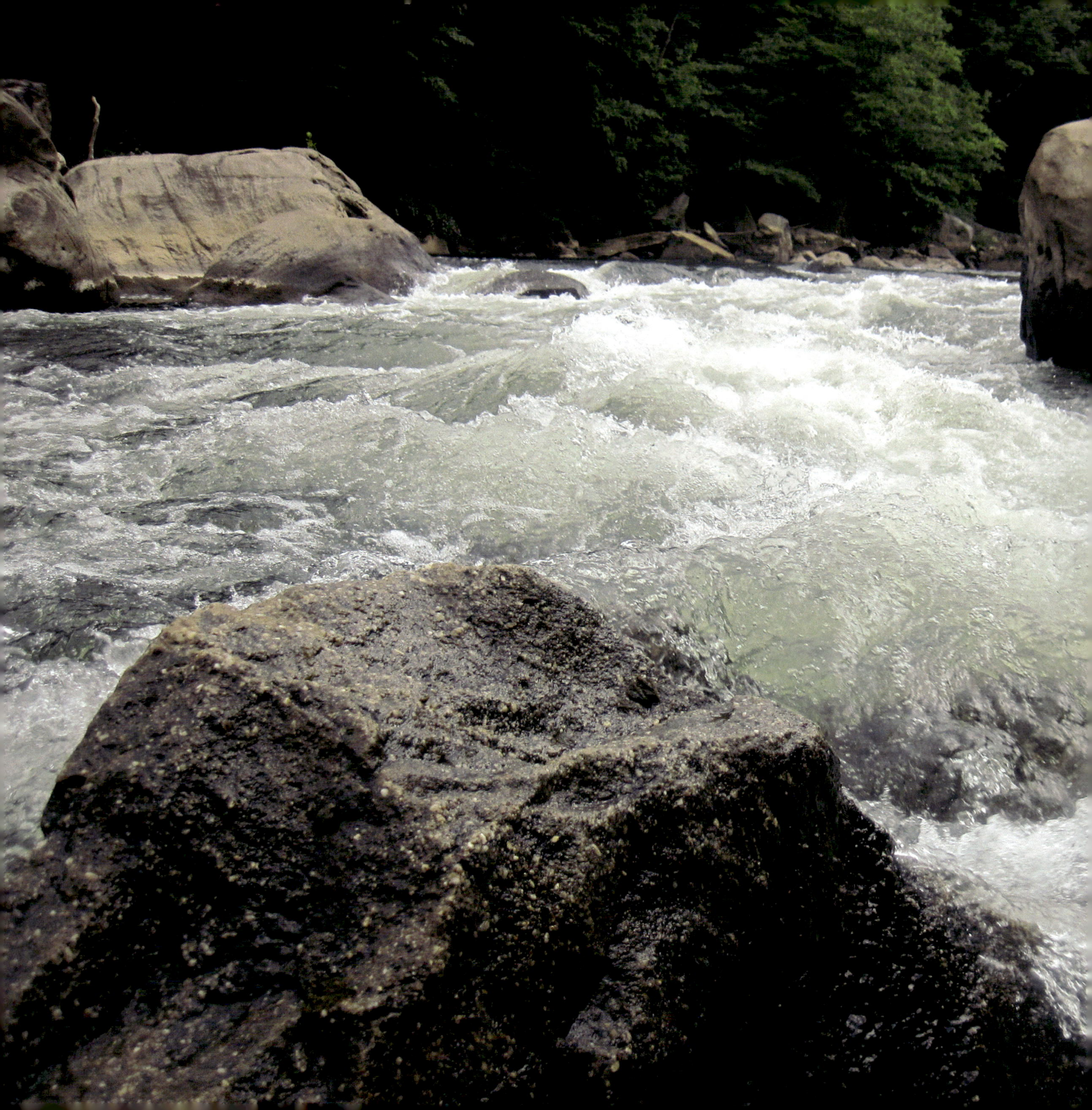

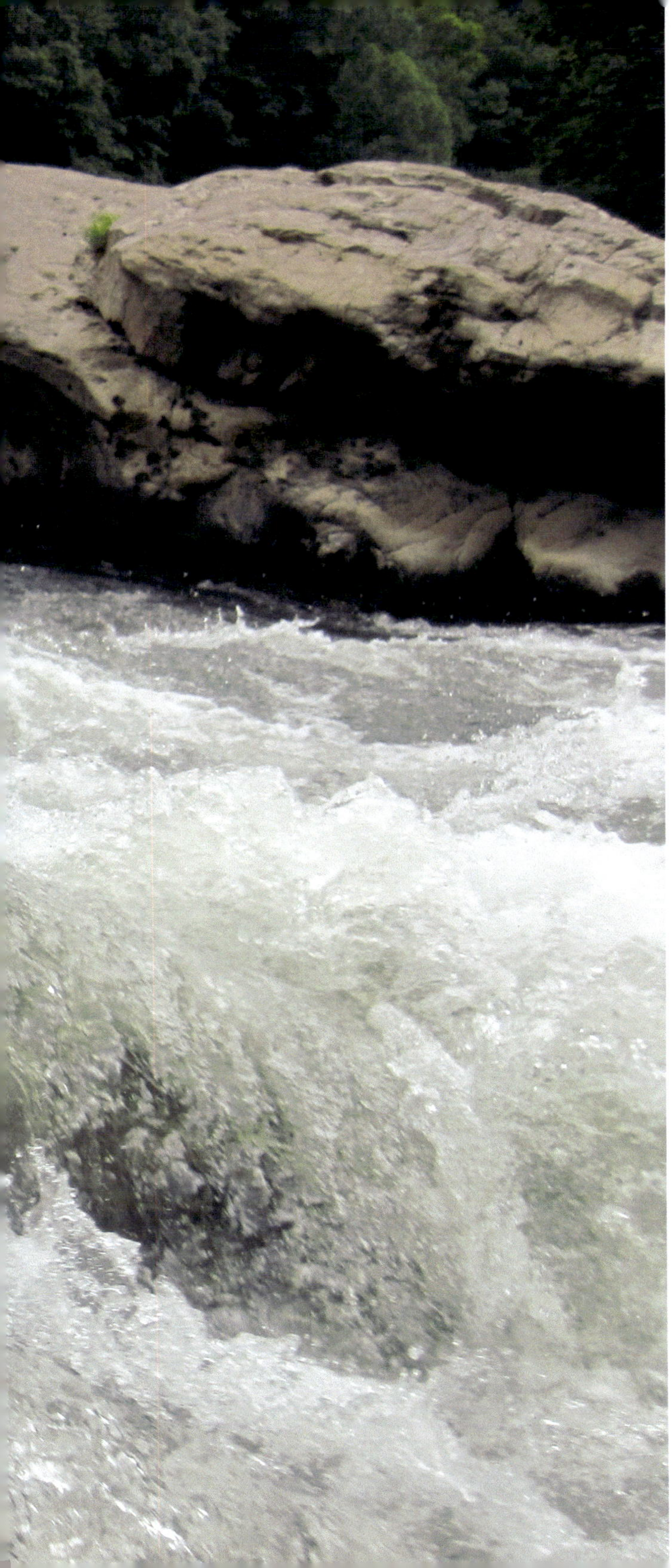

River's End Rapid—named for the horizon line where the water disappears from sight—threads its steep passage among boulders.

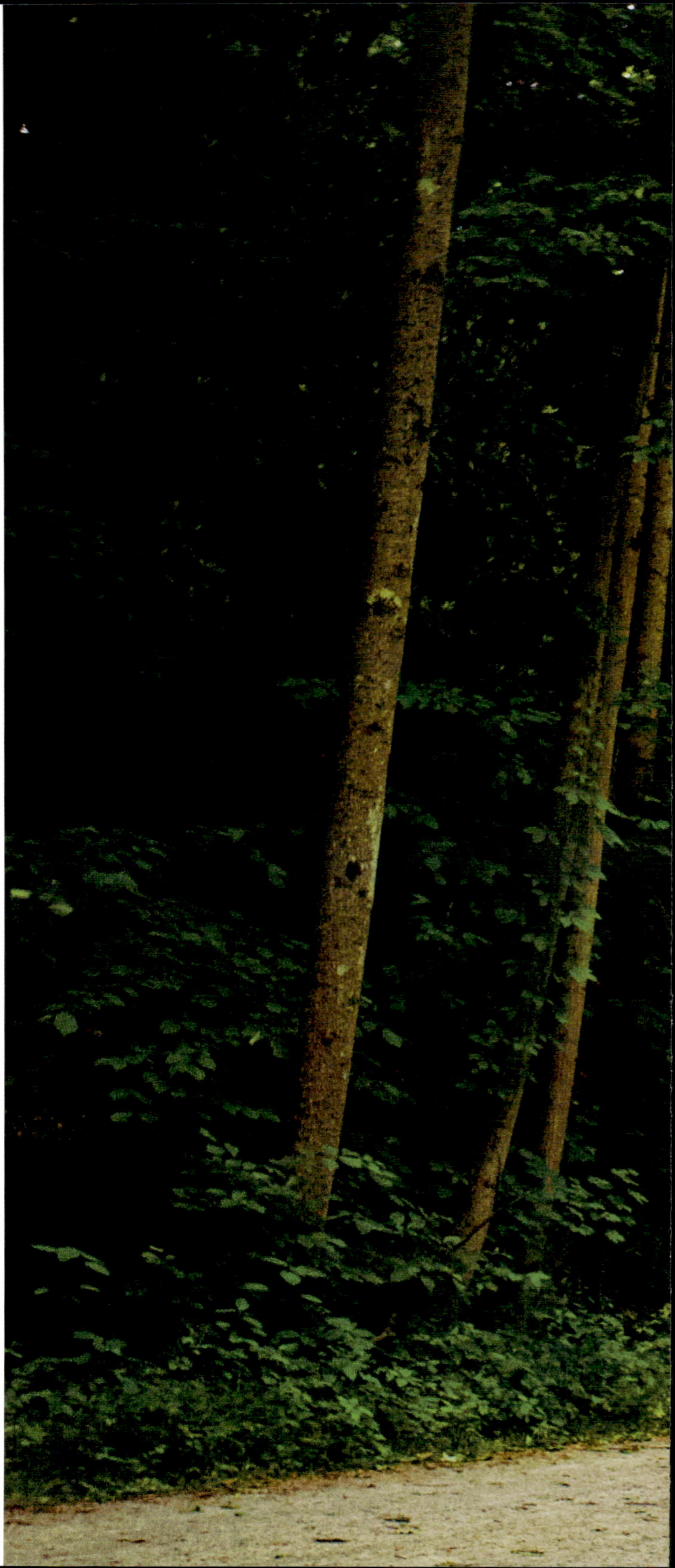

The Great Allegheny Passage trail draws bicyclists through recovering Appalachian forests where tuliptrees tower high and grow larger with each year that they live.

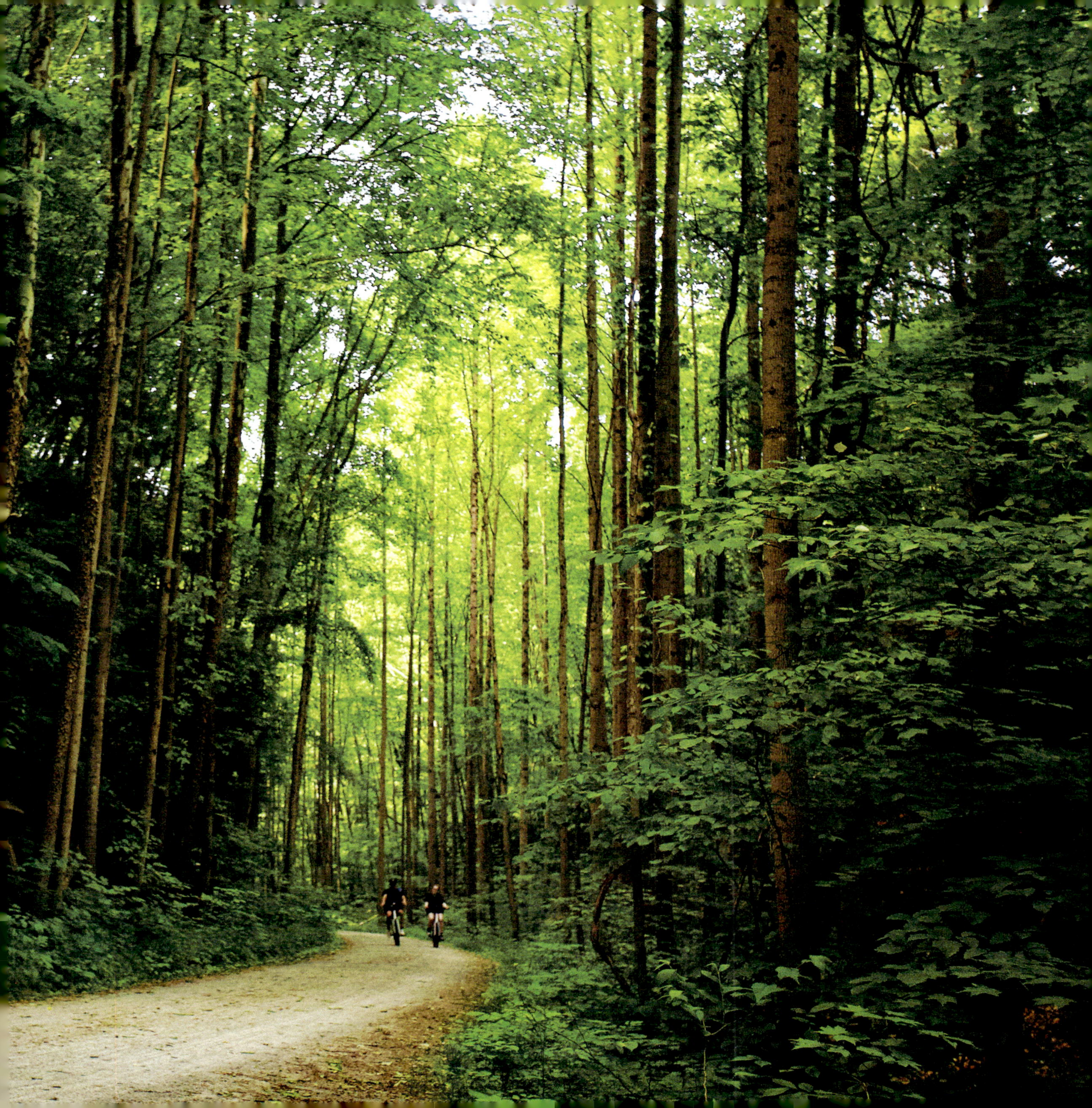

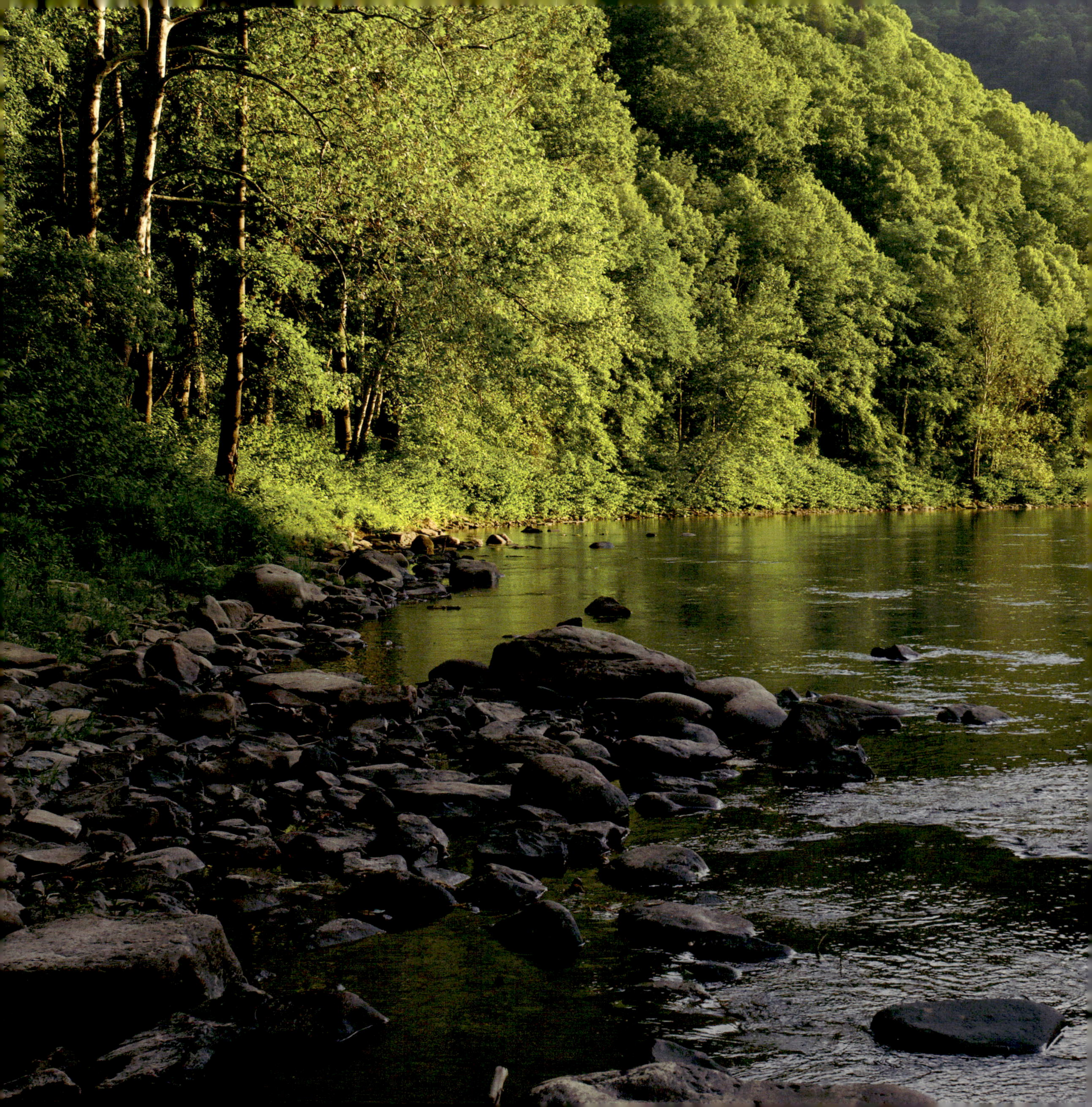

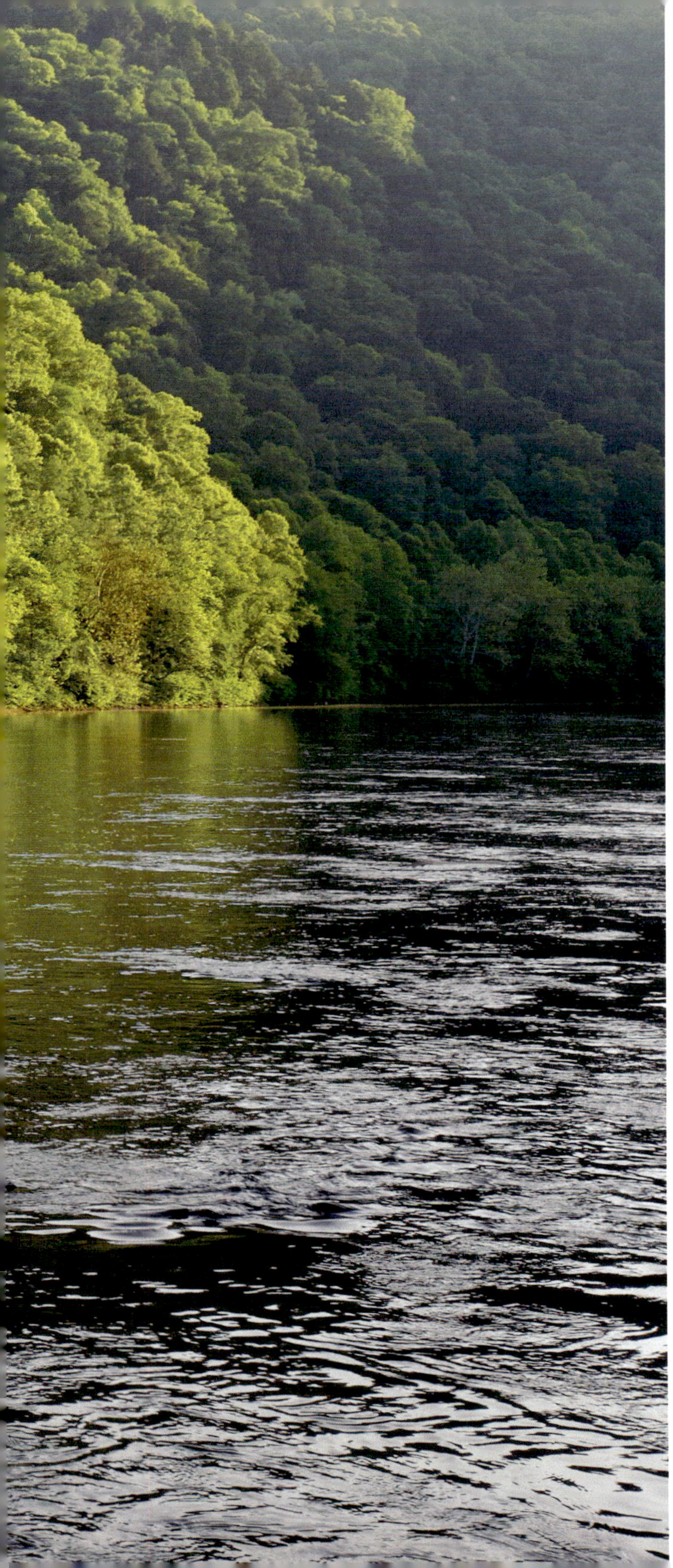

Carving its way through the heart of Chestnut Ridge, the Youghiogheny winds out of the Appalachian complex of mountains and onward to the city of Connellsville.

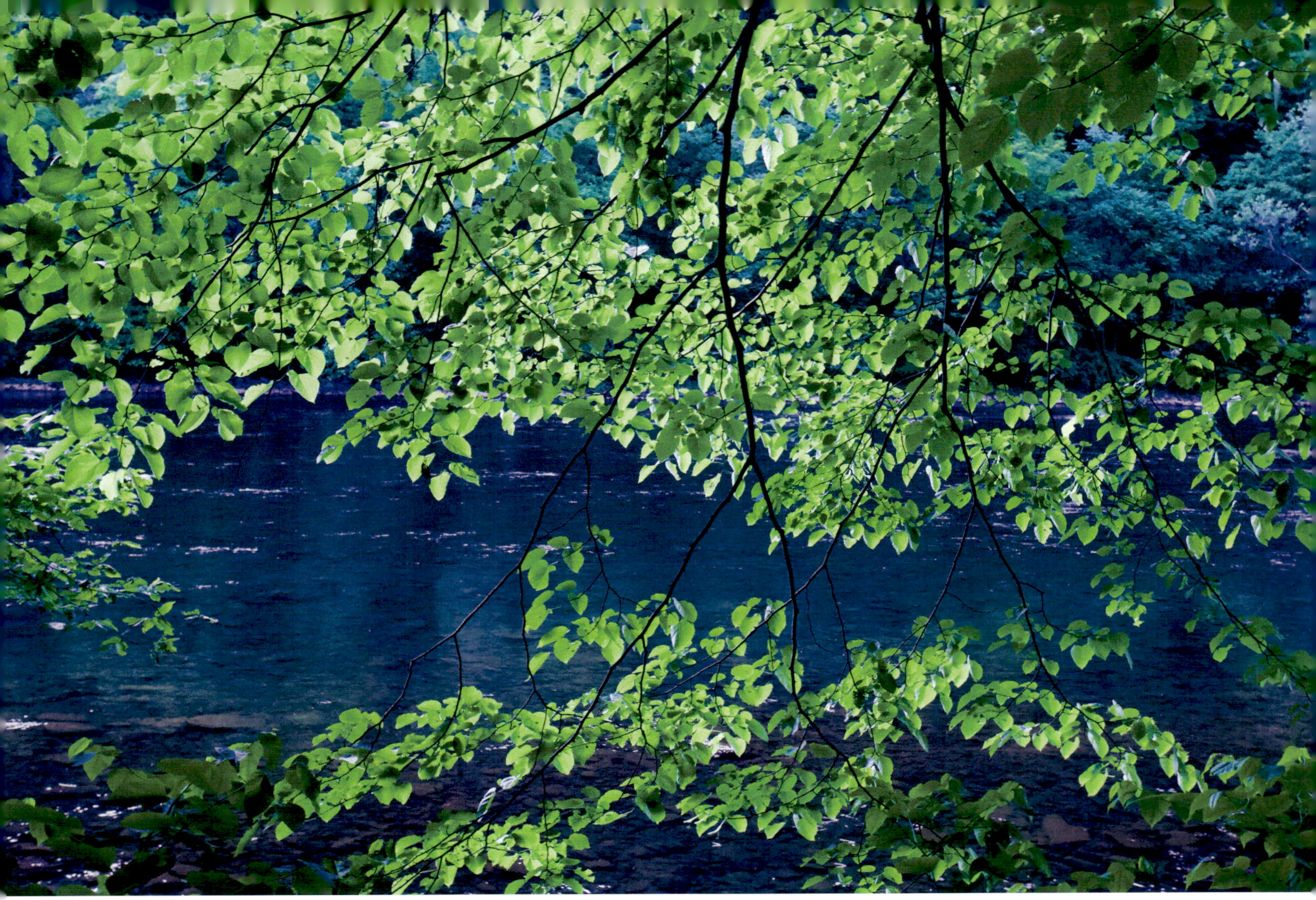

Sweet birch trees shade floodplains in a peaceful stretch of the Youghiogheny as it settles into western foothills of the Appalachian Mountains.

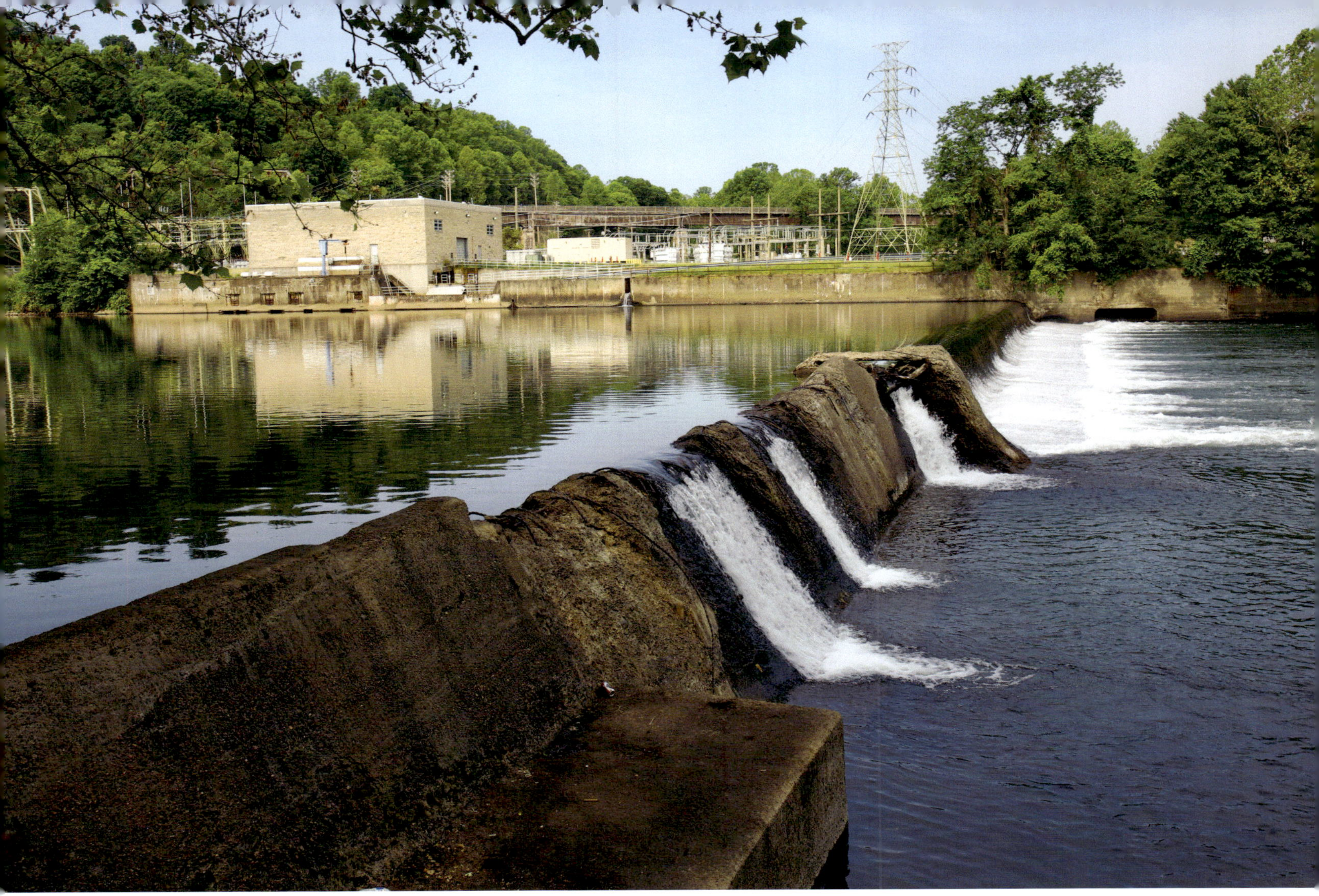

The Youghiogheny's third and final dam blocks the river at the upstream end of Connellsville. Paddlers must portage this unrunnable structure on the right side.

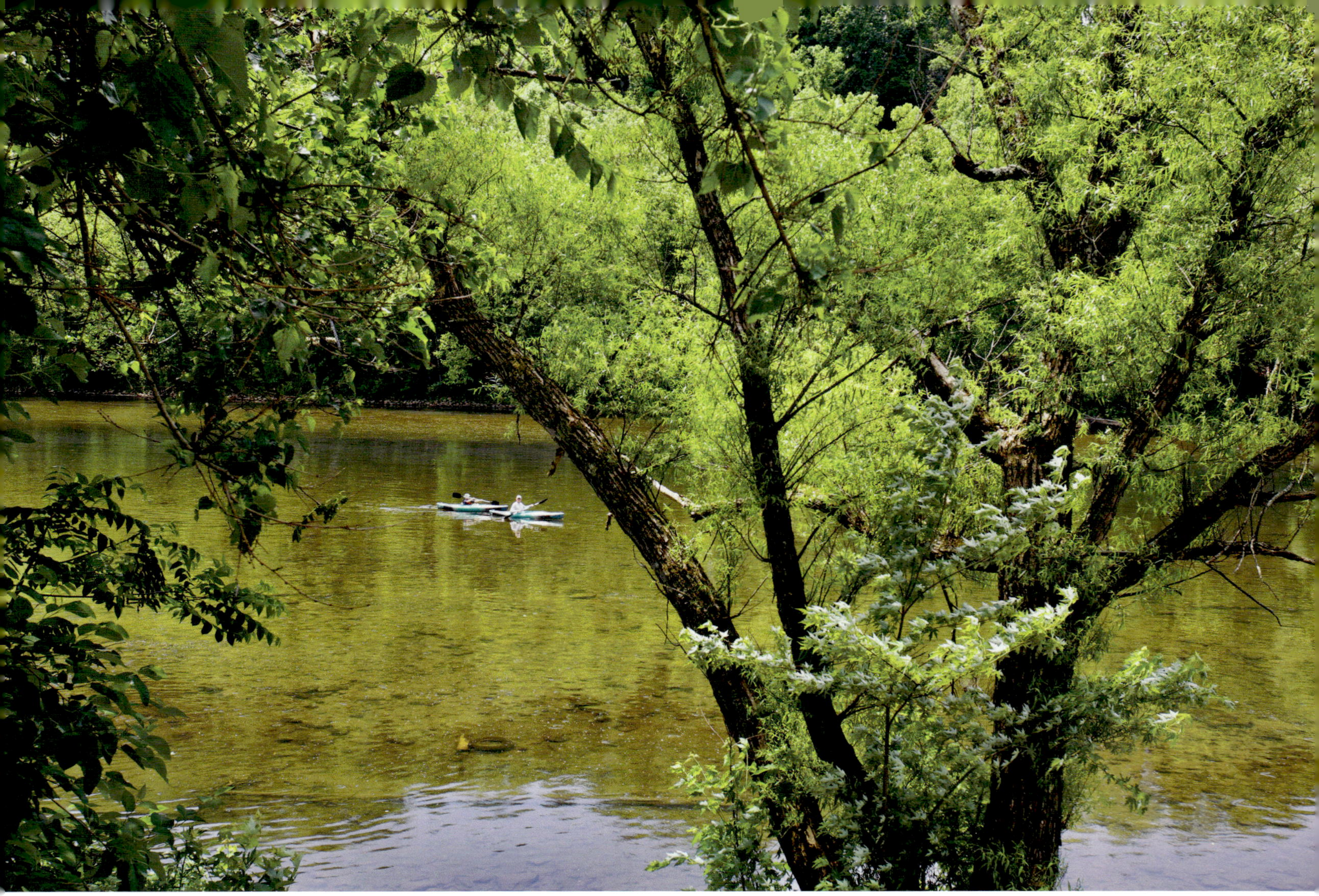

Kayaking and canoeing are becoming popular activities on the gentle lower reaches of the Youghiogheny, including here at the town of Dawson.

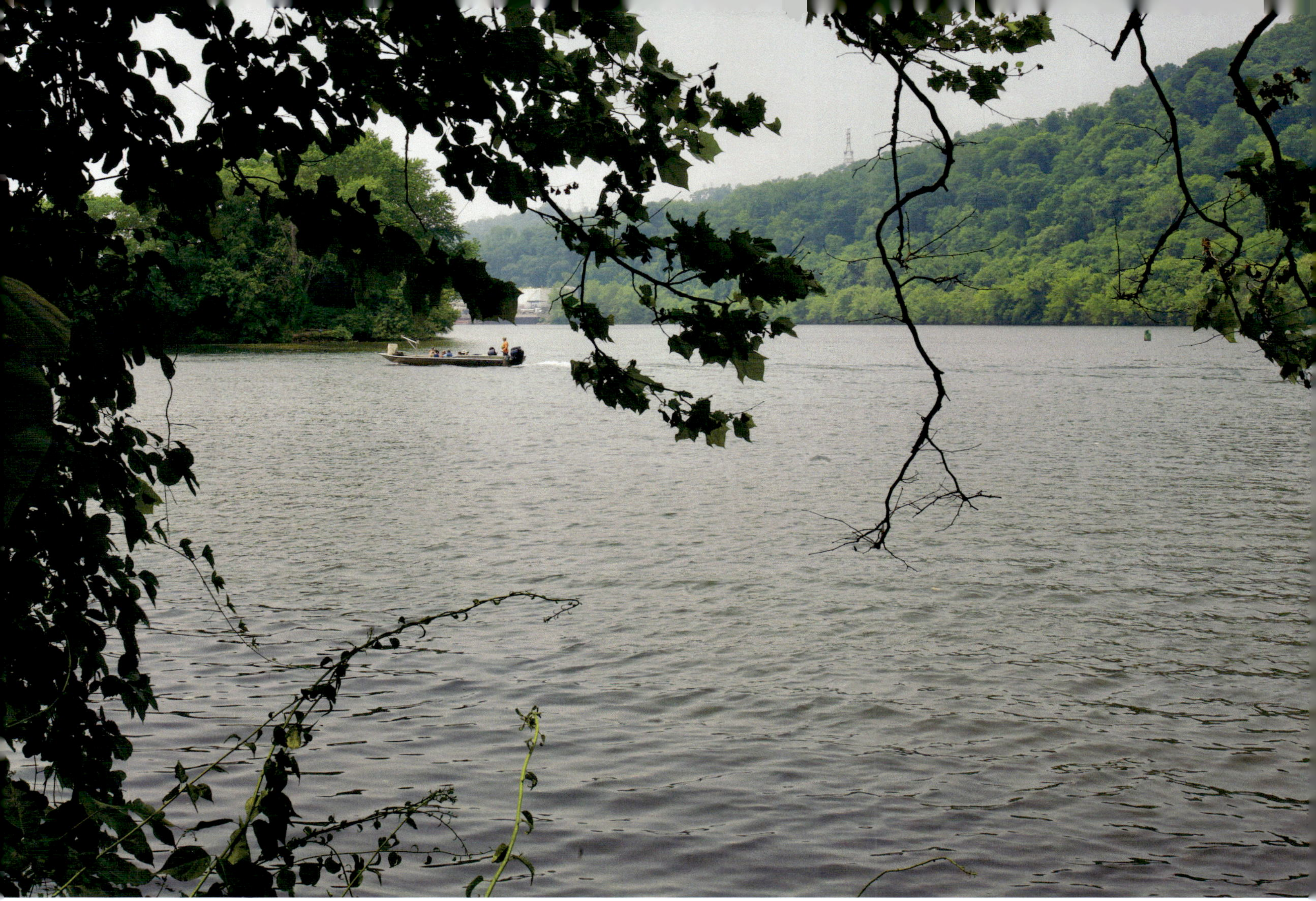

The Youghiogheny ends in flatwater impounded by a dam built by the Army Corps of Engineers for barge navigation downstream on the Monongahela River at Braddock, 10 miles above Pittsburgh. In this photo the Youghiogheny, coming from the left, eases into larger waters of the Mon.

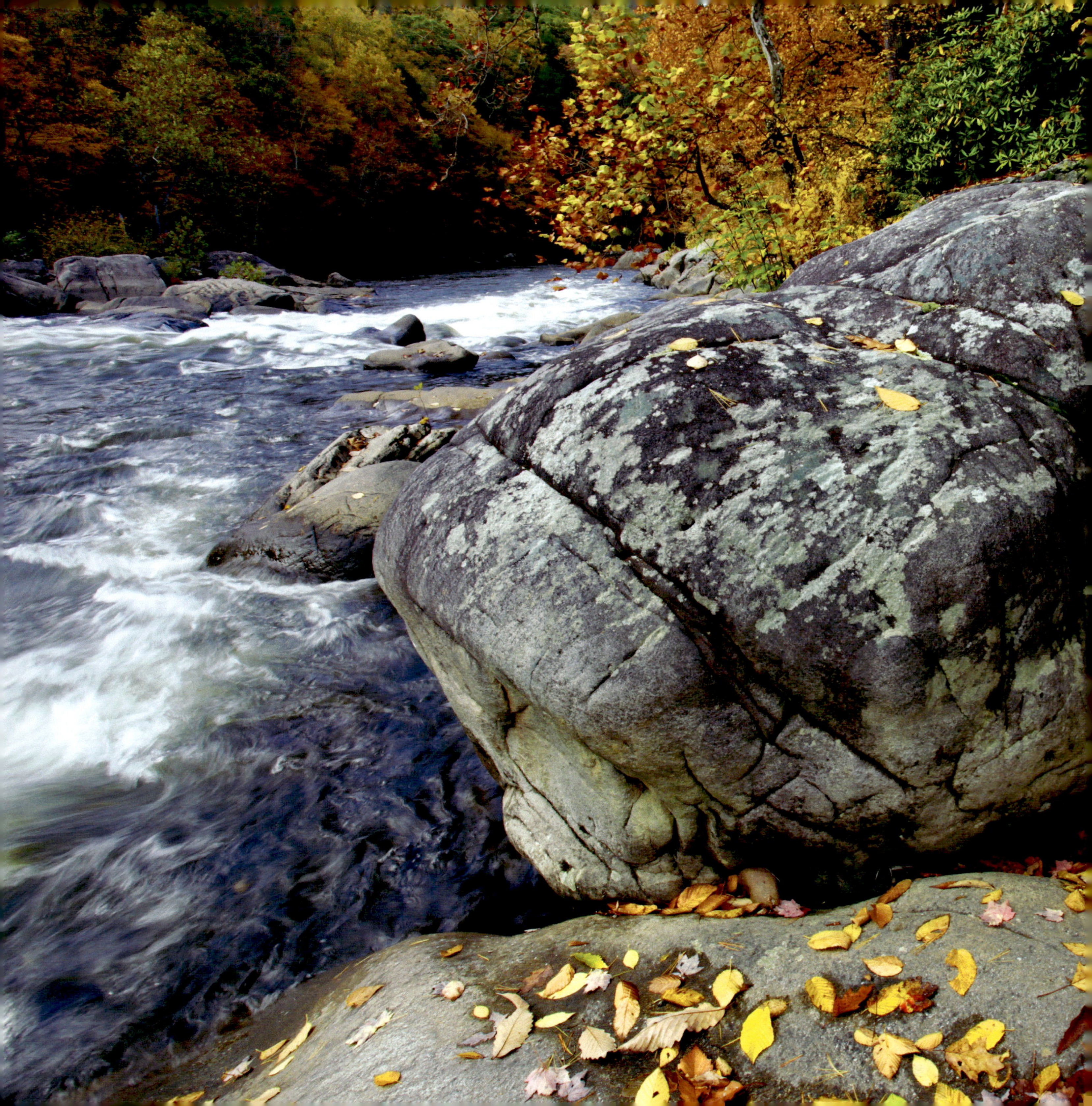

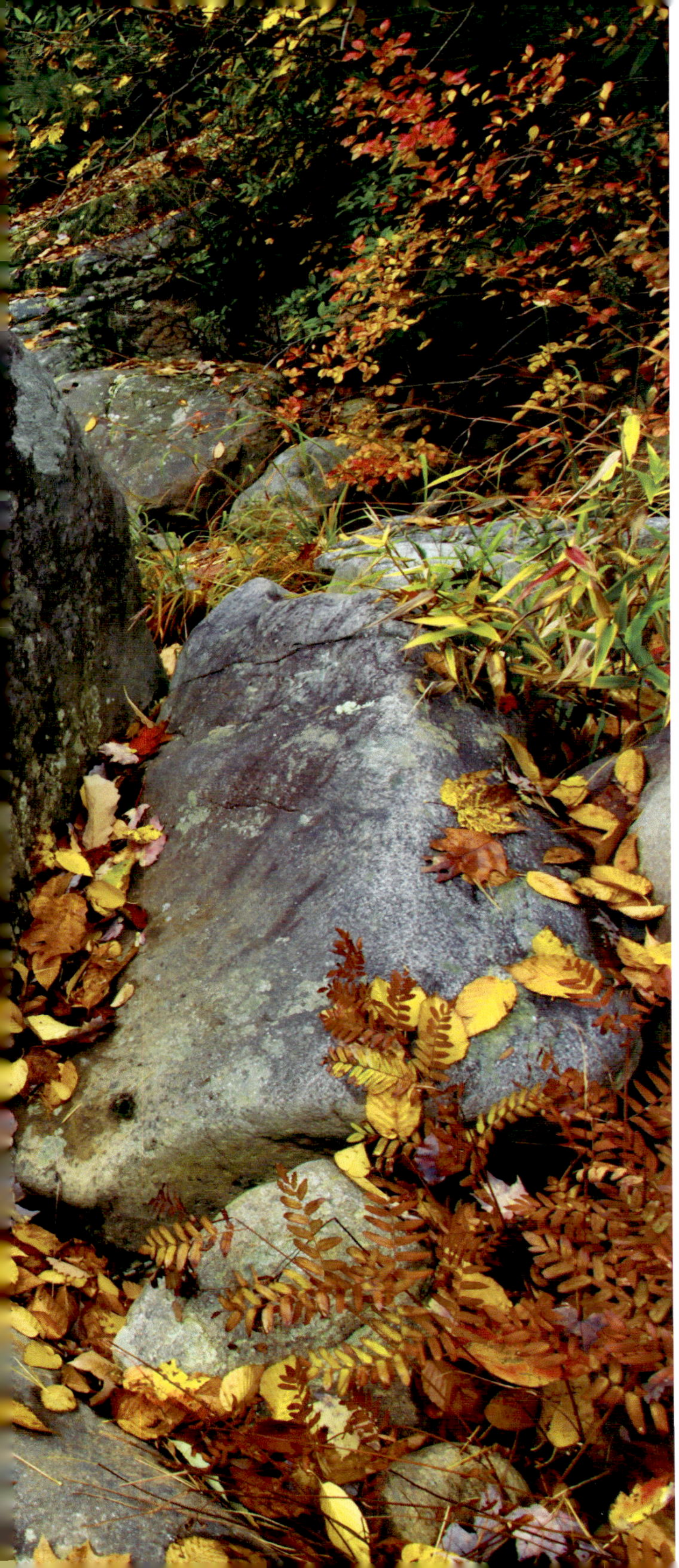

Chapter Five

The More We Look, the More We See

So where do we go to watch the river run?

The short answer is anywhere along the 132-mile course of this freshwater artery, starting with a few acres of headwaters in West Virginia, angling across western Maryland, then 88 miles through Pennsylvania—Appalachian heartland to industrial fringe of Pittsburgh. Perhaps the most helpful advice, if planning only one day at the Youghiogheny, is to go to Ohiopyle State Park in Pennsylvania or Swallow Falls State Park in Maryland. Both in and beyond those preserves, let me highlight the most revealing and reachable sites I've visited as I searched for the scenes pictured in this book.

The Middle Youghiogheny rushes over boulders downstream from Confluence.

Cathedral State Park

The upper 20 miles of the diminutive Youghiogheny—with its first few trickling miles in West Virginia and then running down to the town of Oakland, Maryland—flow almost entirely through private land with the settlement and outbuildings one might expect in rural America, and with fields that farmers cultivate down to the river's edge. But for a public view of the upper-basin Youghiogheny—open for all to see—visit the accessible wonders of Cathedral State Park, southwest of Oakland and just east of West Virginia's Aurora crossroads.

On the north side of Highway 50, turn at the sign and stroll the network of trails through 132 acres of the finest hemlocks remaining in West Virginia. A few five-hundred-year-old trees rise 90 feet, thus far spared the fatal effects of the hemlock woolly adelgid. Large oaks and maples tower over the trails as well, while black tupelos brighten to scarlet in autumn within a forest of thirty tree species. Hosting a short section of Rhine Creek, the Cathedral State Park serves as a rare reminder of what the original Appalachian woodlands and the Youghiogheny's high mountain sources once were.

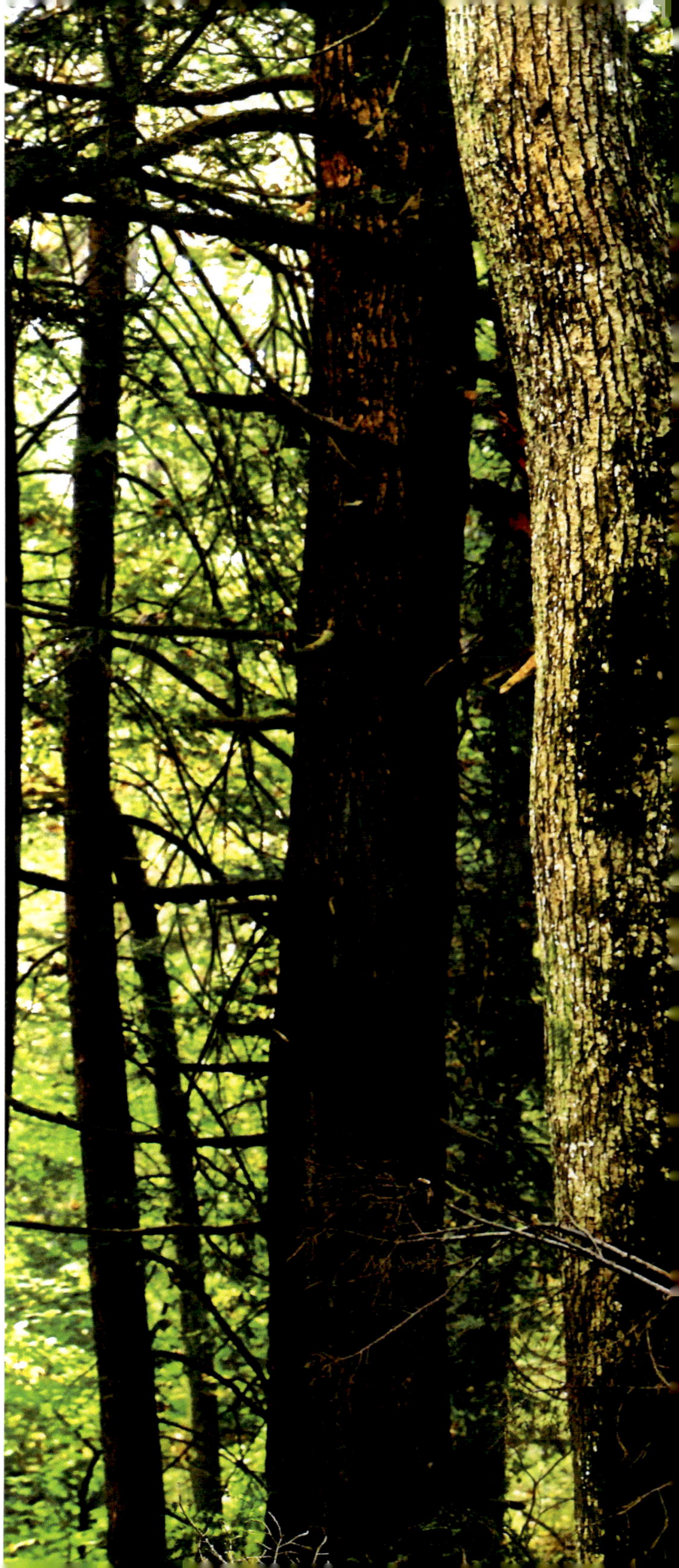

Near the Youghiogheny headwaters, ancient hemlocks rise in Cathedral State Park, West Virginia.

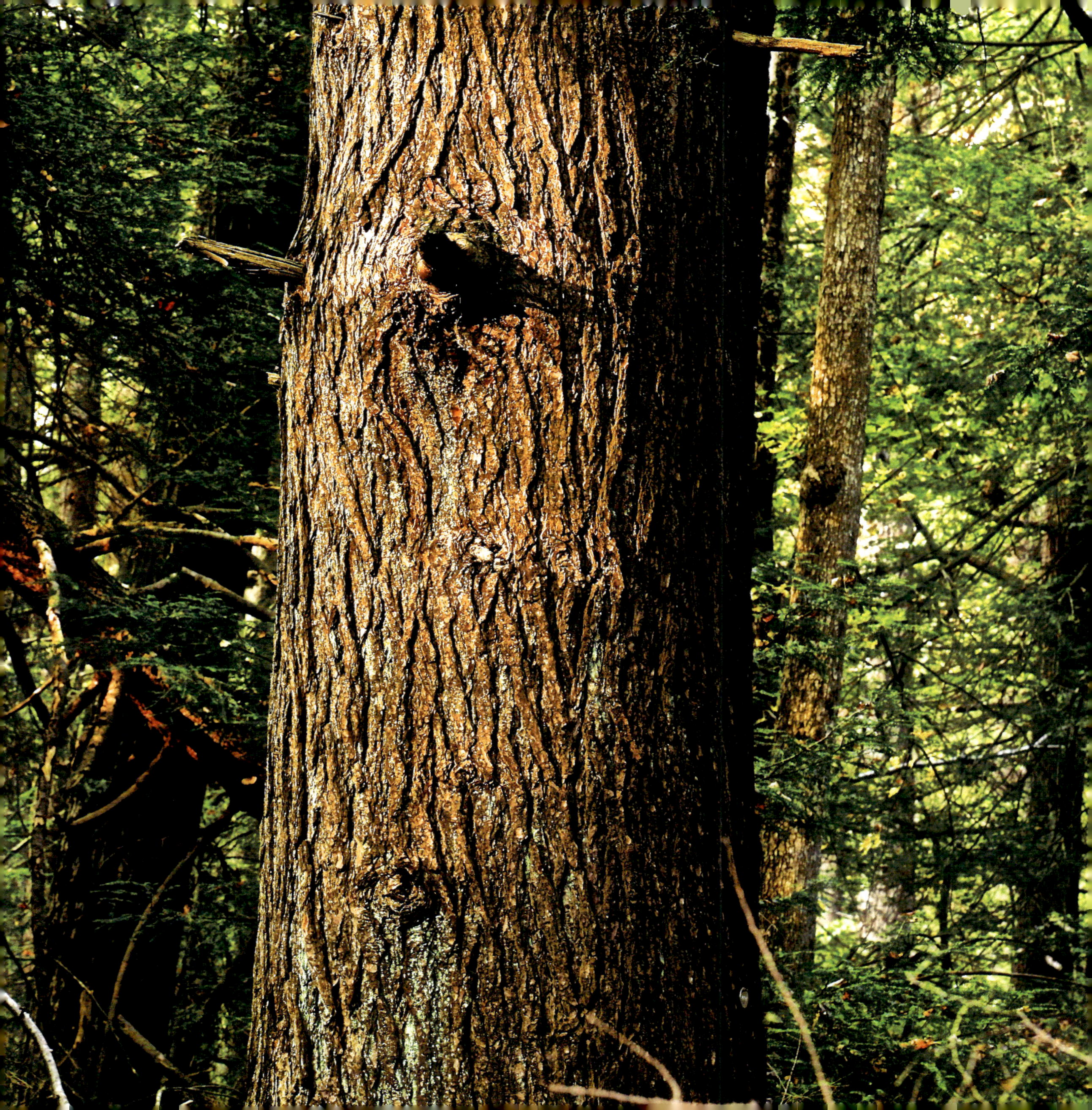

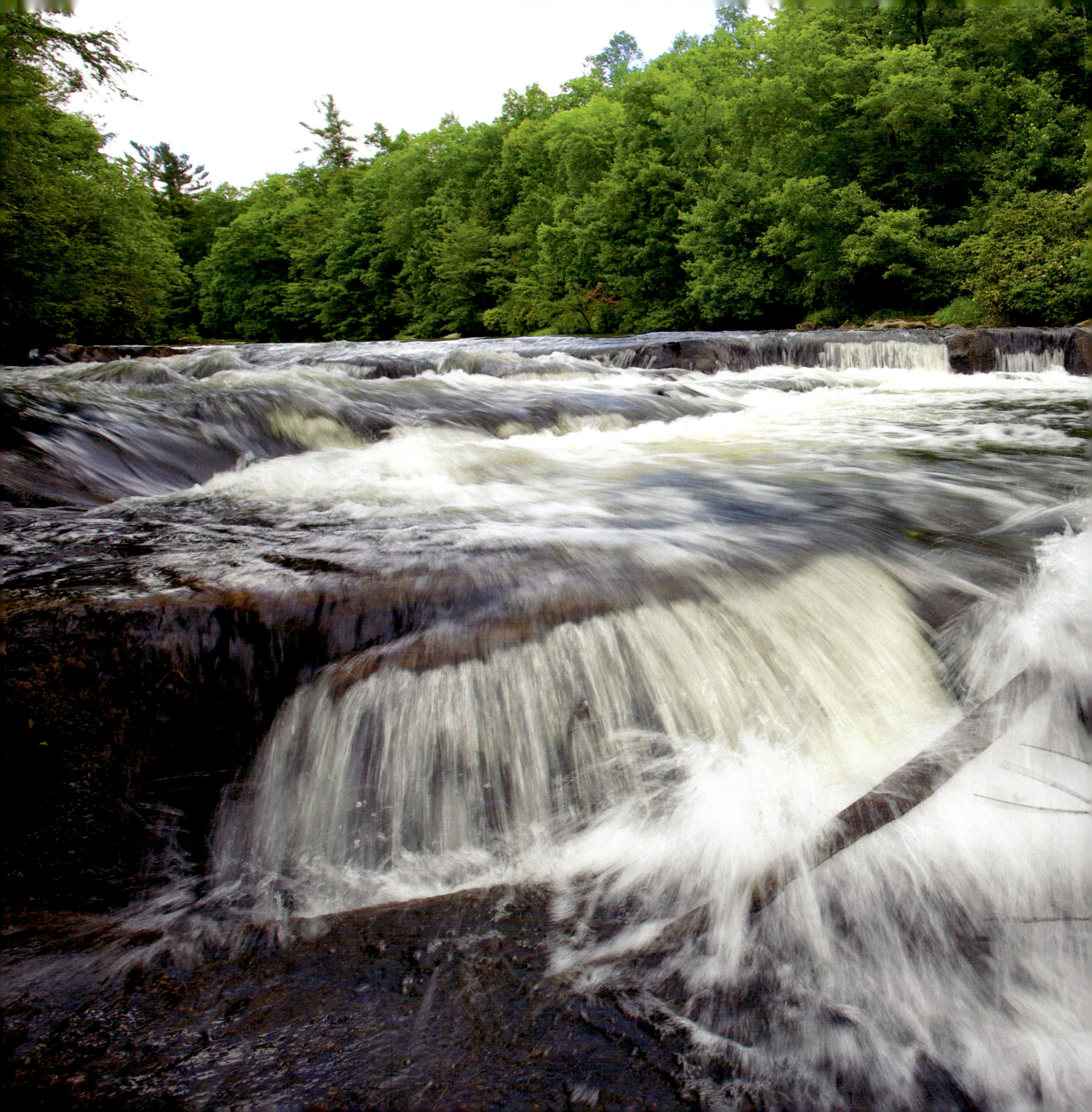

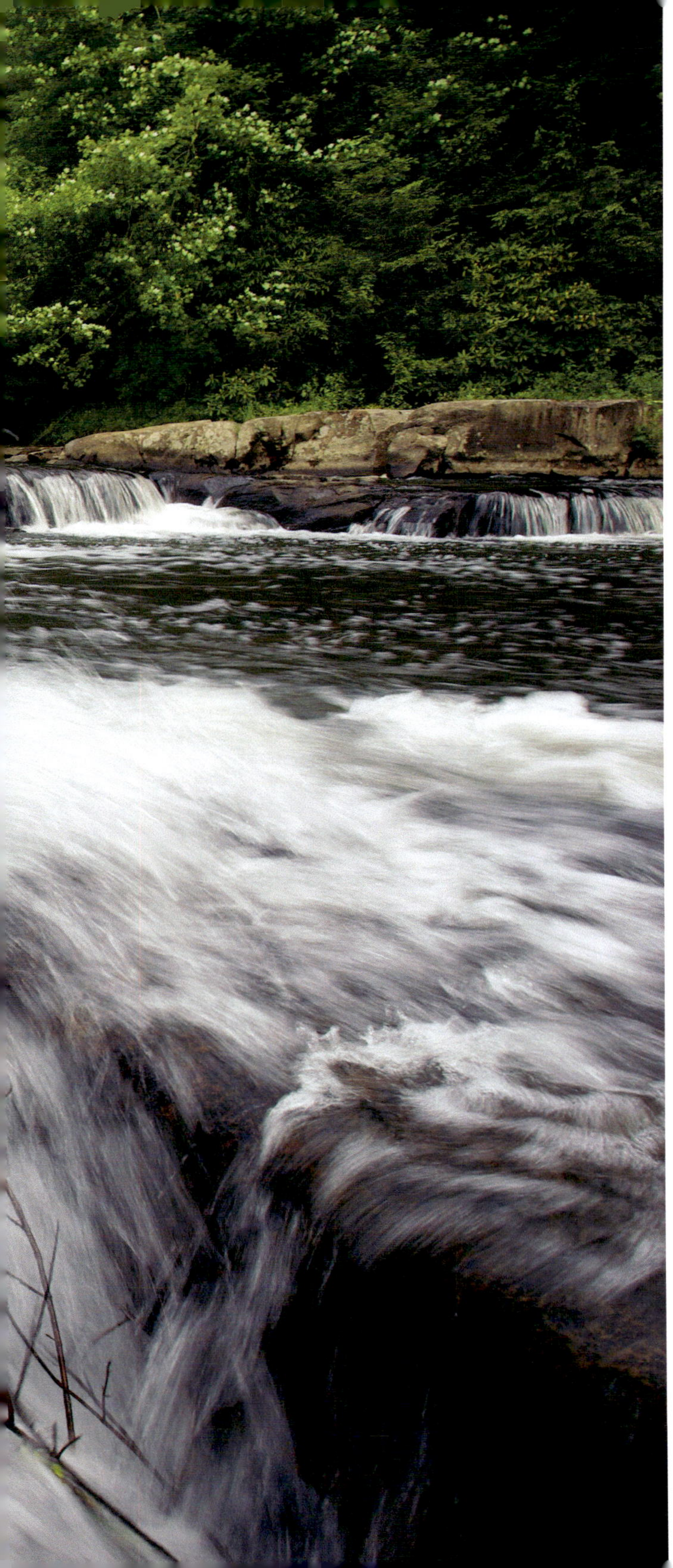

Tippy Top Yough

Nicknamed to set it apart from and upstream of the earlier-named Top Yough, this section of wild river in Maryland aims northward from Millers Run, 2 miles downstream from Oakland, to the Swallow Falls State Park bridge. Expert kayakers during high runoff paddle the first 3 miles of deceptively gentle flow below Millers, then 1.5 miles of challenging Class 2–4 ledge-riddled rapids and congested rock gardens to the Swallow Falls bridge.

To hike along the scenic, publicly accessible lower end of this reach, take Highway 219 north from Oakland 9 miles, turn west on Mayhew Inn Road for 4.5 miles, left on Oakland Sang Run Road for 0.3 miles, right on Swallow Falls Road, and 1.3 miles to the Youghiogheny bridge. Before crossing, park in the lot on the left side of the road. Walk the scenic path upstream until reaching gated private land. Quiet pools for summer swimming alternate with boulder filled rapids.

The "Tippy Top" Yough of rocky whitewater extends from Millers Run, north of Oakland, to the Swallow Falls bridge.

Swallow Falls and the Top Yough

Approaching Maryland's Swallow Falls State Park from Oakland or Highway 219, park your car just before the Youghiogheny bridge in the lot on the left. Instead of walking upstream, as in the previous hike, cross the road and go downstream on river-right. Stroll several hundred yards along the east bank to Swallow Falls' 30-foot sloped cataract. Continue to the rapid at the mouth of Muddy Creek, where that tributary cascades into the Youghiogheny on the other side, and hike onward for several more spectacular miles of the Top Yough's Class 4–5 sandstone gleaming whitewater.

For the standard, more popular view of Swallow Falls, and to see the 257-acre Park's enchanting aged hemlocks, drive across the Youghiogheny bridge to the west side, turn right at the Swallow Falls Park sign, and continue to the paved lot, which requires a modest fee. Hike 1.5 miles clockwise past fifty-three-foot-tall Muddy Creek Falls, Maryland's highest, down to the creek's mouth at the Youghiogheny, then upstream to Swallow Falls before ascending back to the parking lot.

The Top Yough is a Class 4–5 kayak run of 3.5 miles during high flows, for expert paddlers only.

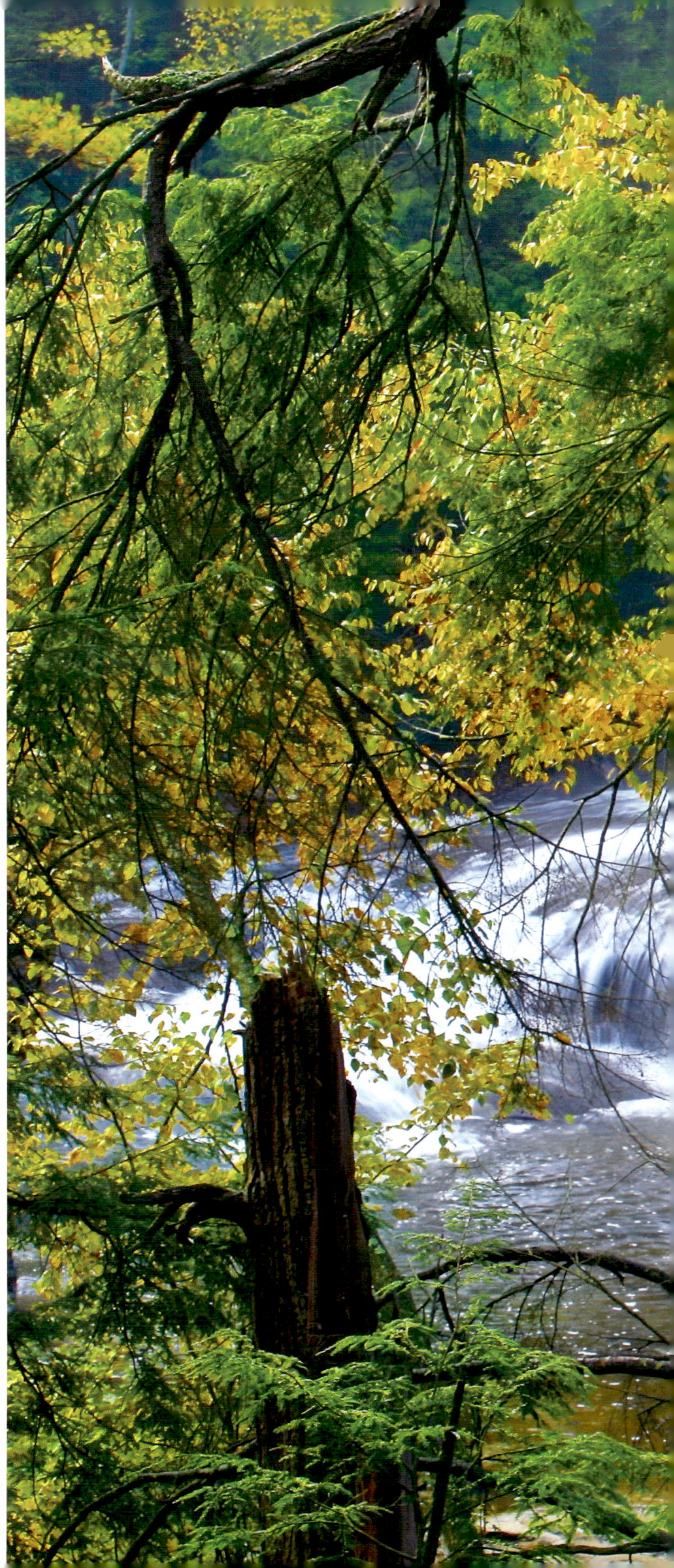

At Swallow Falls State Park in Maryland the Youghiogheny cascades over bedrock ledges. This popular park provides visitors a chance to see the Top Yough reach, with its challenging whitewater passage through stands of aged hemlocks, ending downstream at the Sang Run bridge.

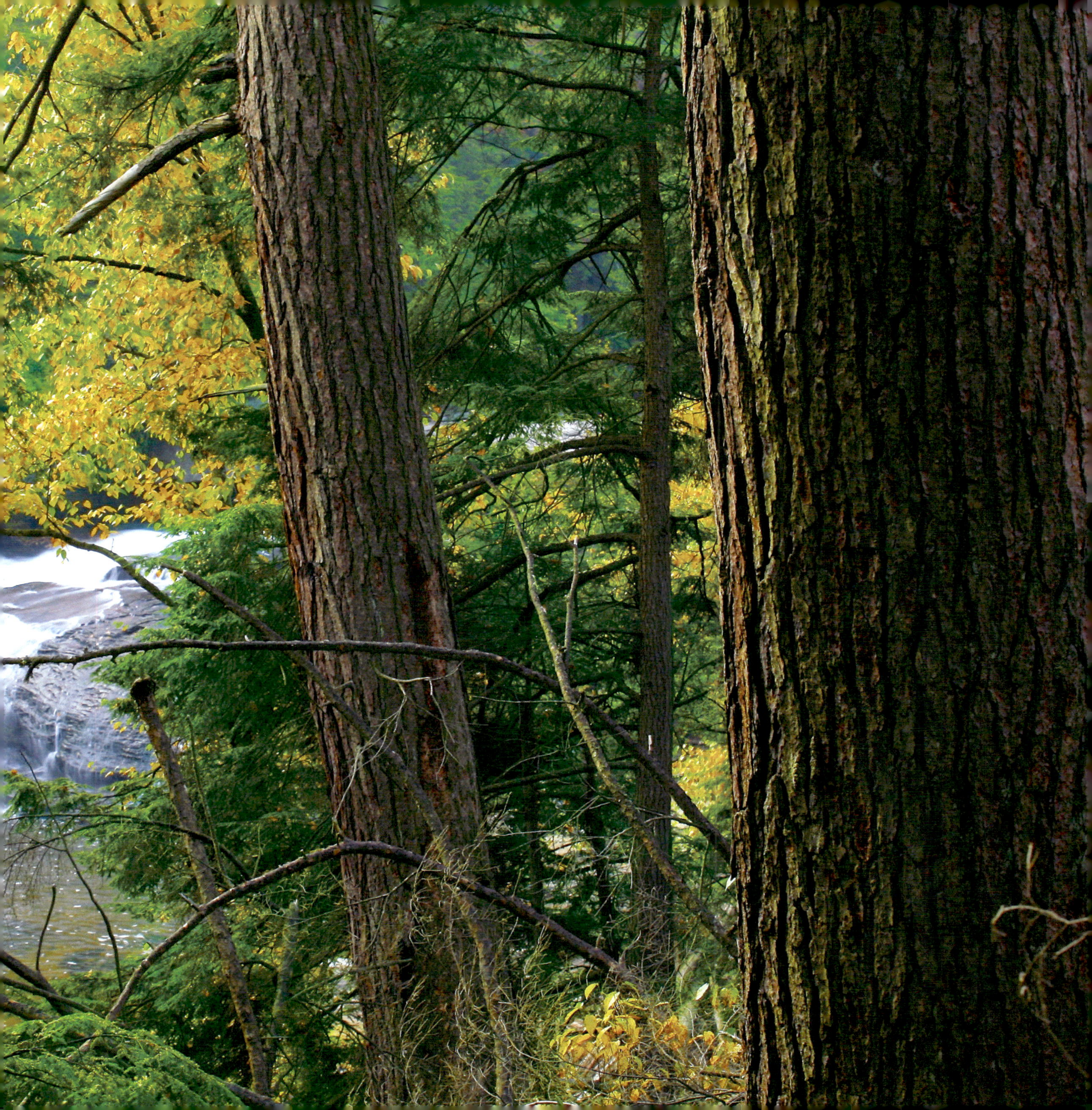

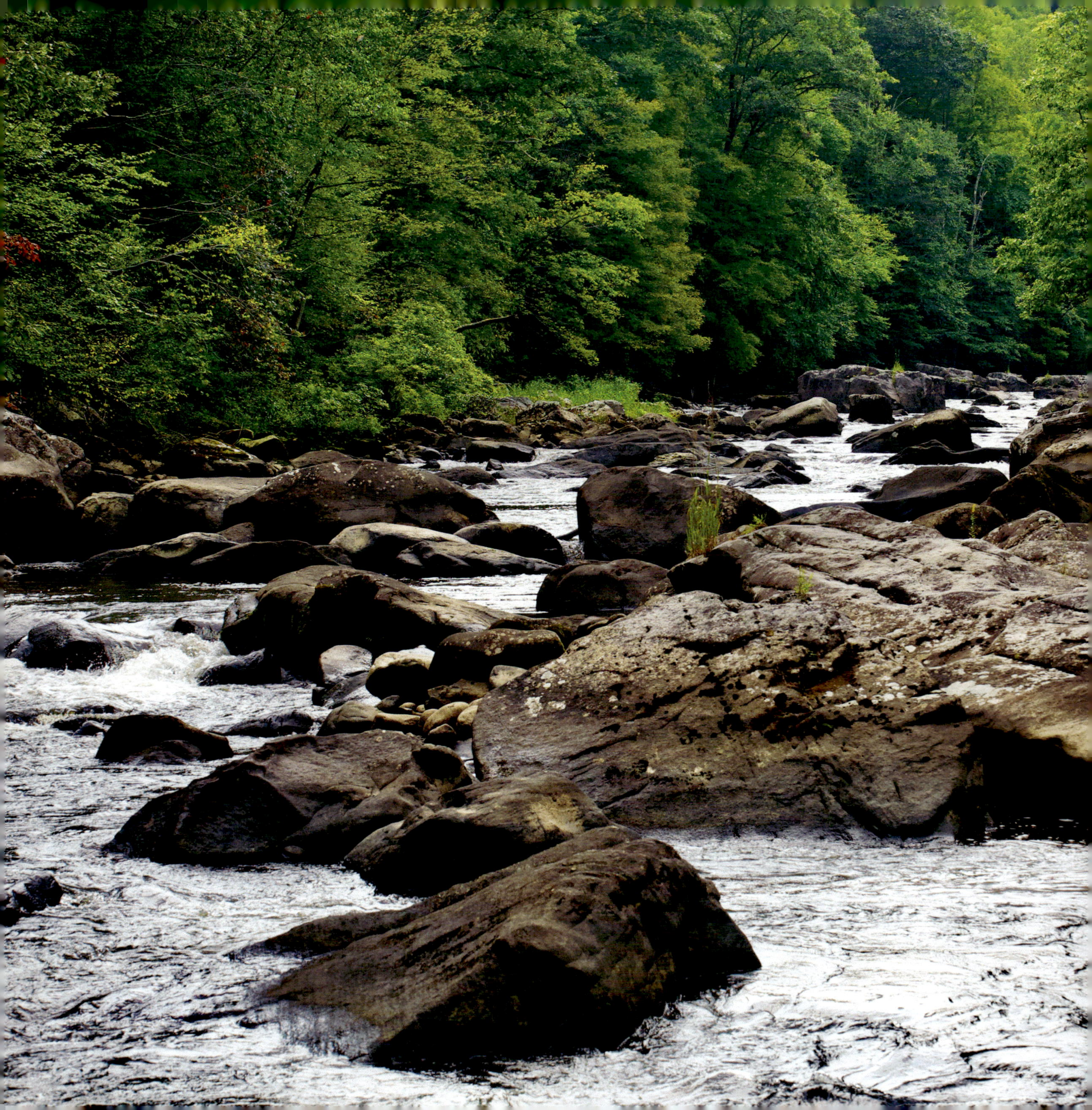

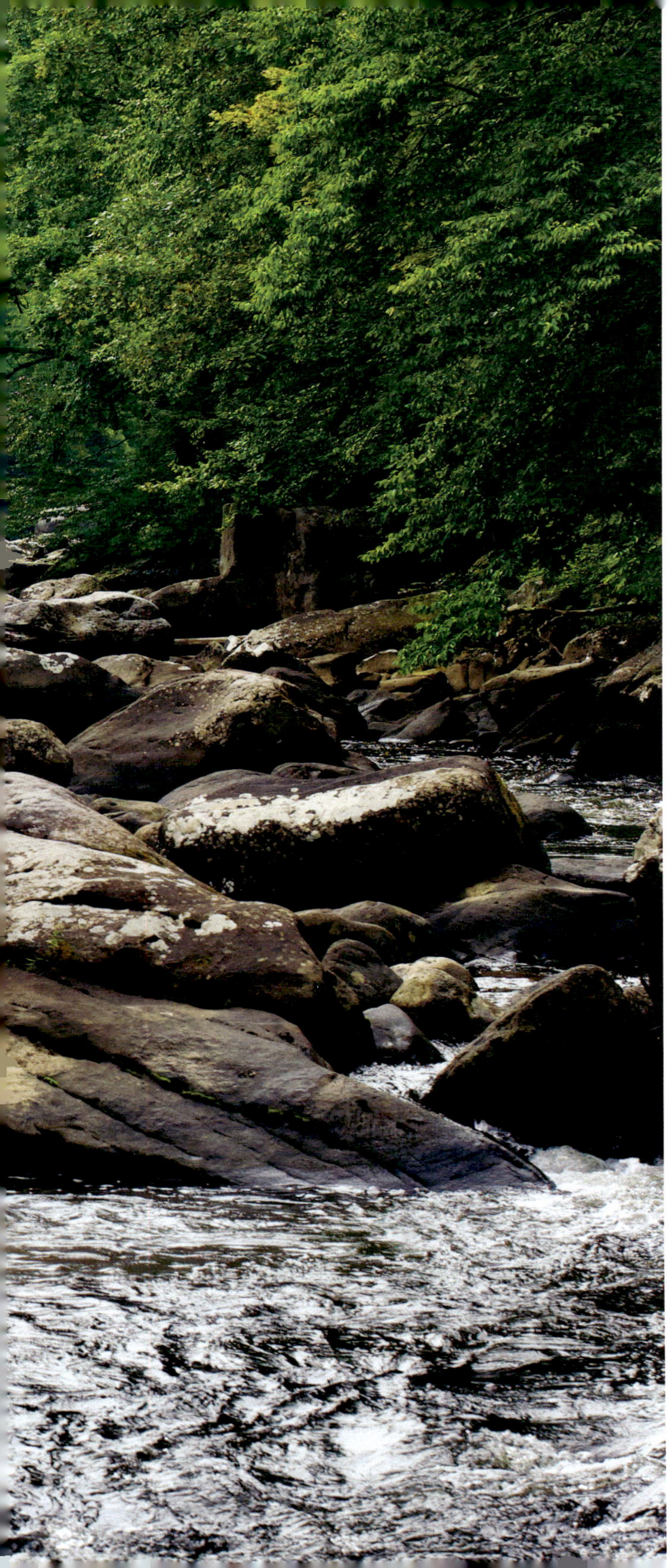

Upper Yough

Lying between Sang Run and Friendsville, the Upper Yough's nationally famous Class 4–5 rapids are not easily reached except by expert paddling or arduous bushwhacking through steep terrain, rhododendron thickets, and rattlesnake haunts—perhaps as it should be for this wildest place in Maryland. But to reach the top, take Highway 219 to the intersection at McHenry, turn west and continue to Sang Run State Park. Much of the corridor below has been acquired by the state for protection. The bridge crosses quiet water, and walking to the major rapids is extremely difficult, though the State Park has 2.5 miles of other trails through woods and meadows. For details check Internet sites for Youghiogheny Wild River Natural Environmental Area.

In contrast, the lowest section of the Upper Yough can easily be reached from Friendsville by hiking upstream on an east-bank (river-right) trail following an abandoned railroad grade with frequent river views. The trail ends at the former coal mine site of Kendall, now reforested. Rapids and rocky congestion increase upstream from there, and walking becomes an athletic—if not ankle twisting—brush entangling escapade.

Friendsville's parking areas for hikers and boaters owe to diligent cooperative efforts among paddlers, American Whitewater, the mayor of Friendsville, and local businesses, which altogether make this small Appalachian town a welcoming place for hikers and capable paddlers.

Above Friendsville several miles of the Upper Yough can be seen from the trail on the river-right or northeast side. Bedrock boulders of this rocky section show during the low-water release from Deep Creek Dam's hydroelectric generator farther upstream. Flows for paddling run higher.

Upstream of Sang Run, a hydropower dam built in 1925 on the tributary Deep Creek releases water to generate electricity during afternoon peaking hours. The dam's federal license recognizes the need to serve whitewater boating at the Upper Yough, with releases planned for two or three afternoons per week during summer boating season. Flows are otherwise low and clogged with rocks. With the put-in park collaboratively managed by American Whitewater and the Maryland Department of Natural Resources, this 9.6-mile Sang Run-to-Friendsville high adventure drops an average of 52 feet per mile but up to a precipitous 120 feet, for expert kayakers and rafters or for physically fit commercial rafting clients only.

Youghiogheny Dam

The dam that backs up Youghiogheny Reservoir 16 miles can be seen by driving from the Pennsylvania Turnpike at Somerset southbound on Highway 281 to Confluence. Cross the Casselman River bridge, in 200 yards cross the Youghiogheny bridge, turn left (south-east) at the sign, and continue up to the public road crossing of the 184-foot-high dam. The spillway as well as a boat ramp, picnic area, and swimming access for the reservoir lie on the opposite (east) side of the dam.

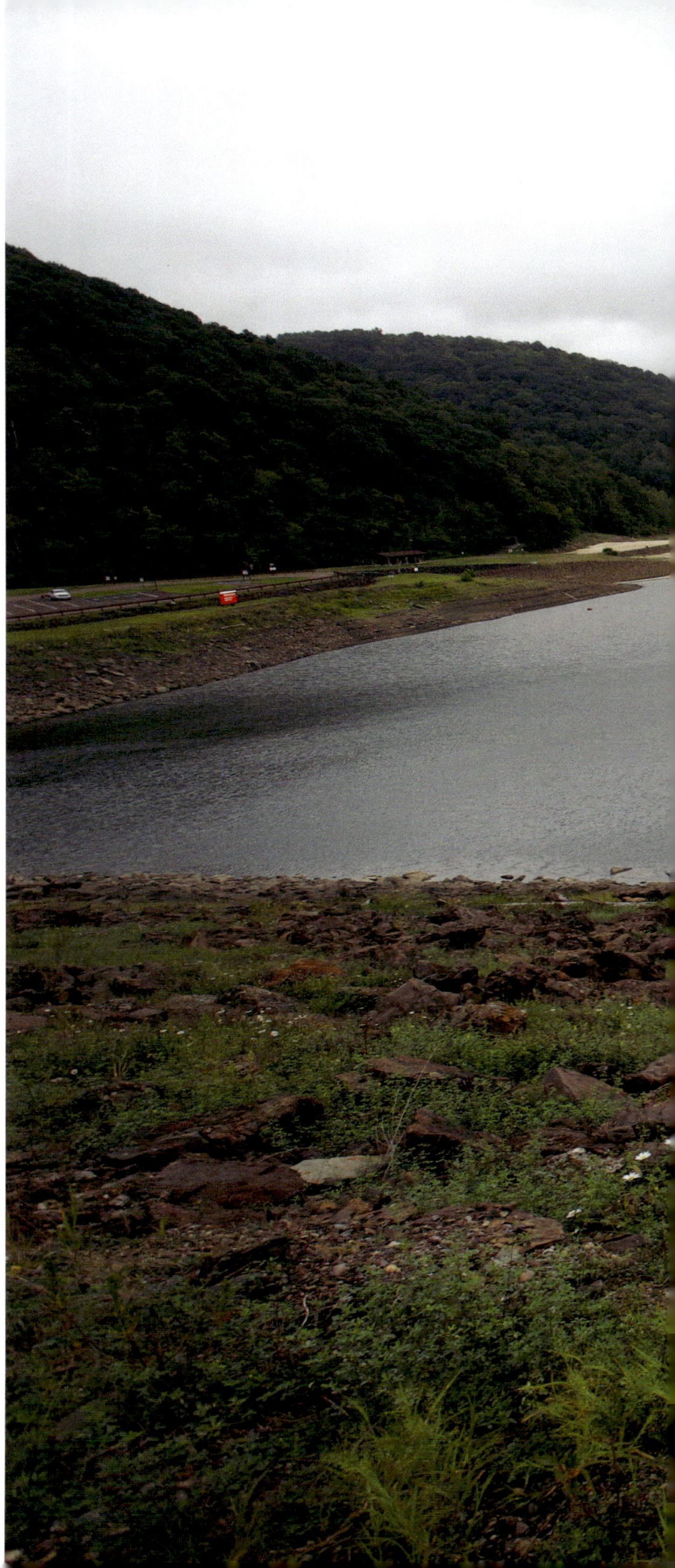

To see Youghiogheny Dam, take Route 281 south from the Youghioheny bridge in Confluence, turn east at the sign for the dam, and drive to the top of the dam and across it.

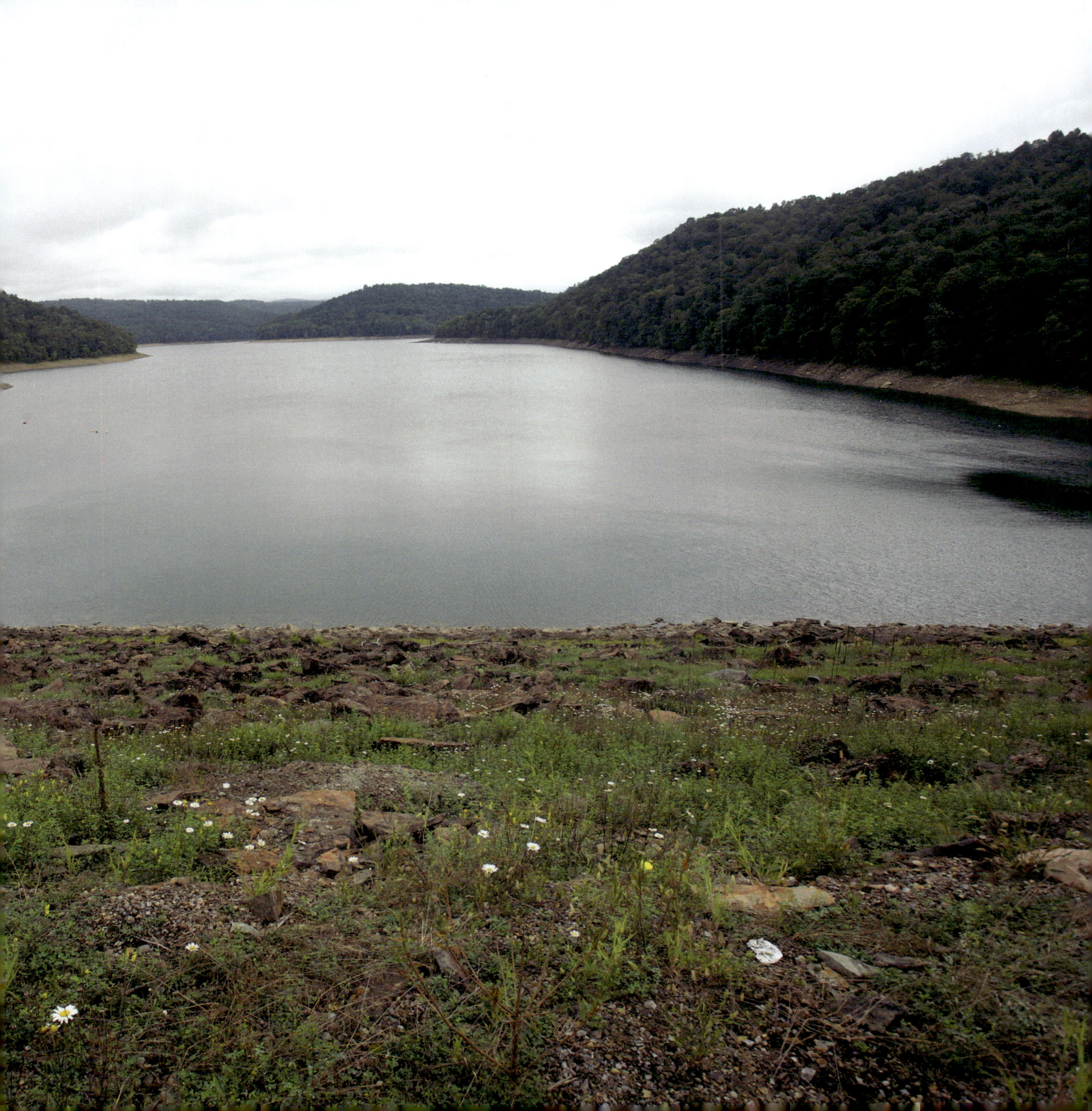

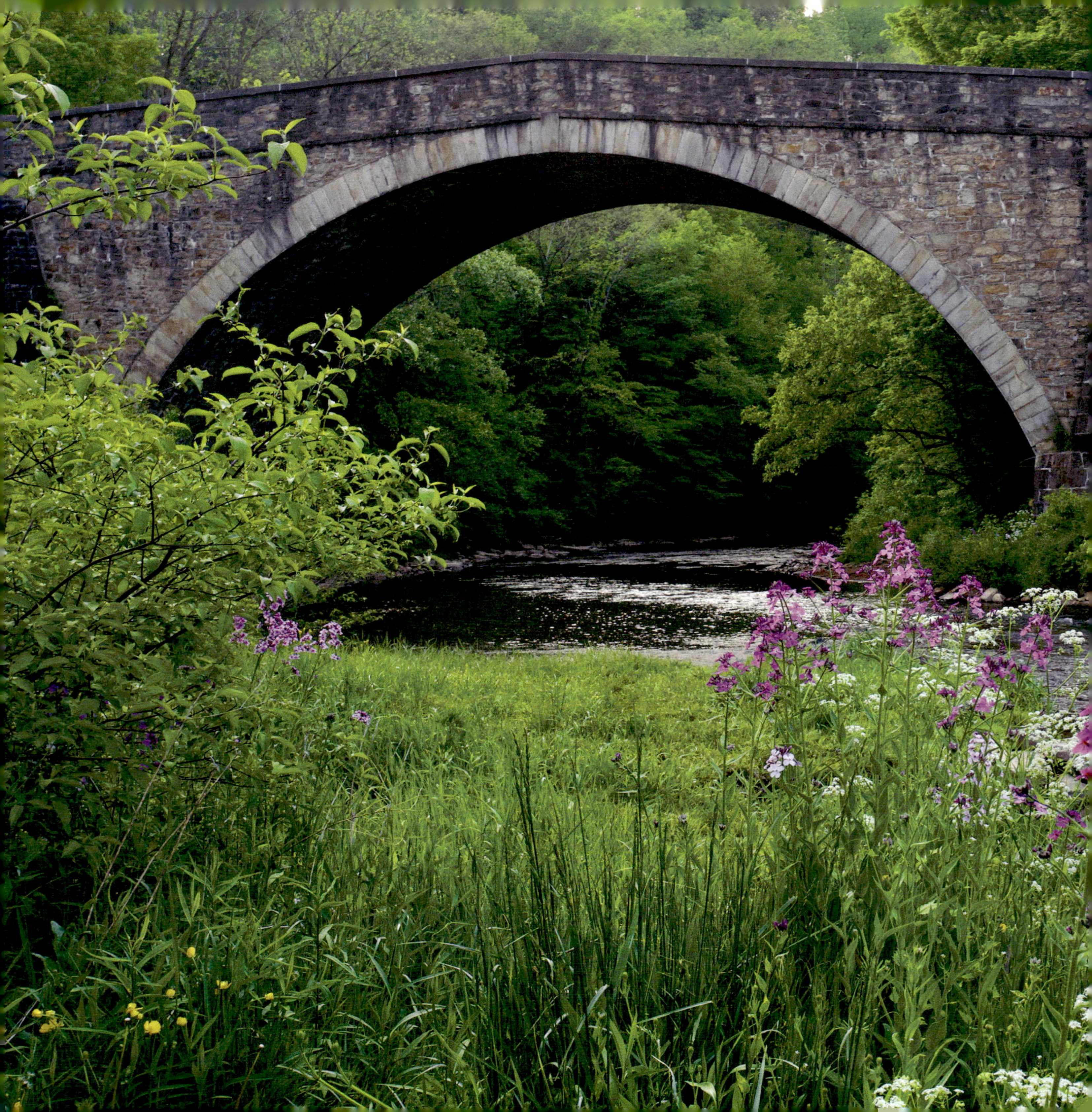

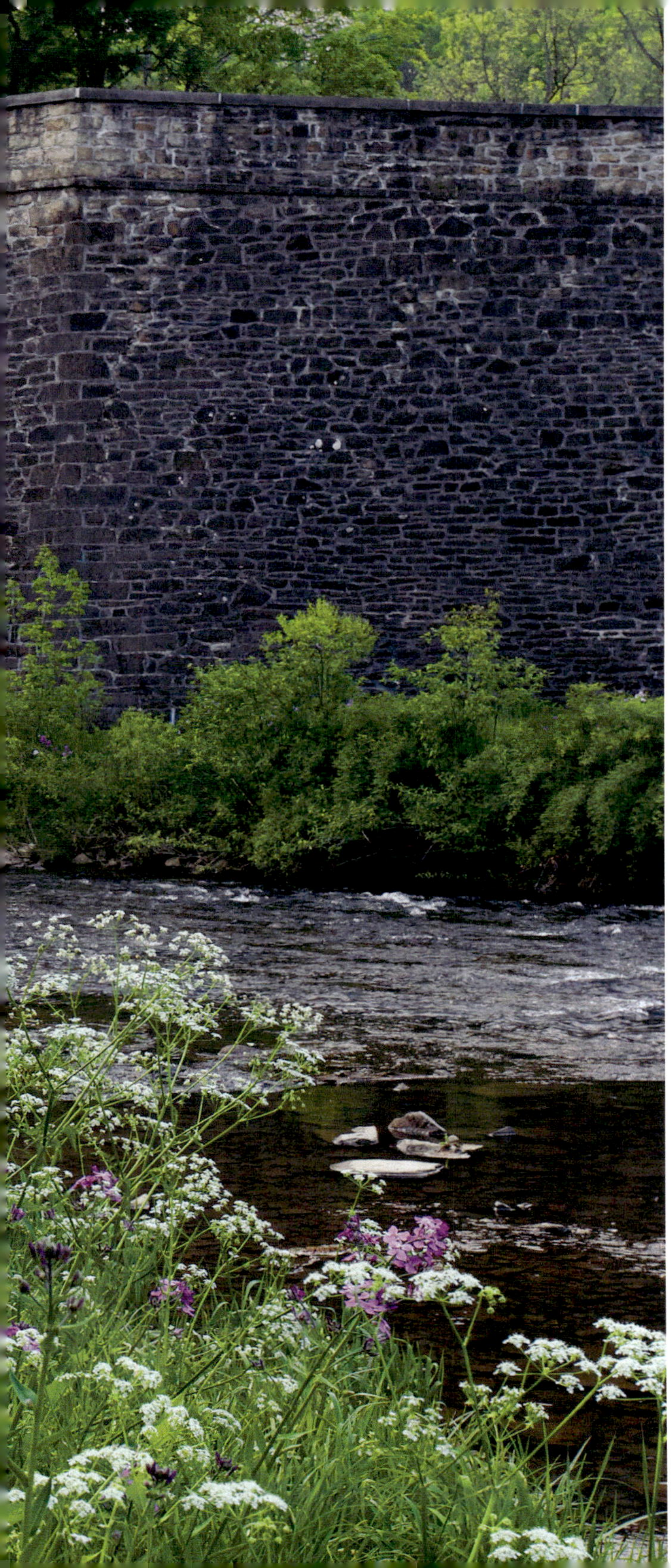

Casselman River

The 55-mile-long Casselman flows from Meadow Mountain northward and riffles under Highway 40 near the Pennsylvania line at Grantsville, Maryland. Just east of that village, one of the nation's oldest bridges was the largest single-span stone arch in America when built in 1817 for the C&O Canal, which was never completed owing to railroad competition. Presidents Monroe, Jackson, Harrison, Polk, and Taylor, as well as young congressman Abraham Lincoln all crossed this bridge, now for pedestrians only.

From there the Casselman riffles through the small towns of Salisbury, Meyersdale, and Garrett, followed by an 8-mile scenic wild gorge to Rockwood. Then the river continues with roadless mileage but bridge access at the settlement of Markleton, where deeply wooded Town Line Run enters on river-left, and also 4 miles farther at the Fort Hill bridge east of Highway 281.

This largest Youghiogheny tributary is not paralleled by any highway, so to see the stream, take the Great Allegheny Passage bike trail. Perfect! Set back in the forest, it follows the river's riffles, rapids, and pools for 30 miles, from Meyersdale down westward to Confluence in a roundabout path with views through the forest and with access at the streamside towns. The 8-mile Garrett-to-Rockwood reach is my favorite, making a fine 16-mile, virtually flat, out-and-back ride through recovering forest.

For this view of the riffling Casselman River and its historic stone bridge, take Highway 40 just east of Grantsville, Maryland, and turn north into the small state park with the historic stone arch bridge.

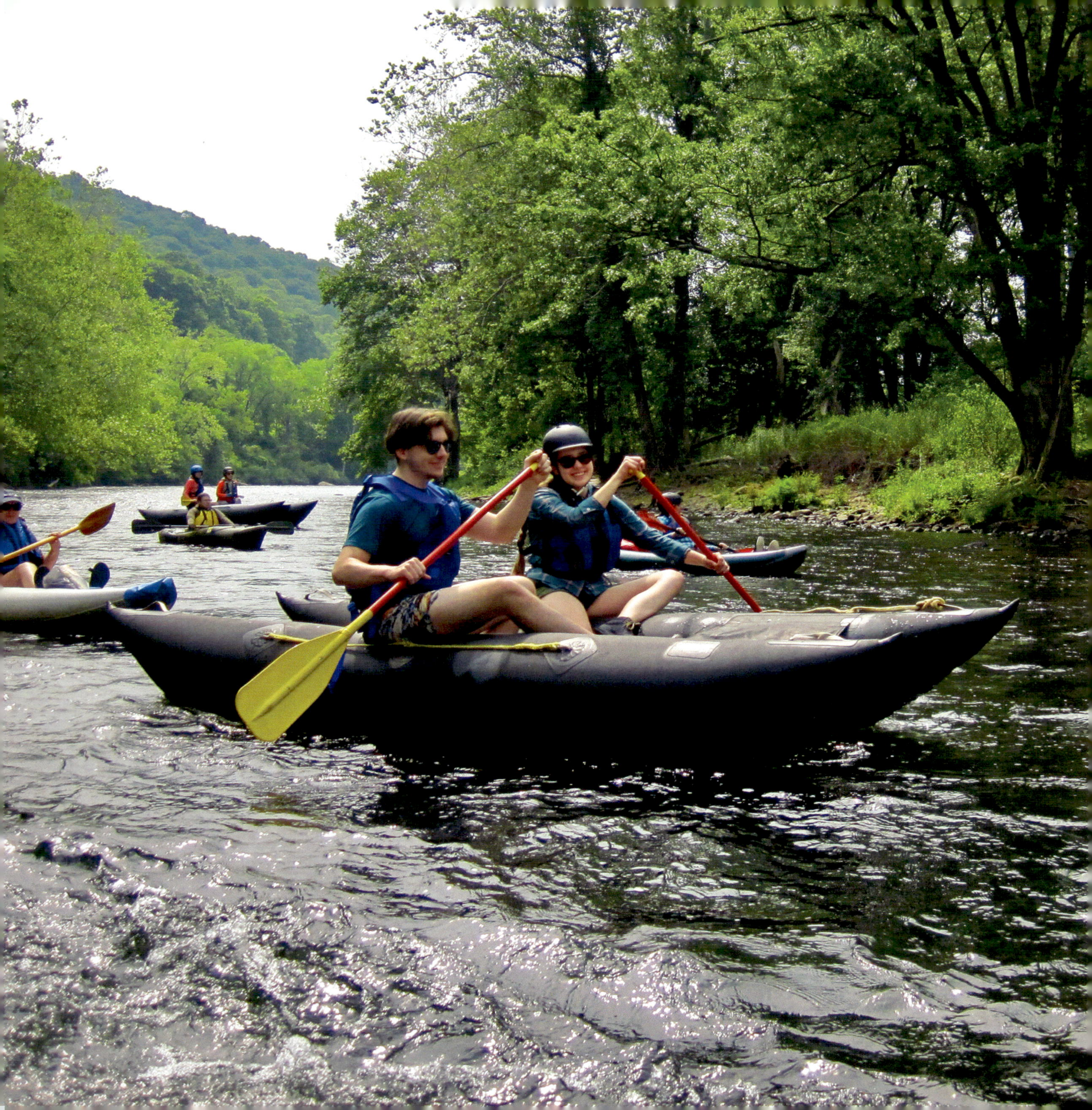

Highlighting this basin, scenic Whites Creek enters the Casselman just above Confluence at the village of Harnedsville. For a short wooded path and fishing access to this shaded little gem, take Route 523 south from Confluence, at the crossroads of Listonburg turn left before the bridge, and continue upstream on Whites Creek Road to angler pulloffs.

Middle Youghiogheny

This stellar 11-mile section of the Youghiogheny, from Confluence to Ohiopyle, is mostly without road access, but it is bikeable via the Great Allegheny Passage trail on the south side. Park on river-left immediately downstream of Highway 281, south of Confluence, or take the narrow paved road 3 miles west from the Youghiogheny bridge to Ohiopyle State Park's Ramcat access for biking, boating, and fishing. From there the trail runs 8 miles to the town of Ohiopyle and onward through Ohiopyle State Park. The entire trail, however, runs 150 miles from its western terminus in Pittsburgh to Cumberland, Maryland. Continuing eastward with the adjoining C&O Canal bike trail along the North Branch and main stem Potomac River, an uninterrupted, road-free bikeway—unequaled in the United States—runs 355 miles, from Pittsburgh to Washington, DC.

The Middle Youghiogheny, from Confluence to Ohiopyle, features moderate Class 2 rapids that are popular among intermediate canoeists and capable novices in kayaks and rafts, including this Shredder—a two-person paddle craft invented on the Youghiogheny by local whitewater legend Tom Love and made in Confluence.

One of the East's premier Class 2 paddling rivers that flows adequately all summer, the Middle Youghiogheny, for 8 miles from Ramcat to Ohiopyle, has an active rail line on the north side but no land development. Riffles, small rapids with rocks to dodge, and relaxing sections of quiet water appeal to novice rafters and kayakers, canoeists experienced in Class 2 whitewater, and standup boarders capable of swift current. Nearing the Ohiopyle left-side takeout, novices should go to the right of Elephant Rock, which rises in the center of the river, visible from far upstream. Beware during high flows, as the takeout lies just below, and Ohiopyle Falls awaits not far downstream.

The Ohiopyle State Park wisely bans Middle Yough access to rental rafters when the gauge at Ohiopyle reaches 6 feet, though the river is pushy, muddy, and forceful at levels several feet less than that.

To reach either end of this popular run, drive the Sugarloaf (Ohiopyle–Confluence) Road. For shuttles, see Ohiopyle's commercial outfitters. Overnight camping may be tempting, but it is not permitted along the river in the Park.

Canoe or raft camping, by the way, is unfortunately not allowed or much suited to any of the Youghiogheny except for the lower reaches below Connellsville, but paddlers of those gentle waters need to be aware of private landownership.

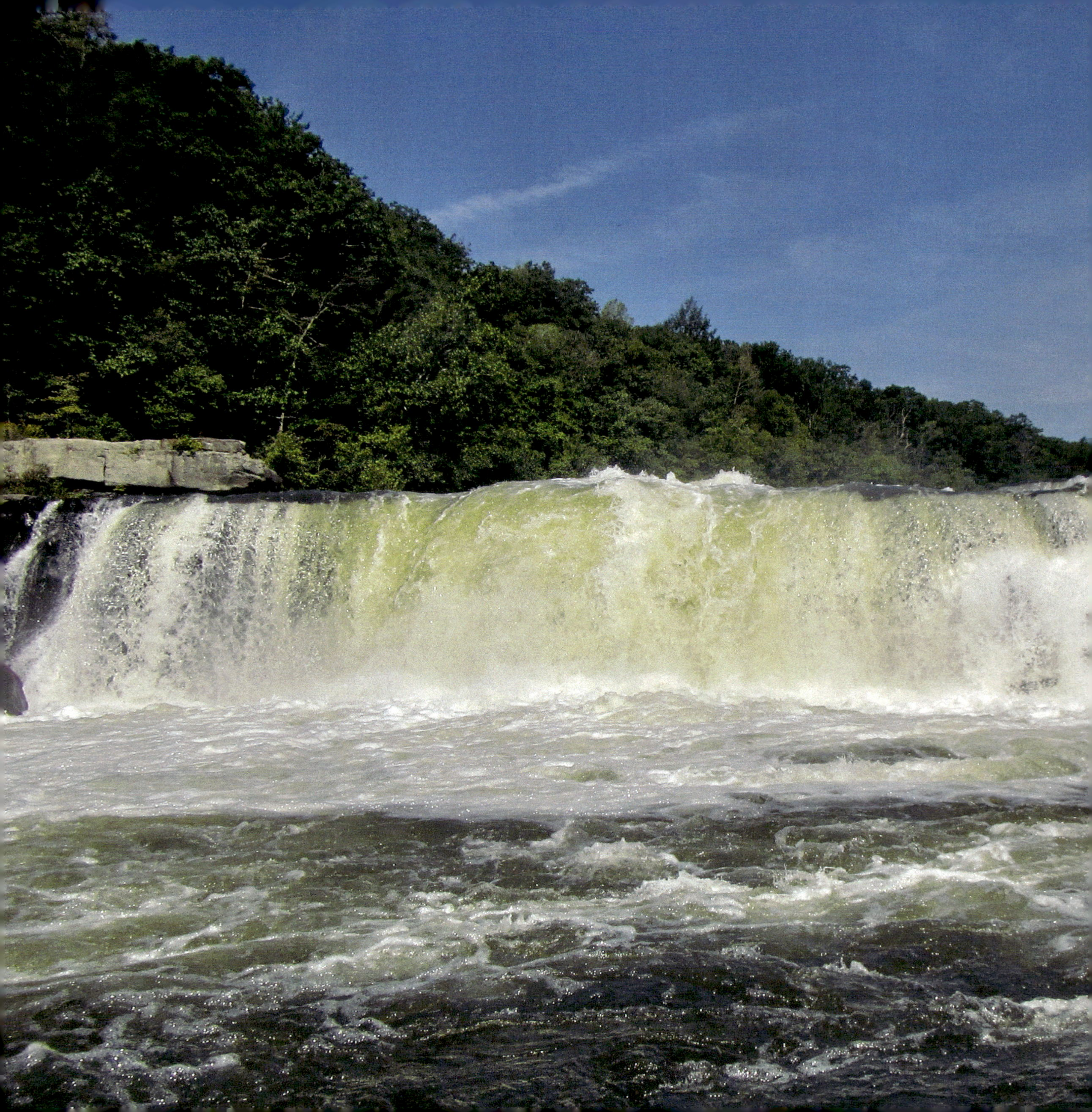

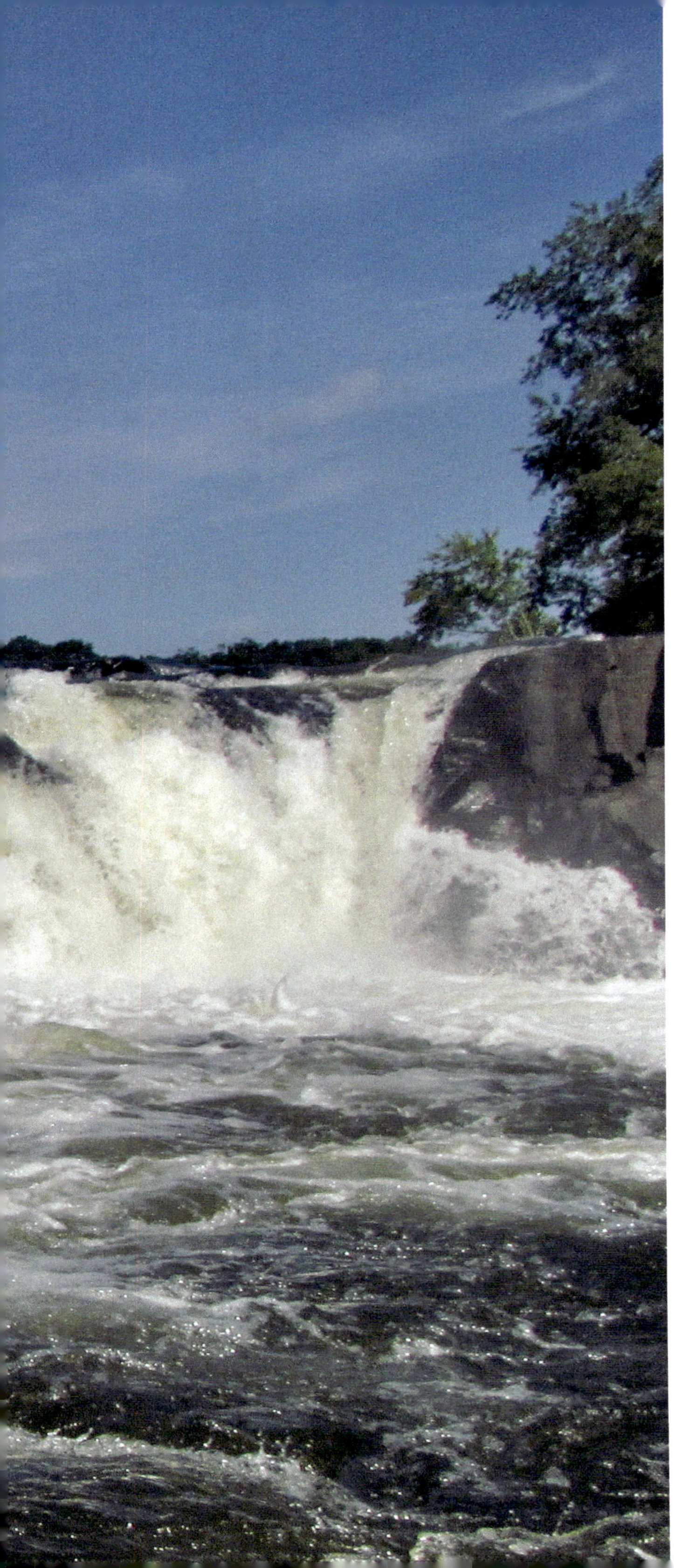

Ohiopyle Falls

This is the big fish, so to speak, of Youghiogheny scenery and river-viewing wonders, so expect crowds and a festive scene, at least on summer weekends. To drive to Ohiopyle from the Pennsylvania Turnpike, exit at Donegal, go east on Highway 31, then quickly south on Highway 381 for 21 miles. At Ohiopyle, cross the Youghiogheny and park on the right in the big lot at the Falls. Viewing platforms overlook the cataract, and paved paths follow the rocky shore between the bridge and the Ohiopyle State Park Visitor Center.

For a more adventurous eyeful, take the quarter-mile walk on the other (river-right) side. When arriving in Ohiopyle southbound on Highway 381, turn right just before the Youghiogheny bridge, and park in the gravel Ferncliff lot. Walk toward the river and downstream on paths across sandstone outcrops. Admire the river's sharp ledge drops leading to the Falls and then the eighteen-foot plunge.

The head-on view of Ohiopyle Falls can only be seen from a raft or kayak immediately below the eighteen-foot drop.

Ferncliff Peninsula

Classic not only for the Youghiogheny but for Pennsylvania and the Appalachians, woodland trails circumnavigate Ferncliff Peninsula where the river cuts its famous clockwise loop below Ohiopyle. Hemlocks, venerable deciduous trees, rocky outcrops, cliff faces, and rapids with large boulders all enchant. From the Ferncliff parking lot (see above), walk toward the river, continue downstream past the Falls, then clockwise around the Peninsula's trail complex for a 2-mile circuit.

Or, from the Ferncliff lot, walk away from the highway and northwest on the bike trail, angle left on the boaters' path before crossing the bike trail's High Bridge, and descend among magnificent tuliptrees to the base of Railroad Rapid. Some of the finest Youghiogheny River views can be reached here by nimble hikers carefully walking the boulders upstream to the top of the rapid. Returning to the main Ferncliff Trail uphill from Railroad Rapid, walk counterclockwise around the Ohiopyle Loop through magnificent forest, past the Falls, and back to the Ferncliff lot.

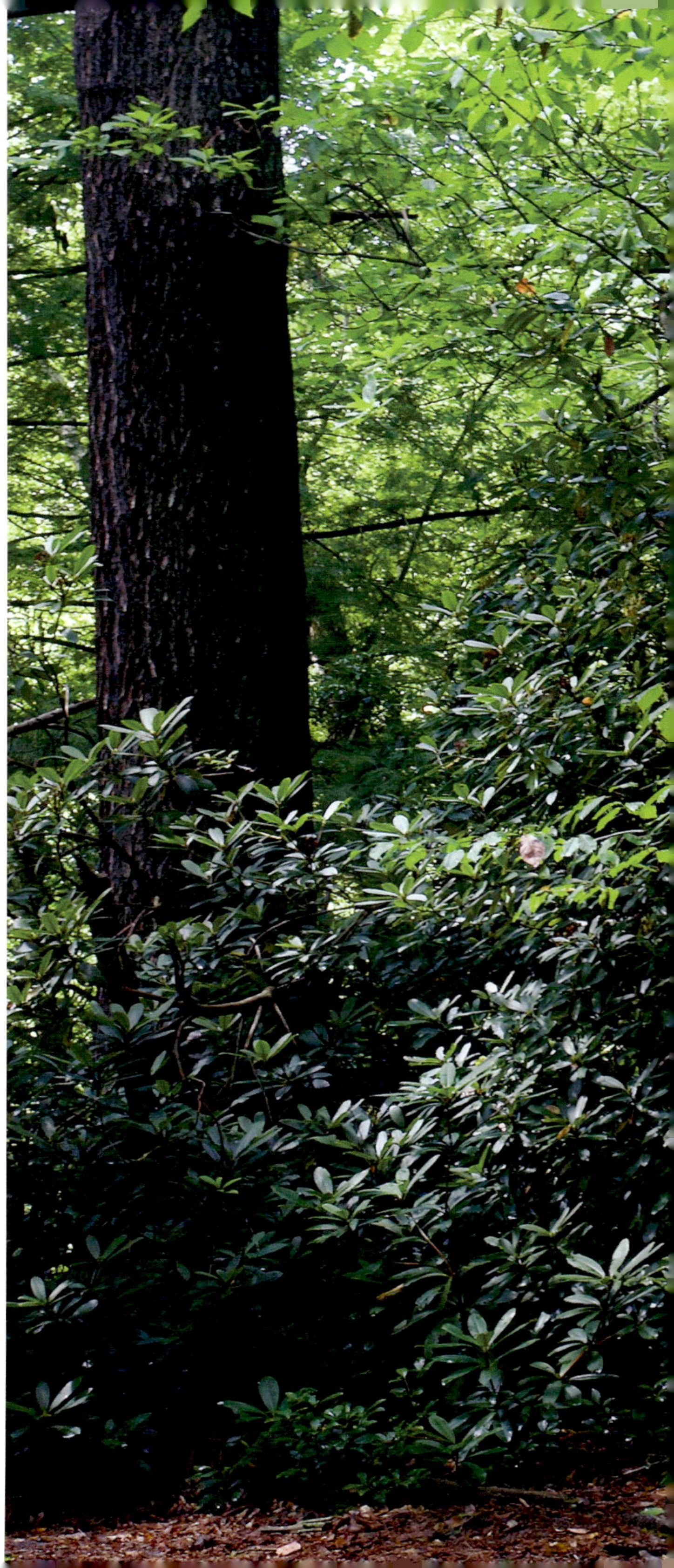

Ferncliff Peninsula's network of trails beautifully tour a recovering Appalachian forest.

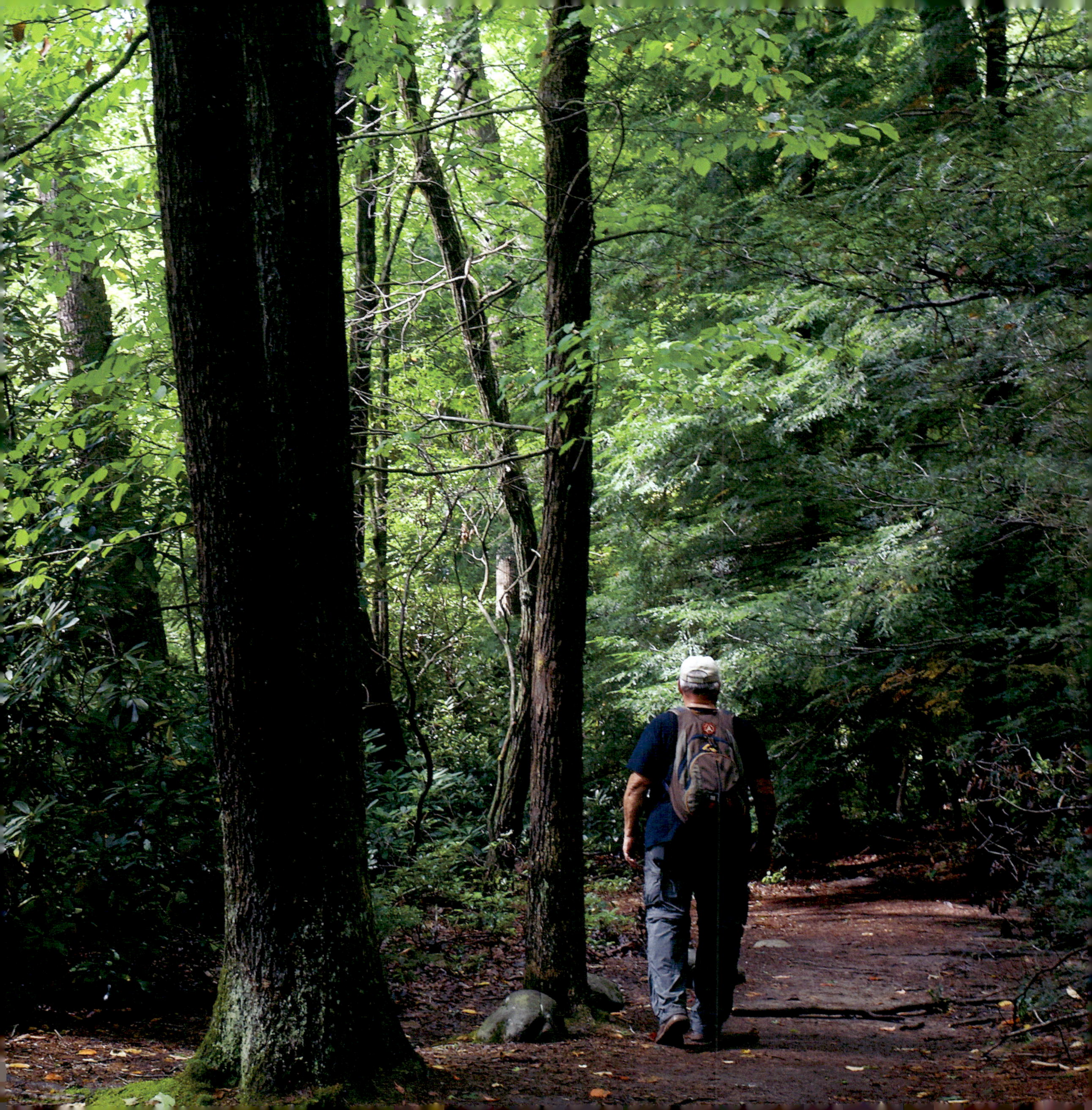

Meadow Run

This largest tributary within Ohiopyle State Park is reached with a quarter-mile walk along the road downstream from the Falls or visitor center, or immediately from a smaller parking lot just south of Ohiopyle at the junction of the Kentuck/Chalk Hill Road and the Highway 381/Farmington Road. Descend the stairs to layered sandstone at Meadow Run's natural flume—a worthwhile destination by itself. Hike upstream through deep forest 1.3 miles to a magnificent flat slab of sandstone, and beyond to the Meadow Run Cascades Waterfall.

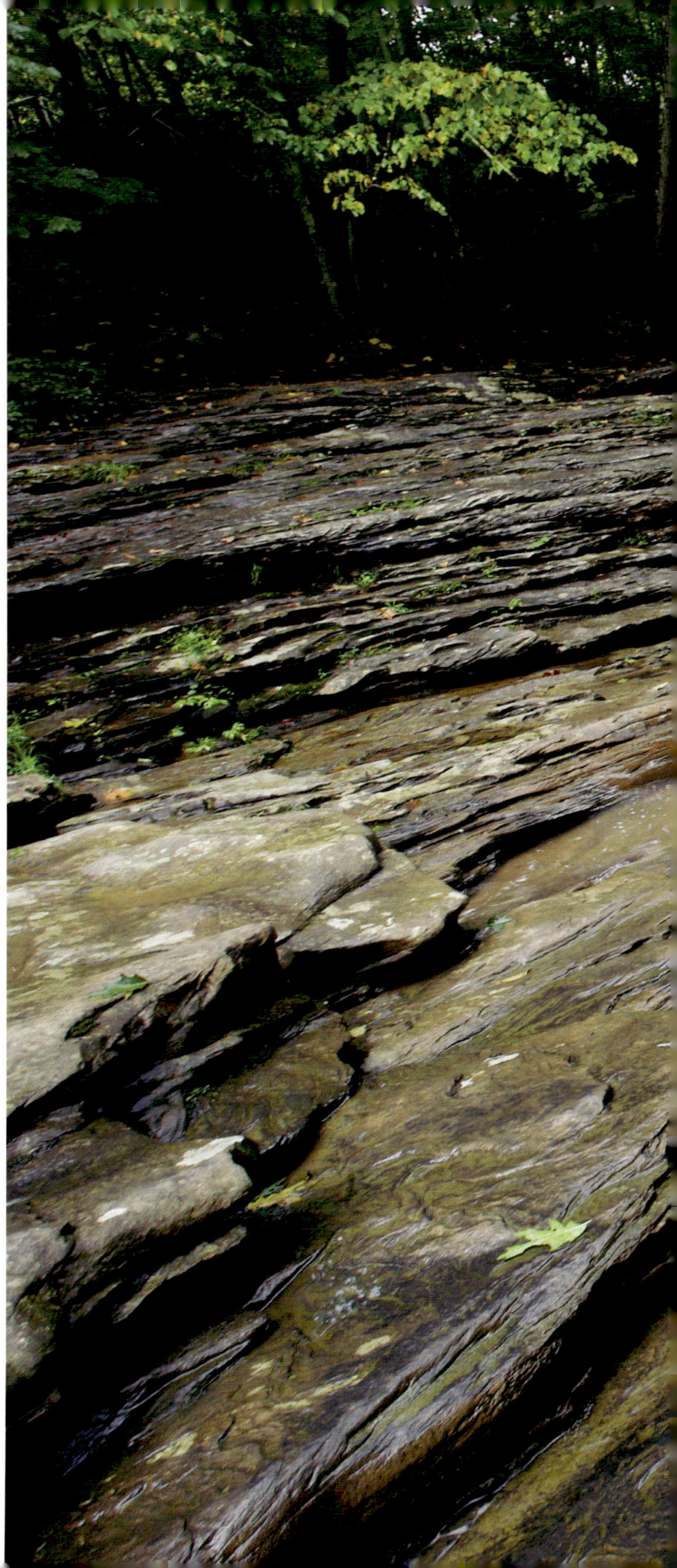

An excellent trail follows the lower section of Meadow Run, a key tributary to the Youghiogheny within Ohiopyle State Park.

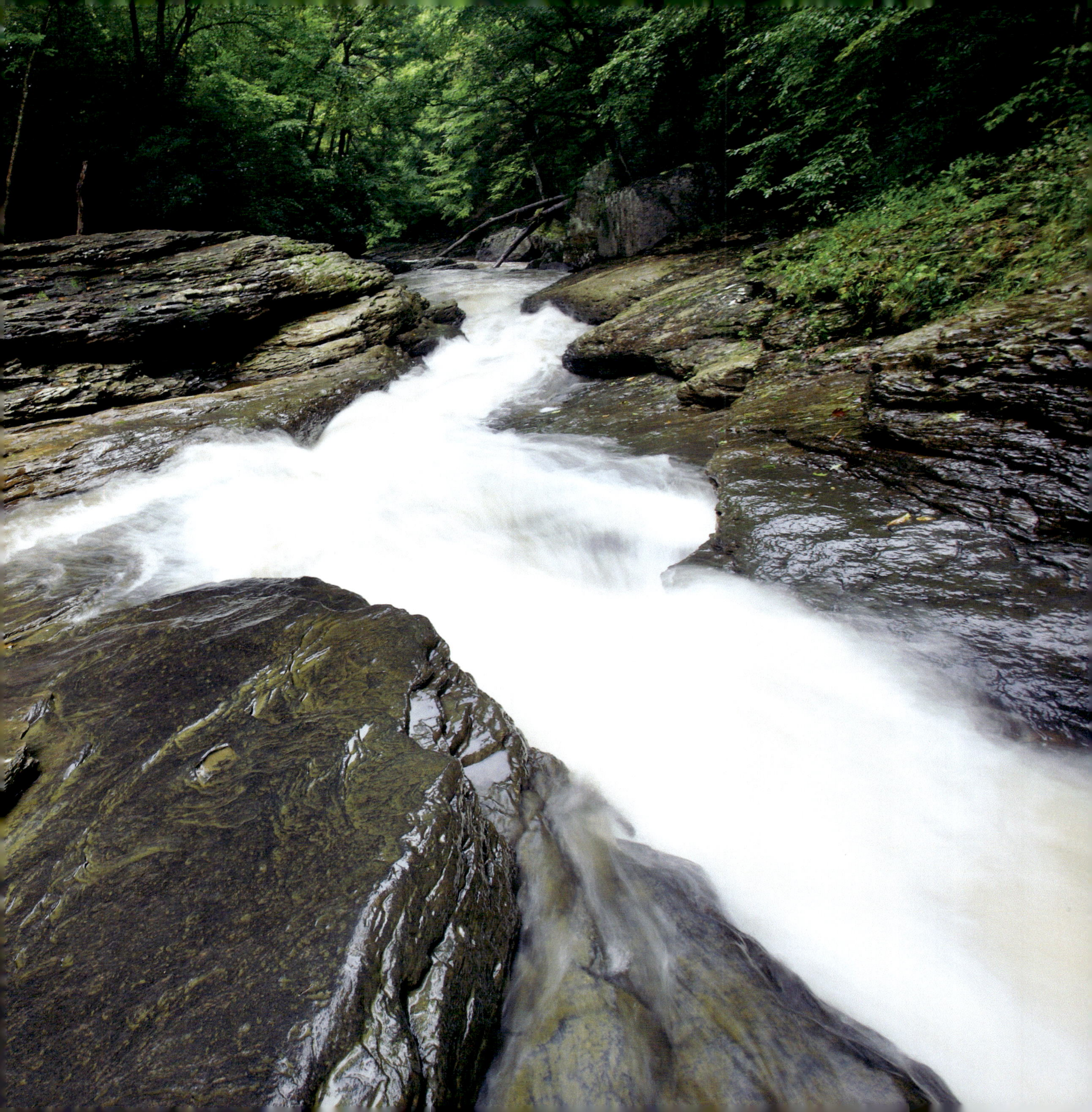

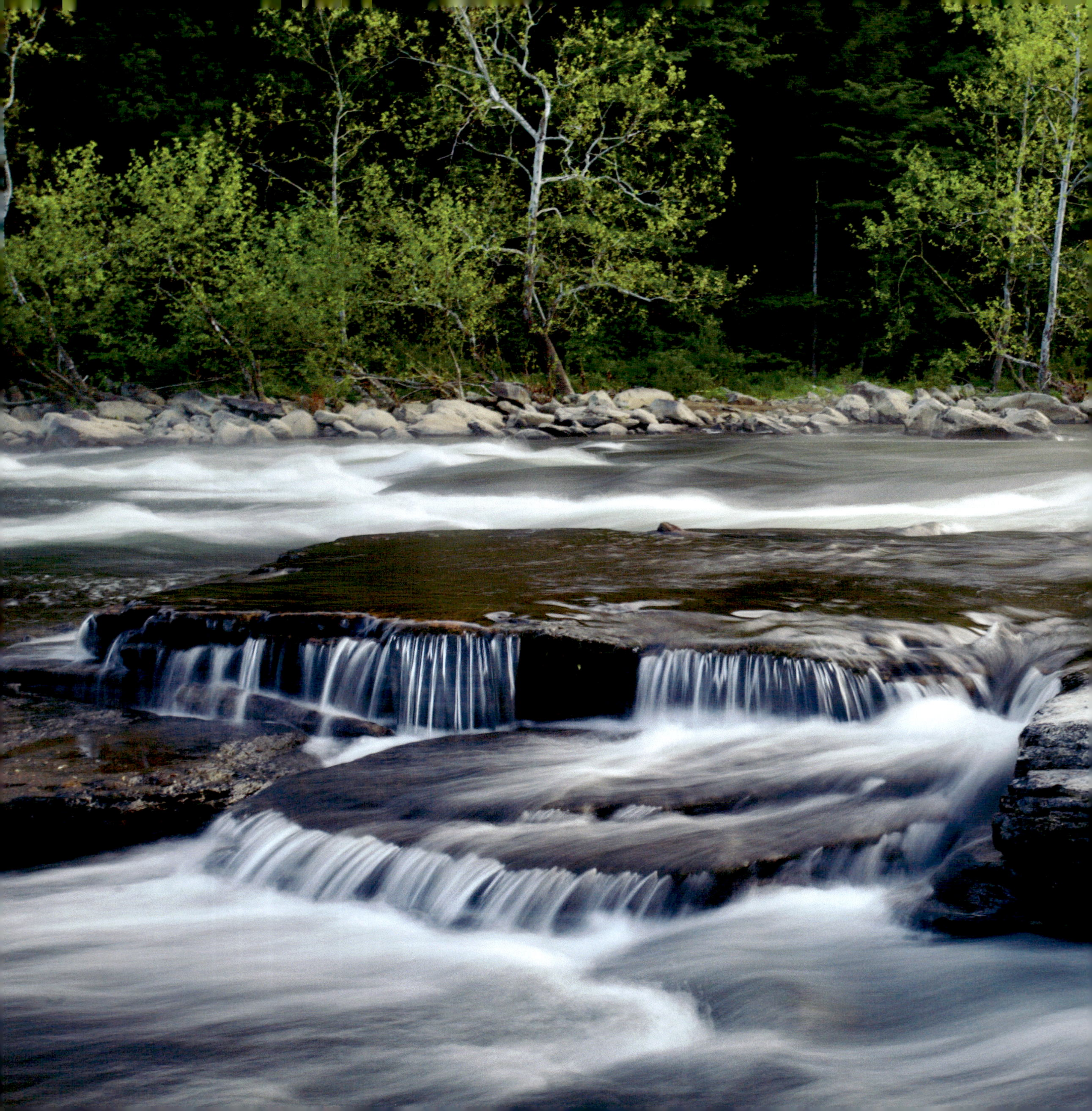

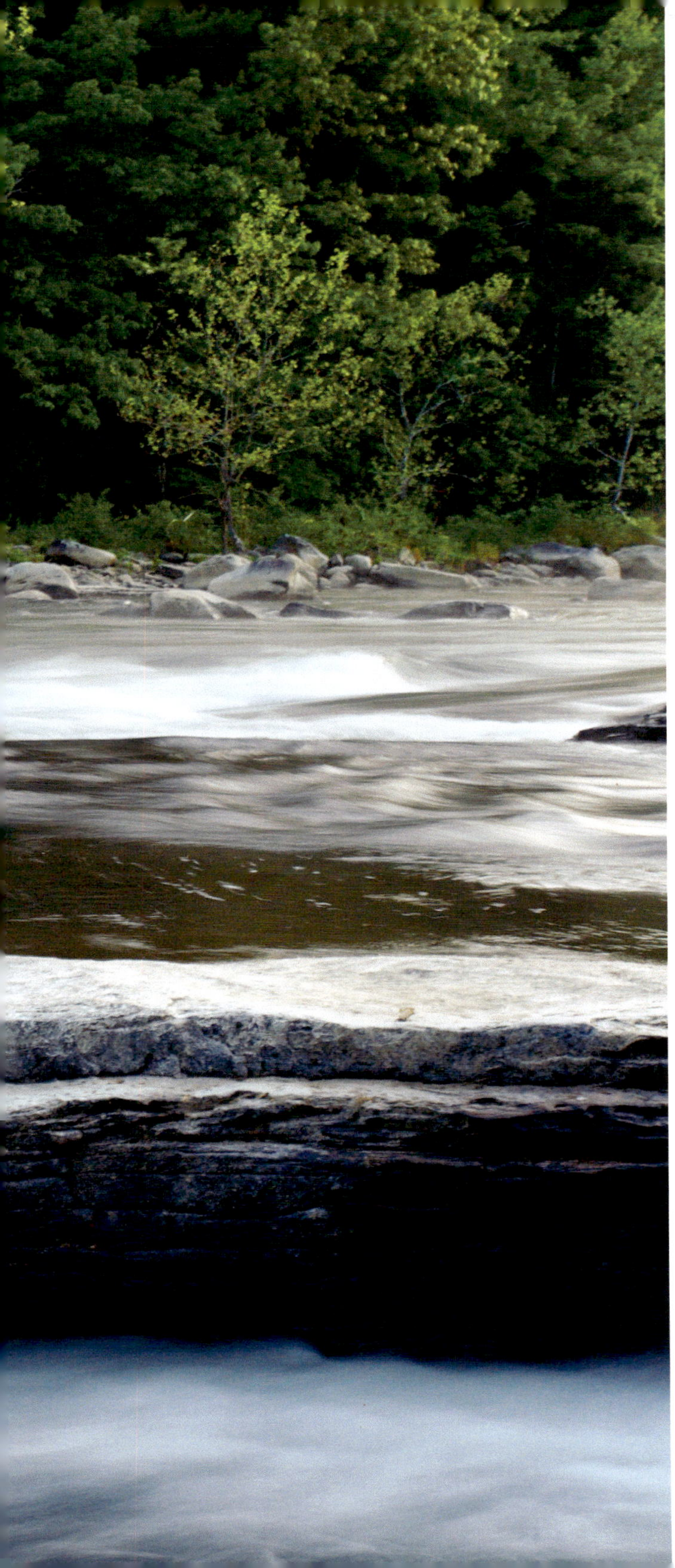

Entrance Rapid, Cucumber Falls, and the Great Gorge Trail

This spectacular hike can be reached from Ohiopyle, Meadow Run, Cucumber Falls, or Ferncliff Park. For a 4- or 5-mile loop, start, for example, at the Ohiopyle Falls parking lot. Walk along the highway downstream, cross the Meadow Run bridge, immediately veer right to the trail down lower Meadow Run and onto impressive bedrock slabs of the Youghiogheny's Entrance Rapid. This riverfront, alone, is well worth the visit with views at the edge of the rapid, upriver to Ohiopyle Falls, and downriver toward Cucumber Rapid and the Kentuck Mountain profile. Explore shoreline intricacies, rock hop all you want, especially at low flows, then continue downstream, where a steep woodland path leads in another quarter mile to the mouth of Cucumber Run.

Take the path up the run a hundred yards to the base of sublime Cucumber Falls, which can also be reached from a small parking lot off the Kentuck/Chalk Hill Road (often filled in summer). After returning back to the river, athletic hikers may want to explore a rugged side path from the mouth of Cucumber Run and downstream along the Youghiogheny to massive rocks edging the boil of Cucumber Rapid. Although it's a rock-scrambling challenge requiring good balance, this is an excellent place to watch paddlers plunging through whitewater.

Not to be alarmist, but whenever walking off-trail, know that the Appalachians are habitat of rattlesnakes and copperheads. That said, in hundreds of days of hiking along the Yough I've rarely encountered them.

The intricacies of Entrance Rapid can be seen along the river-left side, downstream from Ohiopyle and below the mouth of Meadow Run.

But it has happened. So watch where you step, which is not a bad practice anyway, especially off-trail.

For easier walking, before reaching the mouth of Cucumber Run embark down the river-left side of the river on the Great Gorge Trail following an abandoned rail line that once exported coal—and be reminded, once again, of the importance and promise of nature's restoration, if only we give it a chance. The trail continues clockwise high above the river, rounding the outside of the Ohiopyle Loop, all among maturing hardwoods and hemlocks.

After a mile from Cucumber the trail emerges at the northwest side of the High Bridge—a scenic destination in its own right featuring a thrilling view straight down one hundred feet with Railroad Rapid upstream and Lower Railroad Rapid downstream. After crossing the High Bridge to the river-right (eastern) side, walk the bike trail a short leg to the Ferncliff parking lot and back to Ohiopyle.

Lower Youghiogheny Whitewater Run

The Lower Yough, from Ohiopyle to Bruner Run, has for decades been the big paddling attraction and among the most popular Class 3–4 whitewater rivers in the nation. For this exciting and challenging outing, experienced paddlers or those with one of Ohiopyle's four licensed commercial outfitters launch below the Ohiopyle Falls and float 7.5 miles with fifteen major named rapids to Bruner Run.

A classic "pool-drop" river, the Yough has sudden and distinct rapids that feature rocks, ledges, and hydraulic complexities with many options and opportunities for savvy boaters to navigate obscure routes around boulders, to surf, and to paddle back upstream where eddies allow and repeat the drop again. All this by necessity involves the hazards of entrapment and potential boat-pinning against rocks, so paddlers must be experienced and aware. In most cases a relatively calm pool separates each rapid from the next during normal summer flows, the stair-step effect typically allowing for recovery. All this changes at high flows, which create a continuous turbulent descent with significant challenges when recovering after an accident. In cool weather or early season trips, definitely wear a paddler's wet suit or dry suit, and some may prefer a wet suit even throughout summer, just to make sure you're comfortable.

When going with a commercial outfitter, people with reservations park their cars at the outfitters' lot a few miles south of Ohiopyle; Route 381 southbound and left on Dinnerbell Road. The outfitter then buses clients to the put-in and, after the trip, from the takeout back to their cars.

Paddlers running their own trips on weekends from early April to mid-October need to register for a put-in slot and pay a modest fee for the takeout shuttle at the riverbank kiosk adjacent to the Ohiopyle Visitor Center. Advance reservations by phone are recommended for prime weekend hours. On the April-to-mid-October weekends, private boaters shuttle their takeout car from Ohiopyle southwest up the Kentuck/Chalk Hill Road, bearing slight right on Holland Hill Road, then right on Sugar Run Road to the Old Mitchell Place parking lot. After running the river and taking out at Bruner, paddlers board an Ohiopyle Park bus that shuttles them and their gear up Bruner Road and back to their takeout car at Old Mitchell Place. At other times, private boaters drive their cars directly to Bruner takeout and shuttle themselves the whole way.

Although not many people elect to do it, and the safety record is clearly worse than that of floaters

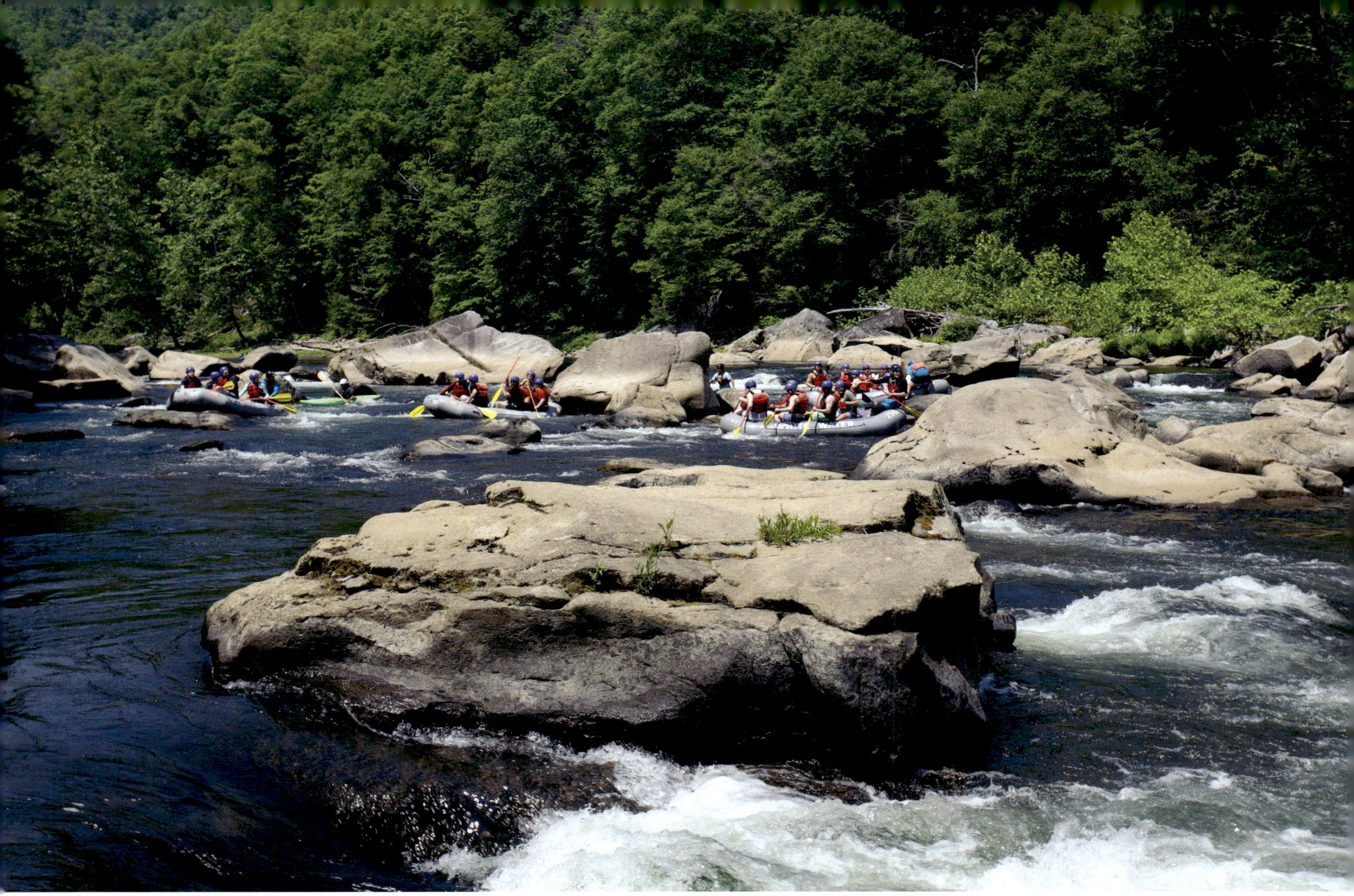

going with commercially guided trips or that of boaters who are competent with their own gear, rental rafts and inflatable kayaks are available for non-guided paddlers in Ohiopyle. The Yough downstream is closed to rented inflatable kayaks at 3 feet on the Ohiopyle gauge and closed to rental rafters at 4 feet. Private boaters with their own high-performance gear are permitted when the gauge is above 4 feet, but they must sign a high-water waiver with the Park. Consider yourself warned. Beware, always, that the power of the river is great, even at lower summer flows, with increasing force as levels rise. Typical summer flows are 1.5 to 2.2 feet on the gauge, or 800 cfs. Within these averages, weekend levels tend to be slightly higher than on weekdays owing to releases from Youghiogheny Dam.

The complex passage of Railroad Rapid can be reached from the Ferncliff parking lot by walking west, or away from Ohiopyle, on the bicycle trail and then veering left on a path before reaching the High Bridge.

For a short option that is open to paddlers with their own boats, and especially kayakers or those with lightweight rafts called Shredders, consider the famous 1.7-mile run around the Ohiopyle Loop and its six distinct rapids, starting at the put-in below the Falls and visitor center and taking out at the Ohiopyle State Park exit sign immediately below Railroad Rapid and above but within sight of the High Bridge west of the Ferncliff parking lot (must-do exit here, or you go the whole way to Bruner Run with no shuttle arrangements). From the takeout, paddlers carry their boats up the steep quarter-mile trail to the Ferncliff lot and then walk or shuttle back through Ohiopyle to the put-in parking area—all doable for the physically fit but a portage I don't recommend otherwise. Launch permits, available at the put-in, are required for the Ohiopyle Loop on weekends.

The Loop is classic among American whitewater runs—nothing quite like it for fabulous rocky rapids and for a stimulating workout of paddling on a section of river where you can walk the shuttle route. It's the closest one will ever come to the mythic whitewater river that runs around in a circle! Short and sweet, and with lots of options for technical paddling that can be extended for hours, it's a favorite late-afternoon or evening run for local boaters and visitors alike.

Ohiopyle to Connellsville

For 16 miles the Great Allegheny Passage bicycle and pedestrian trail follows the west or river-left side

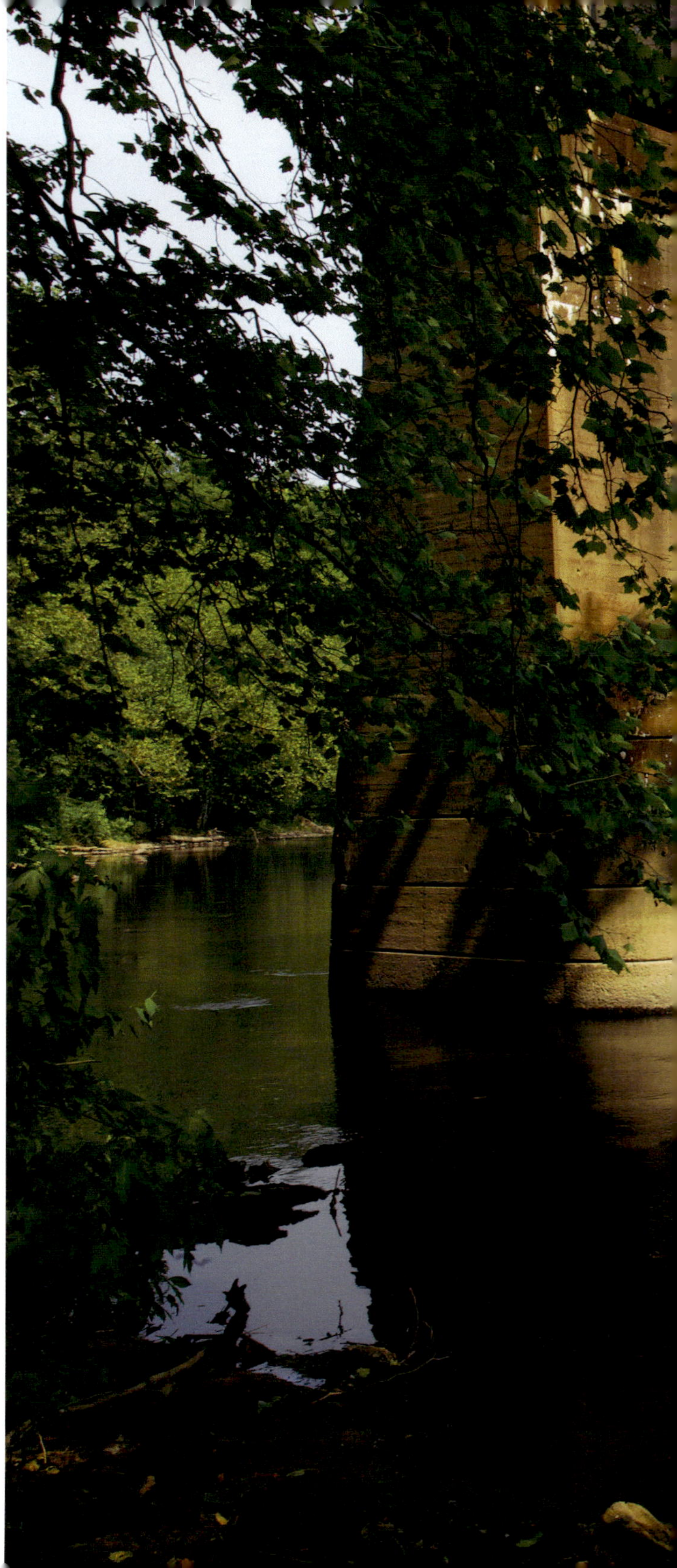

The Great Allegheny Passage bicycle trail follows quiet streets through Connellsville, past relics of the industrial past such as this railroad bridge, through historic towns, and onward to where the Youghiogheny empties into the Monongahela River.

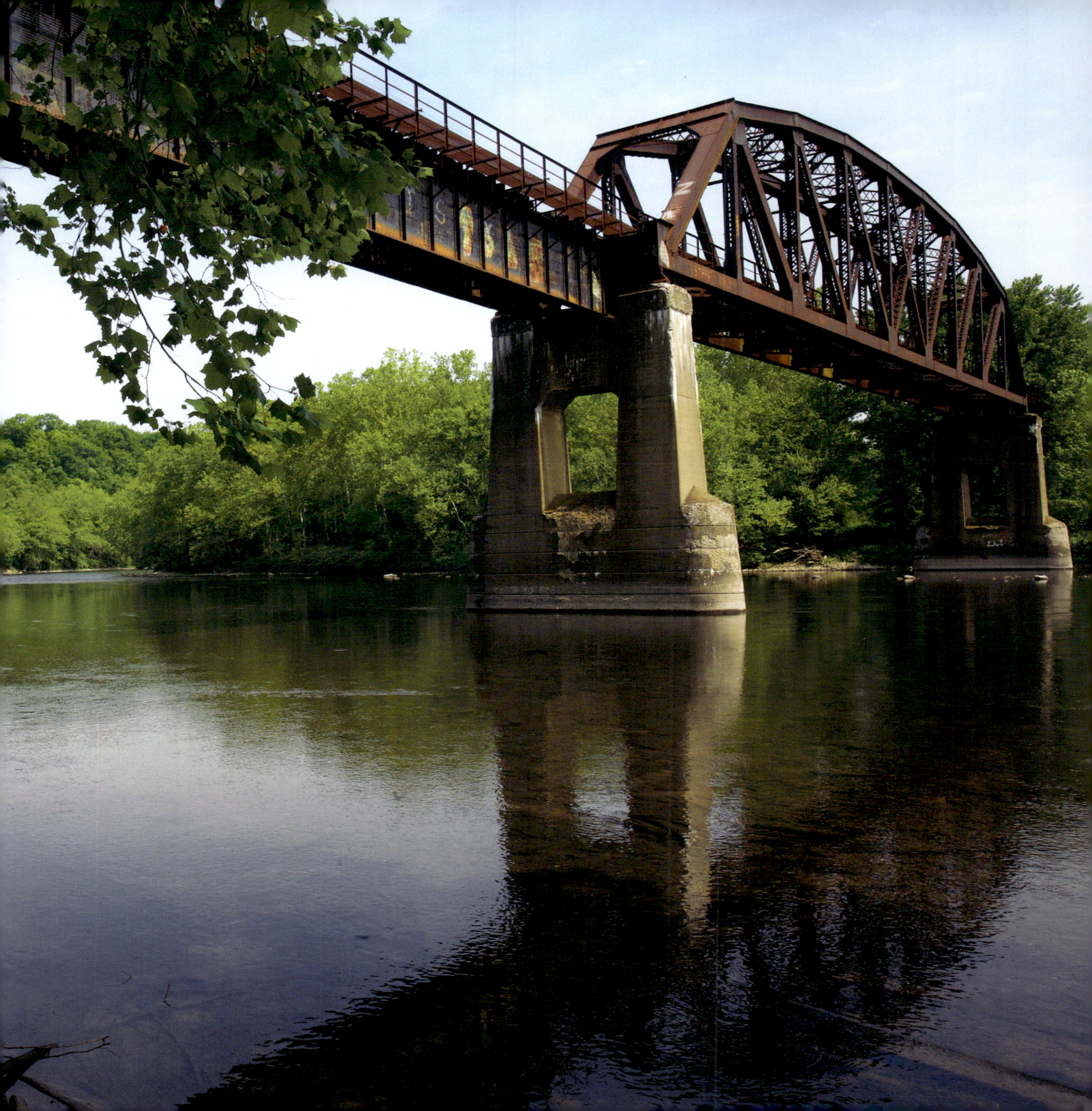

of the Youghiogheny downstream from Ohiopyle to Connellsville. Start at the Ferncliff parking lot and take the trail northwest. Or, from the bottom of this reach in Connellsville, start at River Park off Sixth Street at the downstream end of town, next to the Comfort Inn. The trail passes through forest once cut but now deepening with maturing trees in colonnades along much of the path except for the lower few miles. River views come unfrequently, but the shore can be reached from side paths at Jonathan Run, Bruner Run, and several miles above Connellsville. From either end of this route, an out-and-back trip makes an excellent 32-mile ride. Or do a one-way 16-miler if you want to drive the lengthy and complex 54-mile (both ways) car shuttle.

Paddling a raft, kayak, or canoe down to Connellsville is appealing in concept but a bit vexed in fact, as the only put-in below most of the Lower Youghiogheny's Class 3–4 rapids is the Bruner Run takeout, reached by the Ohiopyle State Park's Bruner Road, which is closed to cars during prime boating hours and then open only to the Park's shuttle bus. Also, beware: the whitewater does *not* end at Bruner Run; several challenging Class 2–3 rapids continue for a few miles below that takeout, making this option difficult for non-experts in kayaks and canoes. For those who elect to resolve shuttle difficulties and to negotiate the rapids below Bruner, much of the 12 miles from there to Connellsville is not rapids, at all, but slow pools and Class 1 riffles—a long day for rafters, especially with the expected upstream wind.

Foremost among complications, the hazardous, unrunnable, eight-foot-high dam just upstream of Connellsville must be portaged. Watch closely for the Westmoreland Municipal Authority water treatment buildings on the left and the cryptic, subtly smooth horizon line at the brink of the dam. The aged and dangerous cement structure is surprisingly hard to see from upstream. Mandatory portage lies on the right, avoiding exposed rusty rebar. Stay off the left shore, where access is aggressively banned by the Westmoreland agency. I know!

At the downstream end of Connellsville, the popular River Park near the Comfort Inn (see above) offers access for paddling farther downriver and parking for fishing and swimming.

Smithton and Cedar Creek Park

Forty-four miles of the Great Allegheny Passage bicycle trail, from Connellsville to McKeesport, follow the river with frequent views to riffles and pools. Among many access points (see *Great Allegheny Passage and C&O Canal Towpath Trail Guide*), Smithton is easily reached by Highway 981 east of the Interstate 70/Highway 51 intersection. Two miles downstream from the bikers' Smithton parking lot the trail enters Westmoreland County's Cedar Creek Park with camping, swimming, and hiking up the deeply wooded lower reach of the creek—perhaps the most natural tributary mileage flowing into the lower river.

Cedar Creek Park, 2 miles downstream from the quaint town of Smithton, offers camping and river access for swimming, fishing, and paddling. Unusual for the lower miles of the Youghiogheny basin, a fine hiking trail tours woodland wonders along the creek.

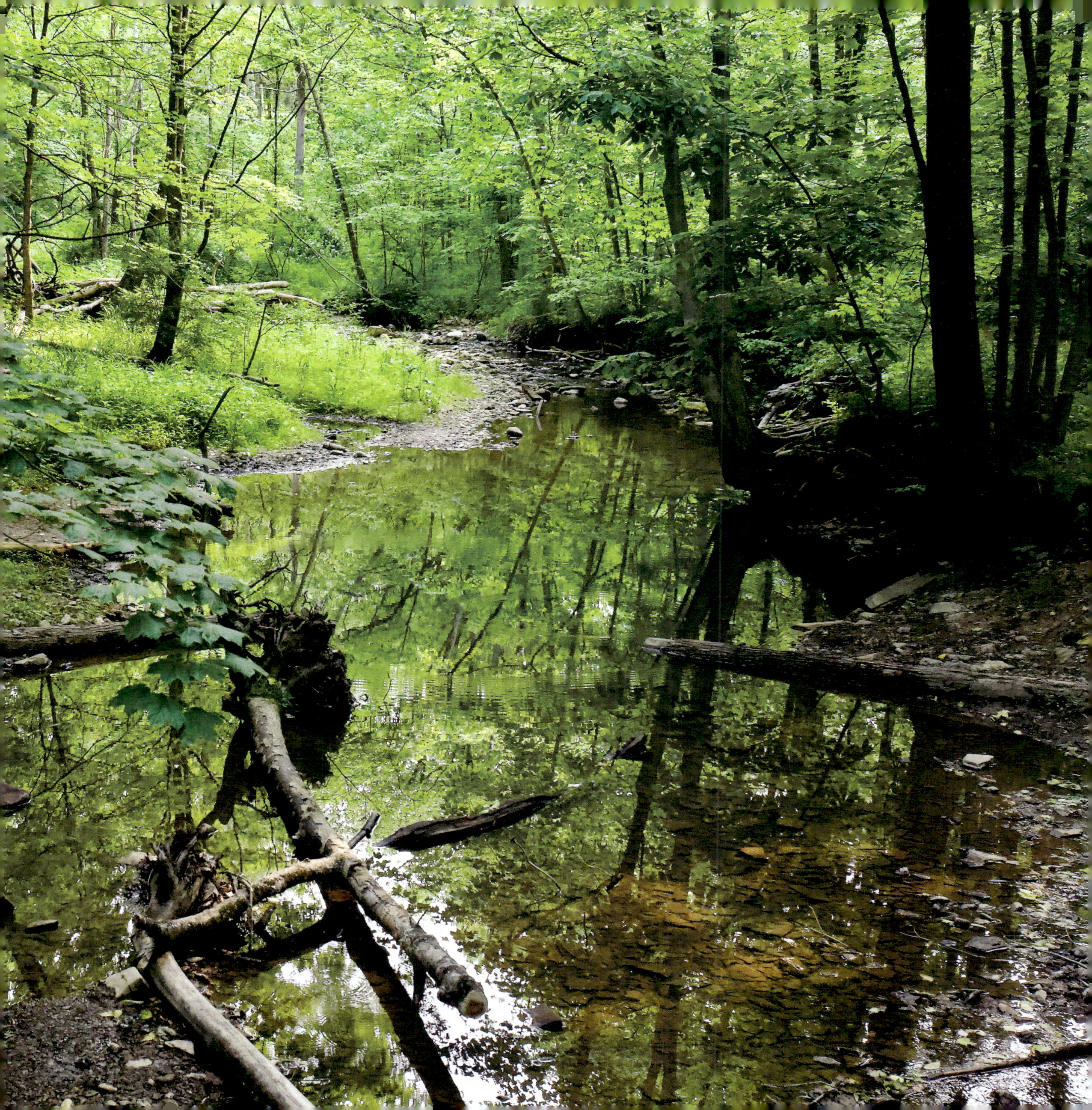

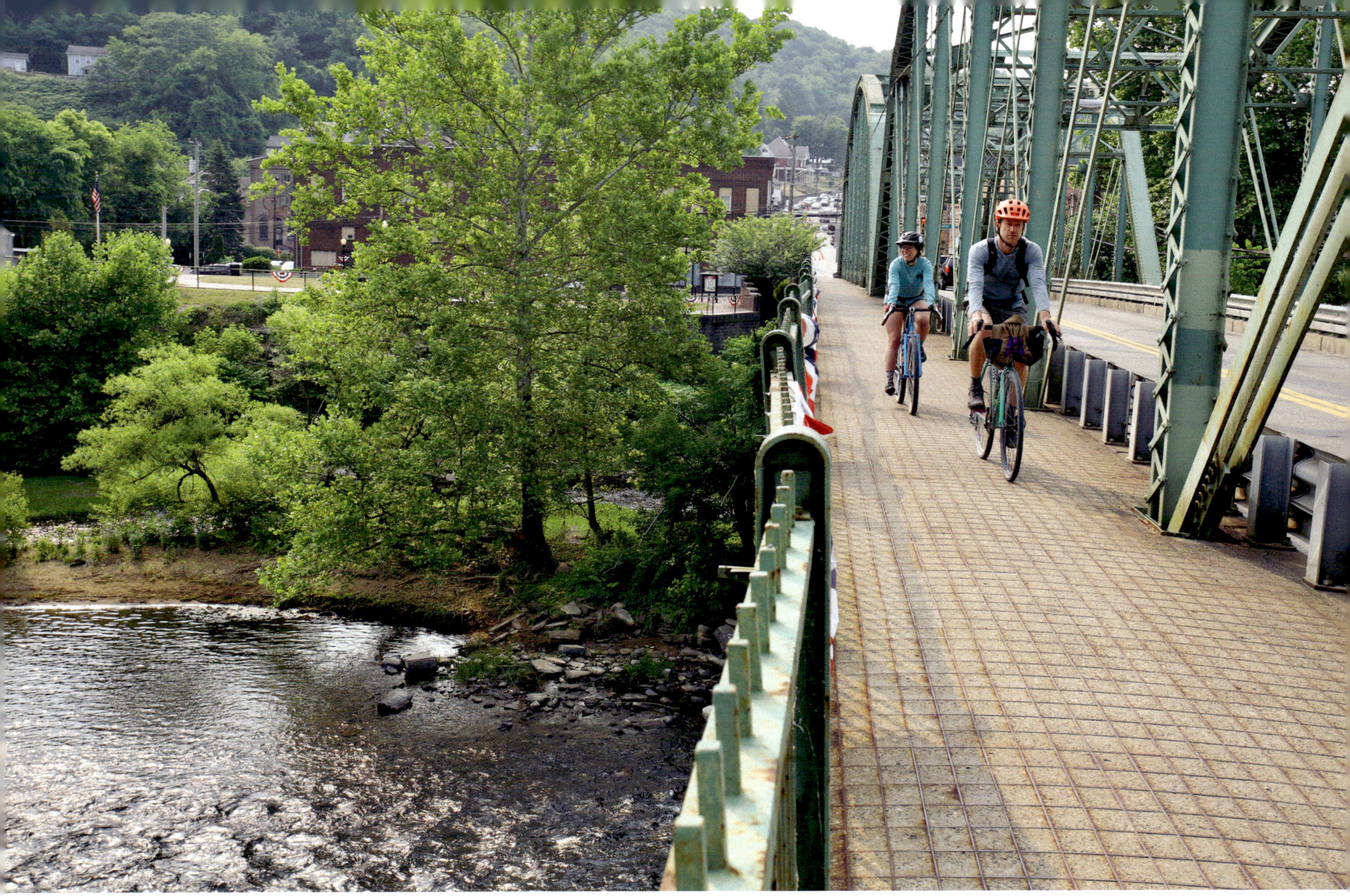

Bicyclists on the Great Allegheny Passage trail can cross the Route 136 bridge over the Youghiogheny and tour downtown West Newton and its historical architecture.

West Newton and Below

The Great Allegheny Passage trail follows the southwest (river-left) side of the lower Youghiogheny. At West Newton the Highway 136 bridge provides easy access to tour the historic downtown with its notable architectural character. Access with parking on river-left is a good place to begin a bike ride or paddle trip down the Youghiogheny's lowest reach to its industrial conclusion and flatwater mouth at McKeesport.

From the late 1800s to the mid 1900s the lowest 20-mile section of the Youghiogheny was a humming corridor of industry, and the river's path today offers windows to the boom-and-bust era of coal, coke, and steel production, evident in occasional lingering buildings and ruins for decades justifying the term

"Rust Belt." As the upside to this, the bike path—from Connellsville down—provides a case study of nature's recovery. Water quality is markedly better than in the 1980s, and the depopulation of once bustling manufacturing centers, in some cases shrinking for the past hundred years, has left us with historic and picturesque small towns that endure. Meanwhile, like a different world, the ridgetops and hills far away from the river—once a rural hinterland—have become the modern corridors of development and commerce along Highway 119, which drops down and crosses the Youghiogheny at Connellsville, and Highway 51 west of the river valley, carrying heavy traffic from Pittsburgh to Uniontown. Thus, regional development here has been inverted from river valley to ridgetop, leaving us with a splendid rewilding corridor along much of the lower Yough. Clustering new building in the lower Yough's towns and preventing overdevelopment along the rest of this newly reclaimed greenway might well be an excellent goal of land trusts, conservation advocates, and supporters of the picturesque and inviting small towns.

Gentle flows and riffles between Connellsville and McKeesport appeal greatly to novice paddlers in kayaks or canoes while riverfront towns including Dawson and Smithton offer convenient access. The lower several miles to McKeesport are impounded by an Army Corps of Engineers dam for barge traffic on the Monongahela River, so don't expect any current there. To avoid the flatwater and the final industrial shorelines, use the takeout ramp at Boston, 5 miles upstream from the mouth.

The bike trail approaching the Monongahela crosses to the northeast (river-right) side and becomes a patchwork with street and road crossings (watch carefully for signs). Then following the Monongahela, the trail concludes in one of America's ultimate urban river scenes at Pittsburgh's Point State Park where the impounded Monongahela and Allegheny merge to form the Ohio River, almost precisely where Lewis and Clark in 1803 launched their pathbreaking journey to the great unknown of the West.

If paddling to Pittsburgh, boats must pass through a lock at the Army Corps dam at Braddock, 3 miles below the mouth of the Youghiogheny. Tall vertical concrete walls inside the lock were made to scale not for small craft but for chains of massive barges. This is a bit spooky, as Mike McCarty and I discovered years ago while kneeling in our canoe within the thirty-foot-deep pit of concrete while the water in the lock drained down, but it's a unique industrial-river experience. For details and schedule confirmation, call the Corps at the lock: (412) 271 1272. Access to the river can be found upstream of the lock at McKeesport and farther down the Monongahela at Homestead, South Side, and Pittsburgh. From there, it's 980 miles of the Ohio River, dammed back-to-back twenty-one times to the Mississippi, with a lone riffle showing directly below the dam at Louisville, Kentucky.

My choice at McKeesport: head back to Backbone Mountain and see the Youghiogheny all over again! I've always found something new and always a view of the river that I've never seen before.

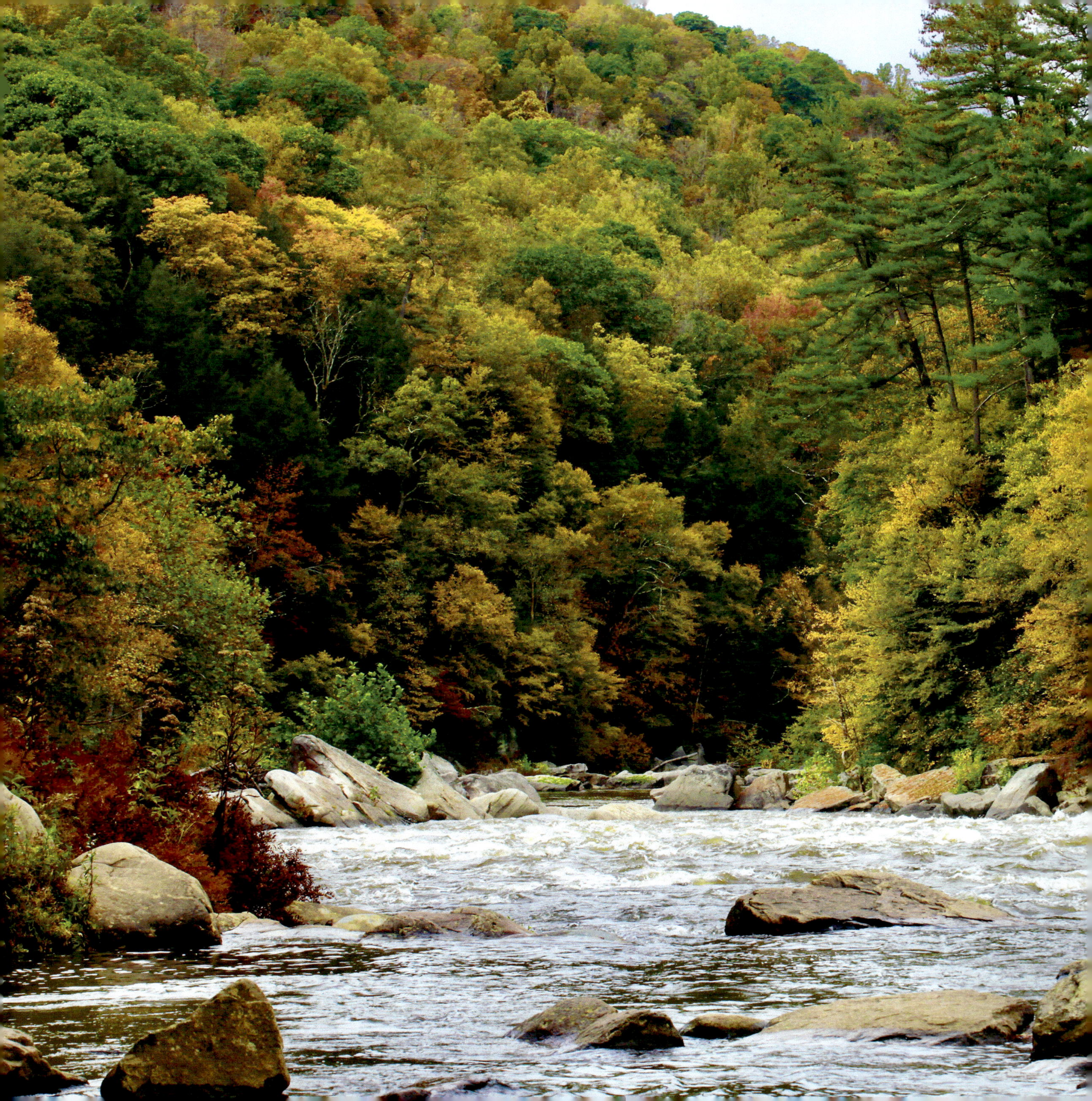

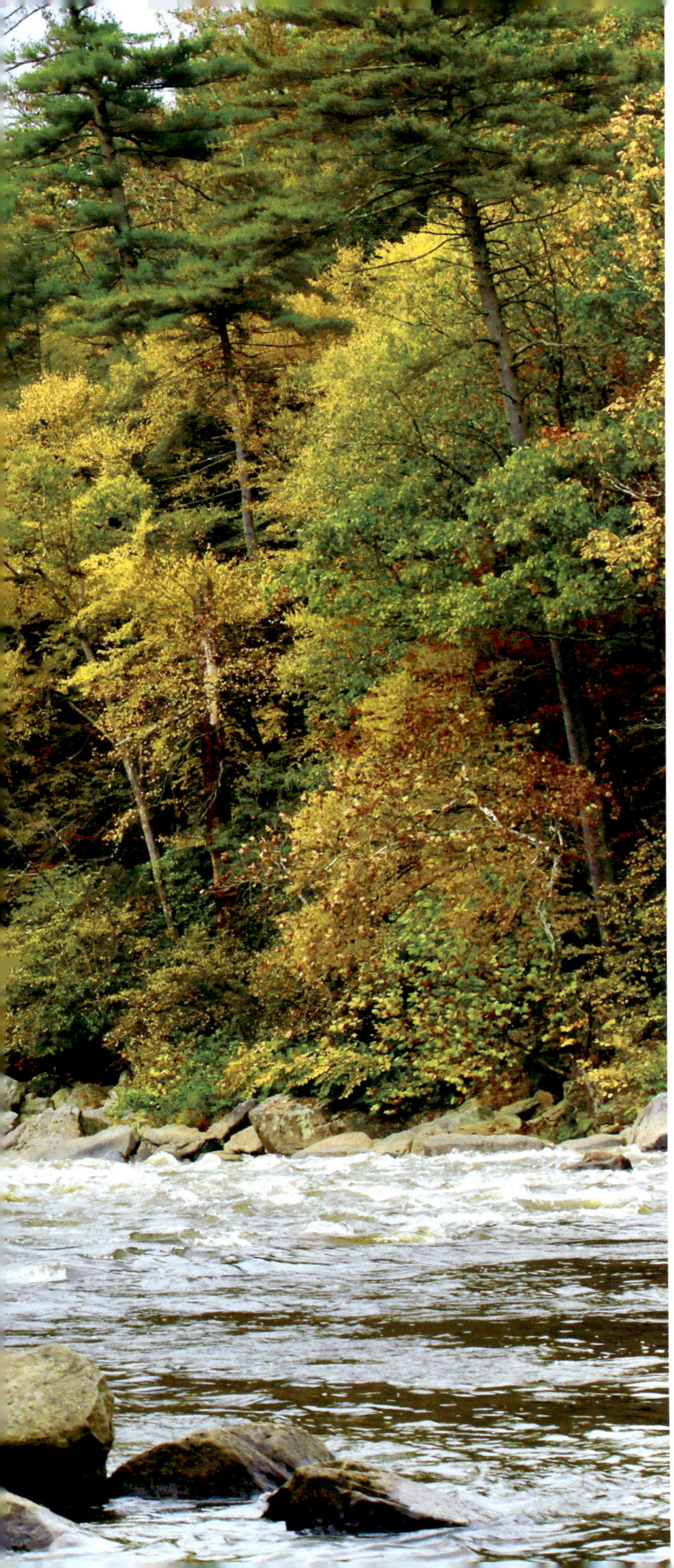

Chapter Six

The Youghiogheny River of the Past and Tomorrow

Less than a lifetime ago the 132-mile-long Youghiogheny River flowed across America's eastern mountains without much notice from the world around it. The town of Ohiopyle remained little known to those beyond the next ridge or two. Few people had even seen this place with the unusual name, adapted from a native language and meaning "white frothy water." I was one of the lucky ones, traveling in the family car to the river each summer to visit relatives living in Ohiopyle, much as earlier ancestors of mine had done since the year 1787.

Recognizing opportunities to protect what remained in a region where nature had otherwise been sacrificed wholesale to logging and mining, pioneers of the Western Pennsylvania Conservancy stepped up to protect special qualities of the river and its leafy green corridor through the scenic gaps of Laurel Hill and Chestnut Ridge. In time, much of the landscape centered around Ohiopyle was acquired from private owners and designated a state park for the protection of nature and for the recreational needs of all Pennsylvanians—in fact, of all Americans.

A state park master plan was drawn up for recreational development and restoration of land that had been clear-cut, farmed,

Downstream from Ohiopyle, the Youghiogheny disappears around its famous 1.7-mile Loop and carves a scenic route through the abrupt rise of Chestnut Ridge.

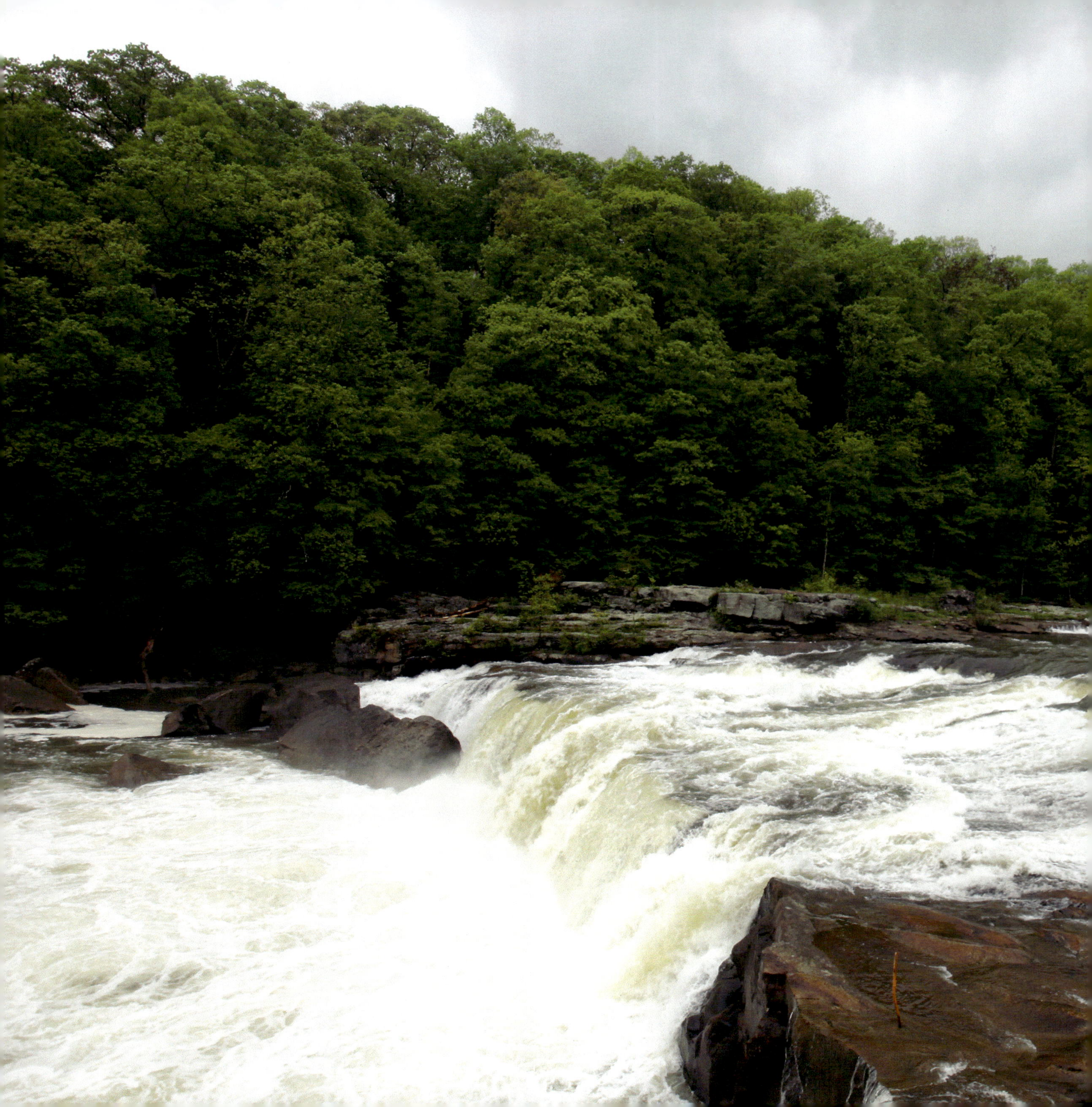

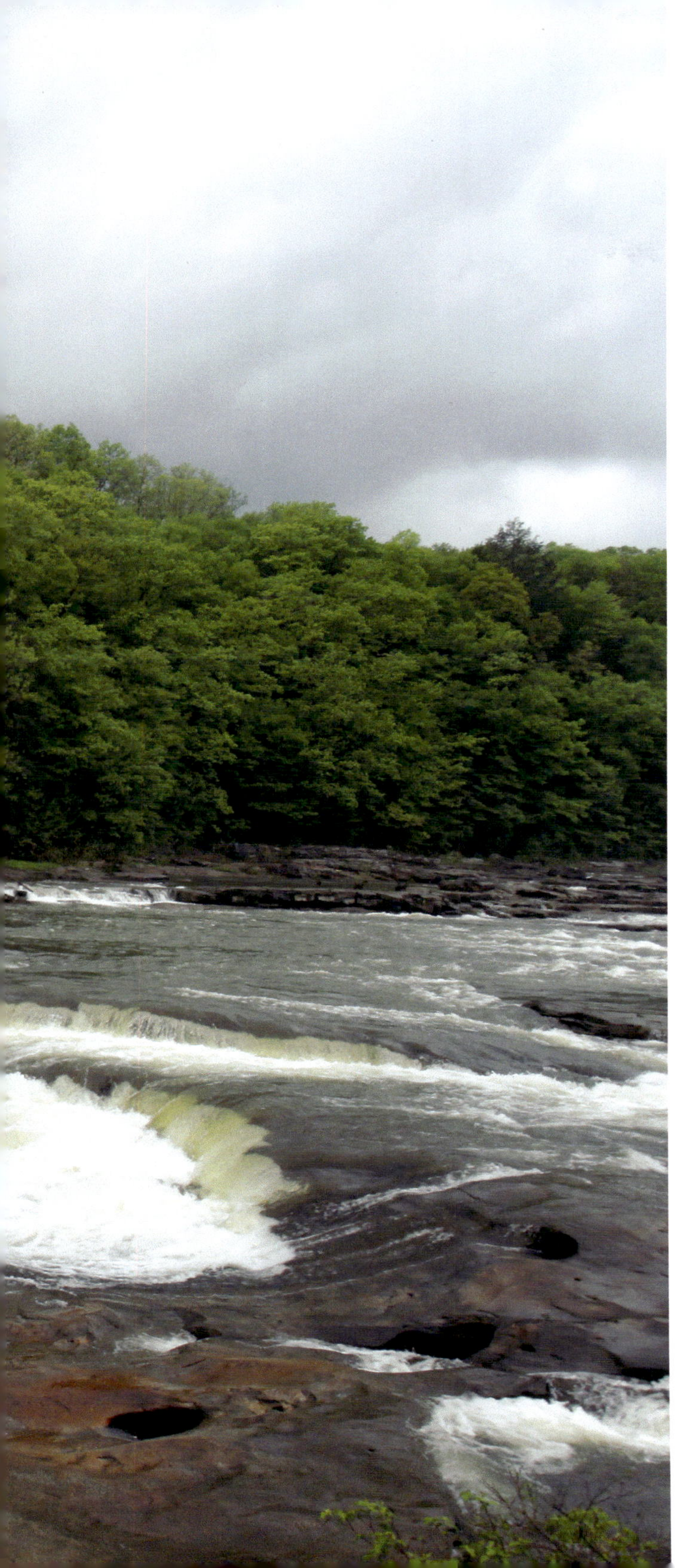

and mined. The state and organizations including the Conservancy, Trout Unlimited, Indian Creek Valley Water Authority, Mountain Watershed Association, Youghiogheny Riverkeeper, and other groups launched heroic efforts to reclaim degraded streams and coal-mined mountainsides for what's now regarded as their highest and best use: the return of nature. Ohiopyle State Park grew to 20,600 acres. Renowned for its scenery and whitewater paddling, and then for bicycling on a nationally known riverfront trail, this became one of the most popular state parks in America, having the prestige and appeal more often associated with a national park.

A premier scenic highlight of Pennsylvania and beyond, Ohiopyle Falls was clearly the signature attraction of the park and the region. Recognizing that, landscape architects designing infrastructure for the Ohiopyle Park crafted a trio of viewing platforms so that visitors can see the Falls and its impressive turbulence without any risk of falling in. Much of the old town's main street was converted from homes and garages to a shaded parking area and visitor center rising up above the river.

Initial iterations of park planning, fortunately dismissed and forgotten in short order, included a golf course, a ski area, and an aerial tram. Facing dilemmas about how much development was enough, the Bureau of State Parks scaled back the options so as to favor protection of riverfronts, streamsides, and mountains, concentrating activity at Ohiopyle. This became the only town within the boundaries of a Pennsylvania state park and a rarity in that respect among state and national parks nationwide. Most adjacent towns, in contrast,

At high flow following summer rain, the Youghiogheny careens over the Ohiopyle Falls.

were left out of their park's boundaries owing to political and business considerations. As a result, they've become commercial strips, built out, lit up, and sprawling without regard to any intrinsic values of their landscapes. See the entrance to the Great Smoky Mountains National Park in Gatlinburg, Tennessee, for a commercially saturated example of what could have been.

Some residents and landowners, including my own grandmother, father, and various uncles, aunts, and cousins on my extended family tree, may have at times been disgruntled about selling their land for the park, or more universally about confronting the myriad changes that cascaded onto the quiet mountain town and its surroundings. But many, including my great-uncle and -aunt, who owned the primary business in Ohiopyle—Holt's Department Store (then truly a department store; now Falls Market serving visitor needs)—recognized, for the long-term future, the value of the Ohiopyle State Park. One of the most profound and unsettling changes was a revamped highway system: a wider bridge, a speedier street through the heart of town, and straighter realignment of roads curving away in all directions, involving the loss of landmarks such as Castle Rock—a sandstone pinnacle that for two centuries had risen up above the road southbound toward Meadow Run.

In many respects, the past has been a tumultuous journey from private property to public park, from neglect to notoriety, and from environmental injury to landscape restoration. Summing up the importance of the place, park manager Ken Bisbee in 2023 noted, "All parks are special, but none are more special than Ohiopyle with this river that runs through. It doesn't get any better than this."

Waters of the Youghiogheny narrow in the turbulence of Cucumber Rapid.

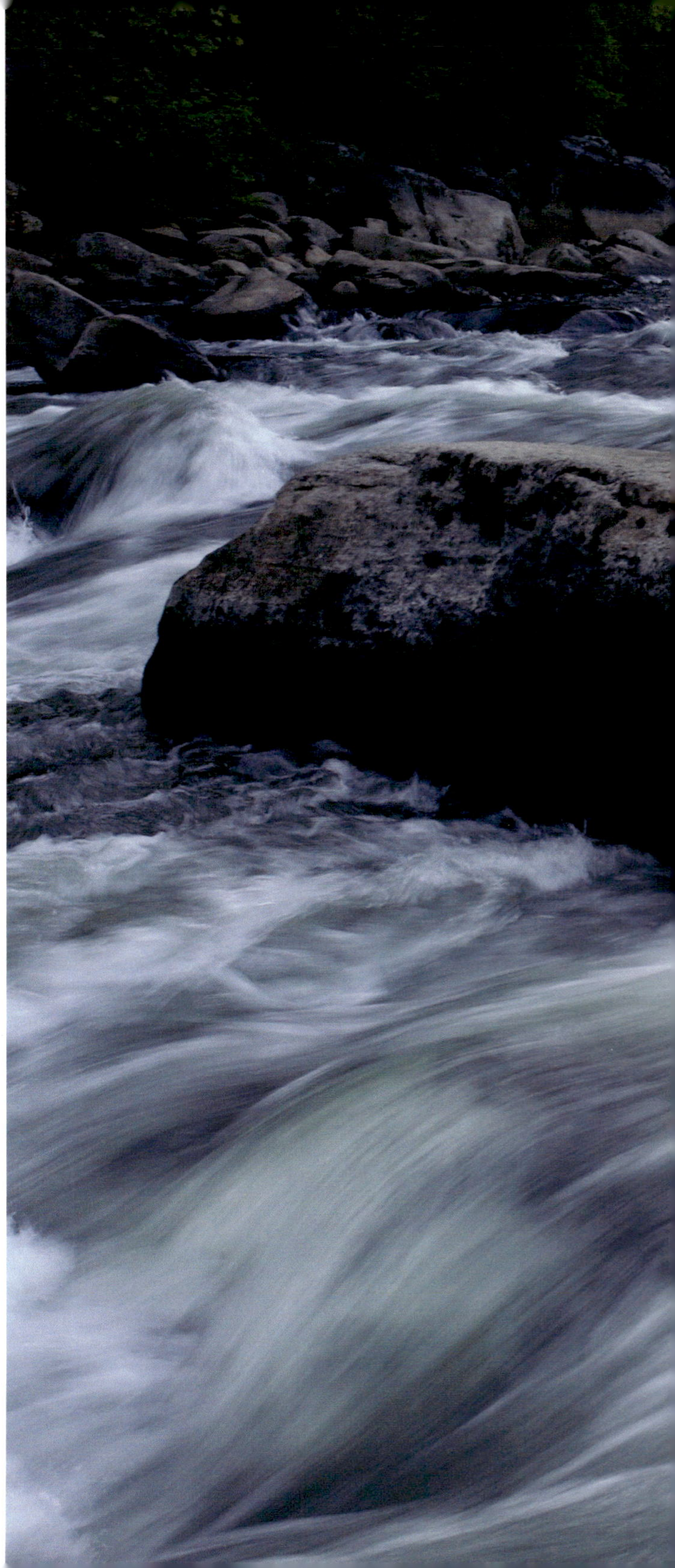

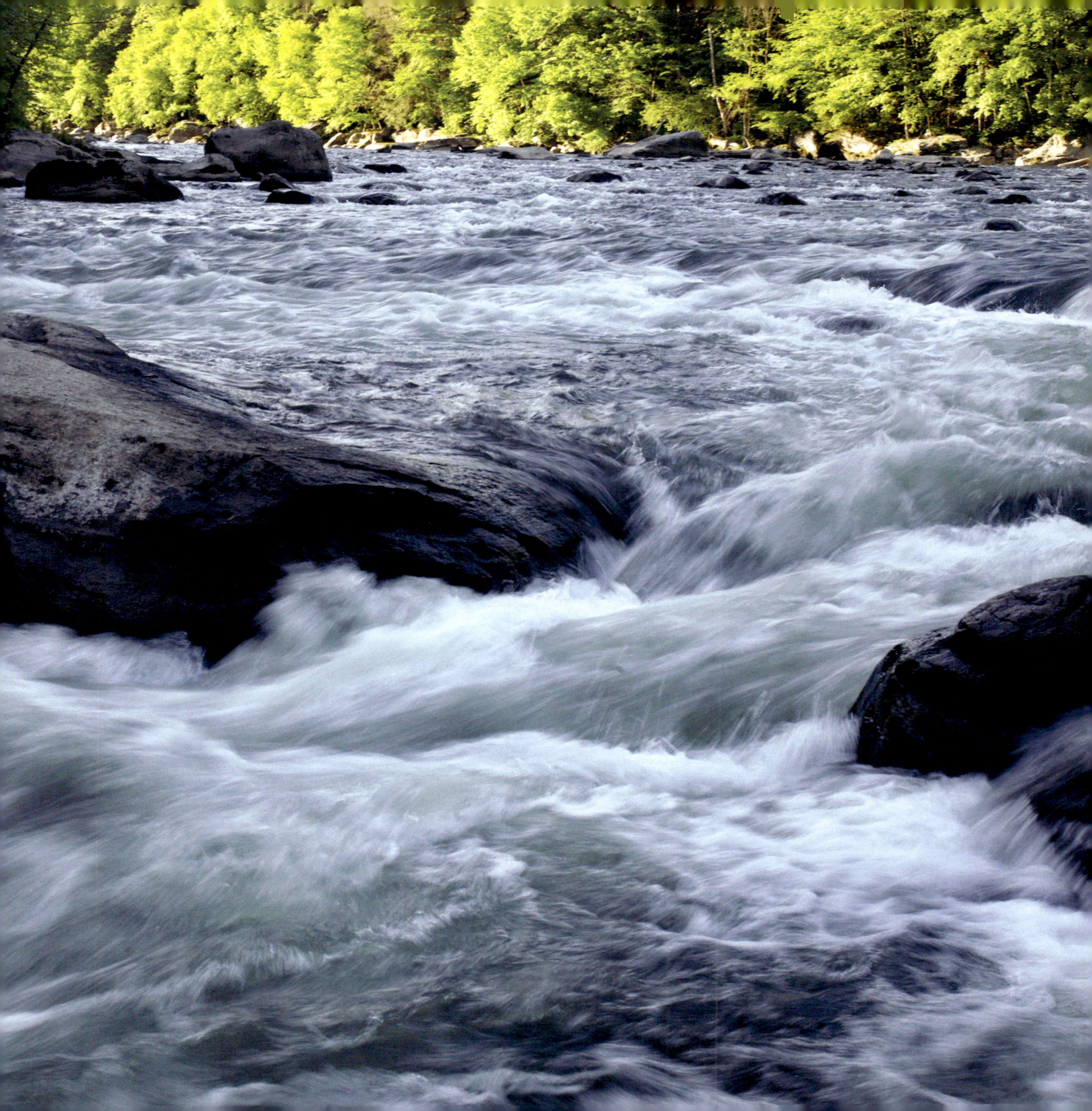

The Youghiogheny basin's three state parks—Ohiopyle in Pennsylvania, the much smaller Swallow Falls in Maryland, and still smaller Cathedral on the tributary Rhine Creek in West Virginia—were all set aside from private development and reserved for public access to the region's greatest scenic attractions. All were destined to become recreational centerpieces within a vast hinterland of mountains, forests, and rivers, enchantingly remote yet eminently reachable from booming cities and the whole East Coast megalopolis. All three parks represented bold, imaginative initiatives to meet needs of the twentieth century. We're now well into the twenty-first century, and the success of those efforts can easily be seen. But also evident are additional needs and opportunities that surface and confront people, communities, and the greater ecosystems of plants and wildlife today.

Through it all flows the river. Whether we like it or not, the Youghiogheny, from Backbone Mountain source to Monongahela confluence, unites everyone who lives, works, or comes for fun, relief, and personal enrichment along the river's path. Many people, for example, need the runoff from the distant summits of Laurel Hill and Chestnut Ridge as drinking water in the cities below, from hometown Connellsville to Pittsburgh's urban edge. While the Ohiopyle Falls might once have seemed isolated on the touched-up postcards of a nostalgic era, the big view of this river now shows that the winding curves of watery current and the probing reach of tributaries connect not only with one another but with all of us. If an abandoned mine along the Casselman River blows out with a sterilizing pulse of acidic water, as one did in 1993 killing most fish for 42 miles downstream, the whole lower Youghiogheny lies at risk. Viewing and seeing the entire river and its branches and comprehending the watershed as one interdependent system of water and life might together bring this larger view to light and this tenuous future to mind.

New linkages now connect by way of the Great Allegheny Passage bicycle trail, which climbs across the Youghiogheny basin, tunnels beneath the eastern Continental Divide, then coasts conclusively down along the Potomac River to nothing less than the nation's capital. What was once only vaguely realized when driving on highways across both named and nameless mountains and bridging both familiar and unknown streams can now be understood with an intimate sojourn by bicycle—a fresh-air mode of muscle-powered travel never failing to reveal more than what can be seen, noticed, or remembered through the windshield of a speeding car.

All this points to greater importance in caring for not just one stream, one town, or one backyard but for the whole place. It's all connected.

This enlarged perspective can be imagined anywhere along the river, and this wider view is also amply evident when comparing the past and the present. Consider, for example, the lower miles of the Youghiogheny. Even after a bustling state park at Ohiopyle had become one of the most visited recreational retreats of mid-Atlantic America, the gently flowing river below Connellsville lingered deep in chilled shadows of its industrial past—one that boasted of thirty thousand coke ovens enabling Pittsburgh's ascent as the steel production capital of the world. For a while. Then came the bust, and for decades the lower river declined and became a corridor plagued by the entrapping bondage of pollution and poverty. But now, cleaner water, a world-class bicycle trail, recovering neighborhoods, fish to catch, and delightful miles of smooth-water paddling have begun to transform the lower Youghiogheny from the industrial boiler room of old into a recreational lifeline showcasing a new economy boosting a whole Appalachian region.

We've seen that Ohiopyle State Park was initiated by wise visionaries of the Western Pennsylvania Conservancy, then boldly created to include more than just the attraction of a thundering waterfall. From it, the park grew to expand recreational possibilities, to encompass neighboring land upstream and down, and to protect the nature of a larger landscape and a longer river, not just for now but for all time and all people. As Pennsylvania Department of Conservation and Natural Resources secretary Cindy Adams Dunn succinctly stated of her mission in 2023, "We're in the forever business." While Swallow Falls State Park, upstream, had set aside a cluster of waterfalls, and Cathedral Park a majestic hemlock grove, efforts at Ohiopyle elevated the parkland purpose to new levels by safeguarding 17 miles of river, by integrating the lower reaches of most tributaries, and by enfolding vital sections of two great mountain massifs—Laurel Hill and Chestnut Ridge.

Beyond the parks, additional and intriguing means of safeguarding what's most valuable about the Youghiogheny have surfaced in recent decades. During an inspired moment in 1976, Maryland's legislature designated 21 miles of the Upper Youghiogheny as a State Scenic River, from Millers Run to Friendsville, mandating protection of scenic and wildland qualities, complete with management guidelines that fortuitously saved the upper Yough from a nonconforming scheme in 2022 to bulldoze a major road-sized trail through the state's finest wilderness. Likewise, a green-forever corridor for the 70-mile-long Laurel Highlands Trail was creatively assembled by—again—the Western Pennsylvania Conservancy, with acquisition of an expansive 11,230 acres of green forest draping Laurel Hill's elegant uplift to the northeast—now one of the Keystone State's stellar long-distance foot paths. Less showy but no less remarkable (or more so, in fact), sewage treatment plants in lower reaches of the river were built in the 1980s, and upgrades have continued to open a welcome new chapter for small towns that had been in decline and hobbled for a century or more by repulsive water quality.

Expansion of state park land, mergers of stewardship with nearby state game lands and state forests, and timely initiatives of communities and trail groups expanding recreational opportunities along the full length of this river, and up its tributaries such as the Casselman along with Indian Creek, Laurel Hill Creek, and others, all support the vision of a whole Youghiogheny as a vibrant and essential system of life.

To step back even further from this waterfront for an even larger view, I see that the state parks and other public lands of the Youghiogheny were established a full half century ago and earlier. In the most recent fifty years, Pennsylvania's population has increased by one million; Maryland's by a whopping 60 percent. But meanwhile, from 1970 to 2024, the collective size of our parks and public land estate has scarcely grown at all. Simply keeping pace with the recreational requirements of an expanding population has failed to occur.

Added to this, we now know that water supplies from a clean river with adequate flows have become more and more essential to basic human needs along

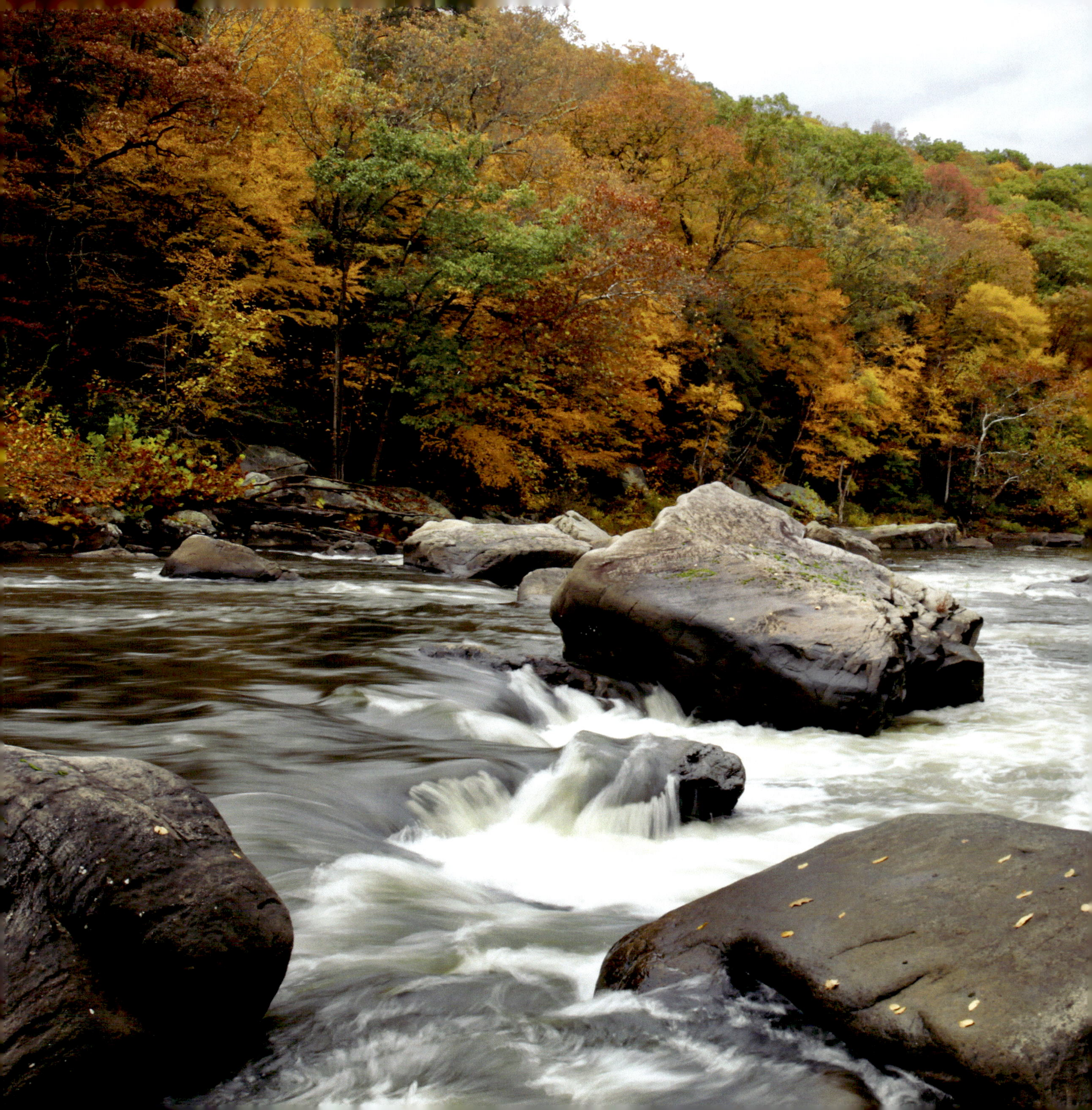

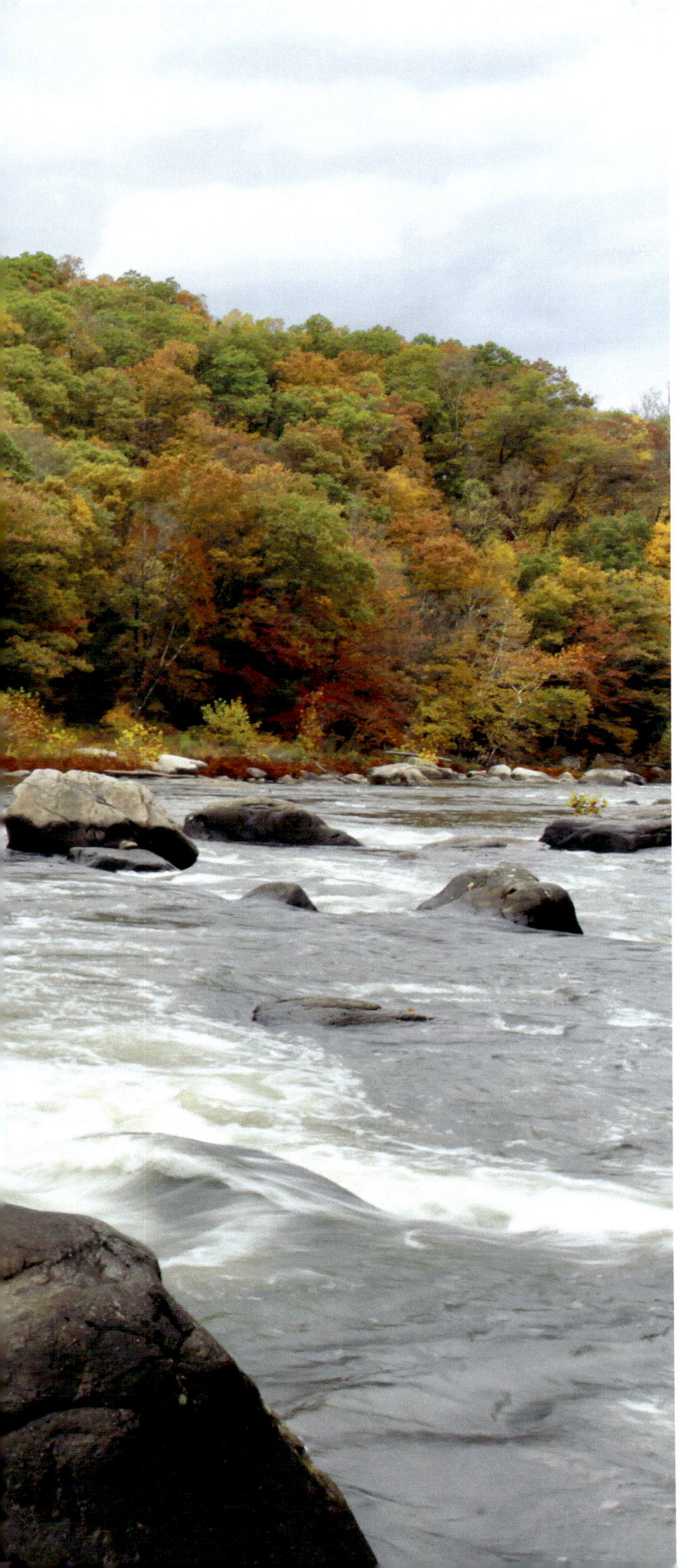

the river. We cannot afford to pollute the river again, no matter what the rewards or profits might appear to be for risks in mining coal or fracking gas. Once taken for granted, native plantlife in this biological meeting ground and outdoor botanical museum face devastating incursions of invasive weeds and exotic insect pests from distant parts of the globe. Suffering multiple stressors, many kinds of wildlife are now forced to retreat from developing fringes of cities, towns, and suburbs. Too often they have no place to go.

On top of that, efforts to curb and reverse global warming require that increasing levels of atmospheric carbon be sequestered in old-growth forests, which through photosynthesis can take the troublesome carbon out of the air and lock it beneficially into the woody trunks of trees, all providing that we let those trees grow large and live long.

In short, the needs for public land, safeguarded watersheds, protected forests, and permanent refuges for native plants and wildlife have grown at rapid rates, but our reservation of land and our dedication of its use for the long-term health of our communities, watersheds, and planet have not kept up with the needs. Innovative and acceptable means of increasing parkland acreages, extending our protected mileage of rivers, and reserving larger tracts of forests so they can grow into carbon-storing reservoirs of life are all essential in the decades and generations ahead. Our care of the Youghiogheny is critical in all these regards, and success here can point to paths that must be followed elsewhere throughout the Appalachian Mountains and beyond.

Autumn colors at Railroad Rapid peak in mid October as another year draws toward its close.

Any view toward a bright and workable future here depends on land and streams that have been set aside for nature and for respectful use. That view begins with seeing this beloved river as nature's gift to the ages. Along the water's edge, a place of wildness and beauty has both endured and been restored. Because of that, we can all come to the Youghiogheny, look into the current, and watch a wild river run.

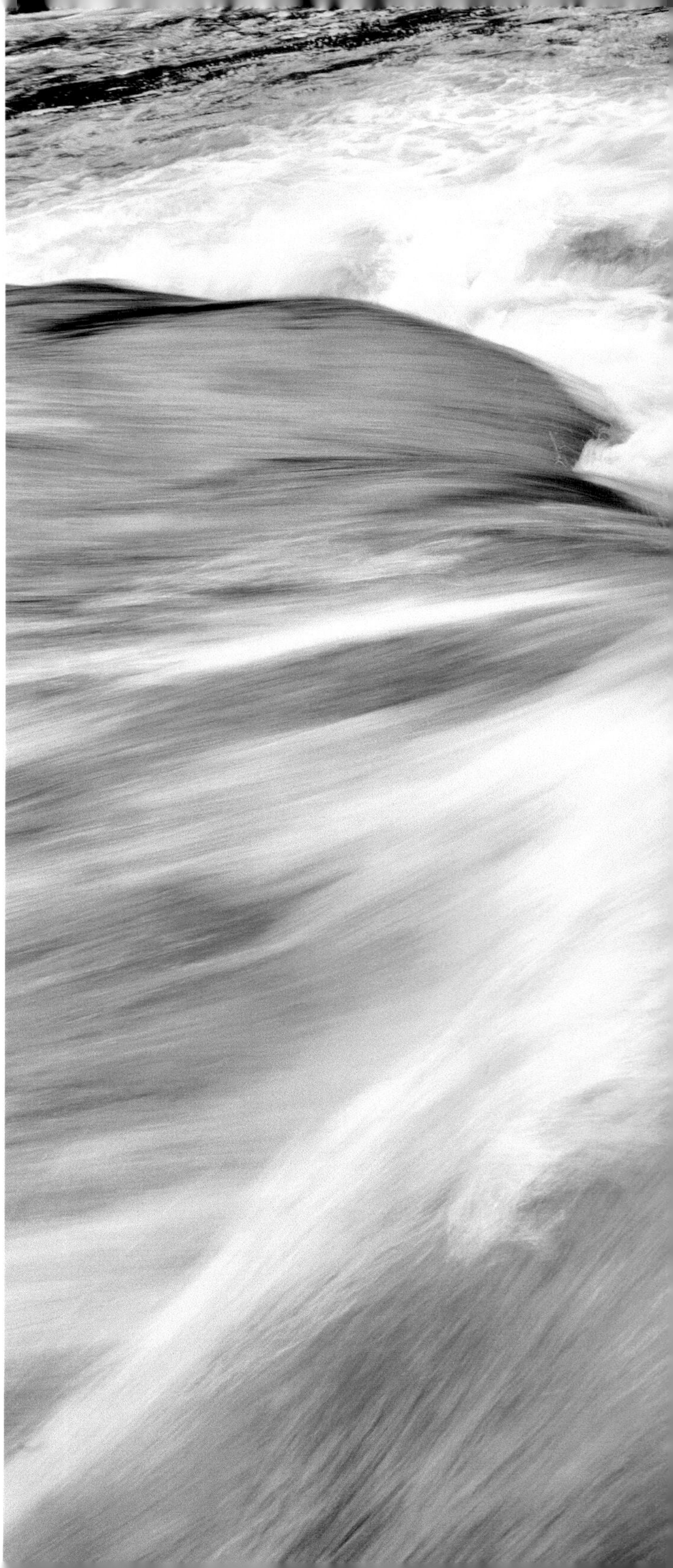

The Youghiogheny continues to flow in timeless patterns seen by all who come to look among the Appalachian Mountains of Pennsylvania, Maryland, and West Virginia.

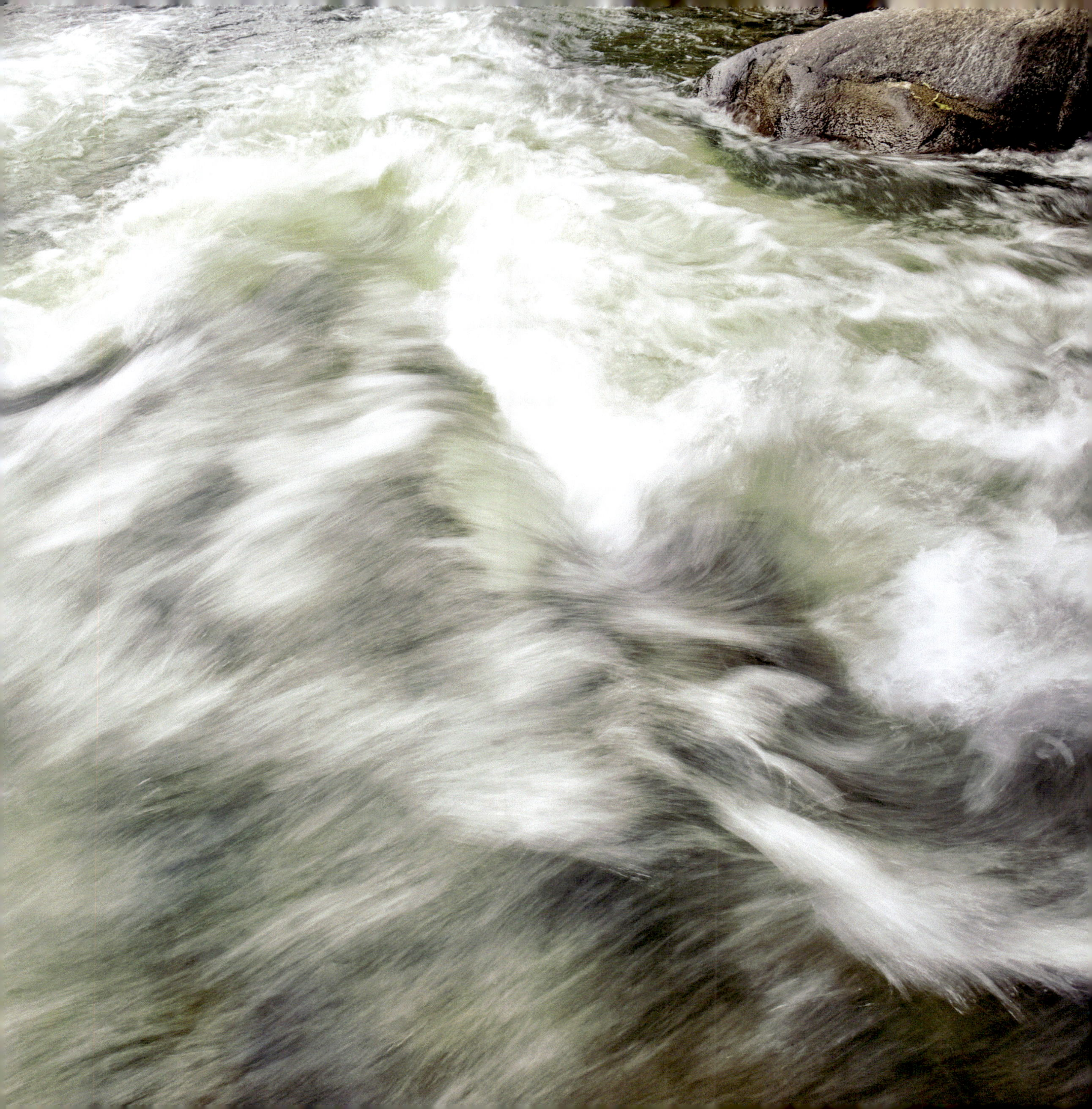

Acknowledgments

First of all, thanks to my wife, Ann Vileisis, for her unending support of my writing, my photography, and all of my work. I simply could not do what I do without her. My brother, Jim Palmer, helped with everything from lodging at his cabin near the Youghiogheny to his paddling partnership spanning many years, including a few critical tows of me and my overturned canoe into eddies. My sister, Becky Schmitz, likewise helped with her timeless love and encouragement in countless essential ways.

University of Pittsburgh Press staff made this book possible, going back to director Fred Hetzel's original enthusiasm for my first Youghiogheny book and continuing with directors Cynthia Miller and Peter Kracht. I especially appreciate my editor Abby McAllister, along with the marketing director and Youghiogheny aficionado John Fagan and his capable crew, Lesley Rains, Caleb Gill, and Kelly Thomas. Alex Wolfe designed the book with expert artistic skill.

Colleen O'Neil and Eric Harder of Mountain Watershed Association and Youghiogheny Riverkeeper provided vital information, congenial company while biking the lower river, partnership in planning festive outreach events, and most importantly soul-reaching inspiration that—because of people like them—we can all work toward a bright rather than a bleak future for our most cherished places. Steve Storck, ever thoughtful about natural resource and recreation issues, especially along the upper river, guided me on foot to his section of that wild waterway. Brook Lenker of the Keystone Trails Association, and formerly of FracTracker Alliance, helped with information about gas drilling. Robin Morris welcomed me to Paddler's Lane—her beautiful campground and exquisite riverfront resort near Confluence, which I happily regard as one of my very best homes-away-from-home.

Paul Wiegman's impressive lifetime of knowledge about all things natural was crucial to my understandings of this place. Amos Ludwig, Barbara Wallace, and Ken Bisbee shared wise insights about the park that they work diligently to manage and protect, and they each patiently answered questions from me that trickled into their email for a year or more. Former park manager Larry Adams shared invaluable knowledge and revealing observations as only a real pro can do. Outfitters Eric Martin and Mark McCarty regaled me with stories and sharp reflections of paddling and commercial rafting, past, present, and future. Bert Kerstetter pitched in with pivotal financial support for the printing of color photos that make this book uncommonly elegant, if I do say so myself. Cindy Bower generously read the manuscript and offered excellent suggestions. My timeless old pal John Lichter once again guided me, informed me, and joined me on the water in our shared spirit celebrating a place that's special to us and to many others.

Sources

For a key source that informed and guided me throughout this effort, see my earlier book, published in 1984 and updated in 2023, *Youghiogheny: Appalachian River*. I'll list just a few important additional references here.

A trove of data and inventory is available from the Pennsylvania Department of Conservation and Natural Resources' Natural Heritage Program, *Natural Resources Inventory and Management Recommendations for Ohiopyle State Park*, December 31, 2010. To flag just a few citations there, in roughly the order that they appear, see coverage of brook trout in Jonathan Run on page 54 of the *Inventory* and Ramcat Run on page 114. The lack of fish in Cucumber Run is noted on page 56. Gas deposits on Sugarloaf Mountain are mentioned on page 114. For the hemlock woolly adelgid, see page 26; for plant diversity, page 5; for Barbara's buttons, page 7; for ice scour, pages 8 and 33. The policy of favoring native species to combat exotic invaders appears on page 47, and the recommendation favoring management for old-growth forest can be found on page 88.

For geologic information, including Lake Monongahela, I consulted a white paper by the Pennsylvania Bureau of Topographic and Geologic Survey, *State Park Geology Day, October 4, 2014*.

My 2018 book, *Twilight of the Hemlocks and Beeches*, details and illustrates the problems that hemlock and beech trees suffer from the exotic blights threatening those trees along the Youghiogheny and throughout their ranges.

A reference for fish by Michael L. Hendricks and others, *The Zoogeography of the Fishes of the Youghiogheny River System,*

Pennsylvania, Maryland and West Virginia, was published in the *American Midland Naturalist*, July 1983.

For the landmark study of fish by David G. Argent and William G. Kimmel in 2012, see "Physiographic and Anthropogenic Factors Influencing Fish Community Composition in Tributaries to the Youghiogheny River in Pennsylvania." *Northeastern Naturalist* 19(3): 431–44, 1012.

The sequestration rate of carbon in old-growth forests is documented by Mark G. Anderson in *Analysis of Sequestration in Old Forests*, Northeast Wilderness Trust, vol. 1.1, 2021.

The data on fracking wells comes from *FracTracker's Comprehensive New Pennsylvania Map*, FracTracker Alliance, January 20, 2022, https://www.fractracker.org/2022/01/introducing-fractrackers-comprehensive-new-pennsylvania-map/. The increase in water use for fracking is from Jake Bolster, "A Fracker in Pennsylvania Wants to Take 1.5 Million Gallons a Day," *Inside Climate News*, September 17, 2023.

For the Upper Youghiogheny in Maryland, see *Maryland Scenic and Wild Rivers: The Youghiogheny*, by the Youghiogheny Scenic and Wild River Local Advisory Board (1996), on file with the Maryland Department of Natural Resources.

A wealth of natural and cultural data is revealed in the compendium *Youghiogheny River Conservation Plan: Part 1, Headwaters to Dunbar* Creek, prepared for the Pennsylvania Department of Conservation and Natural Resources by the Mountain Watershed Association, 2023.

About the Author and Photographer

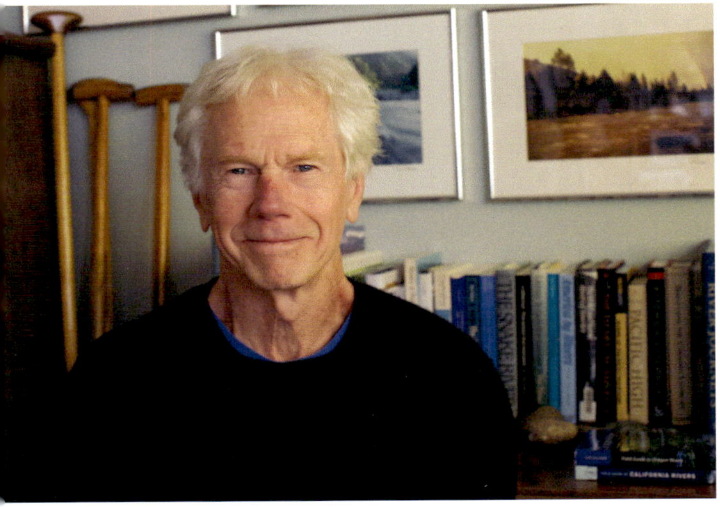

Author and photographer Tim Palmer, photo by Ann Vileisis.

Tim Palmer is the author and photographer of thirty-four books about rivers, the environment, and adventure travel. After graduating in landscape architecture at Penn State University and working as a county land-use planner in Pennsylvania for ten years, he began his full-time career as author and photographer. In 1984 he wrote *Youghiogheny: Appalachian River*, which the *Harrisburg Patriot-News* credited as "the best book ever written about a Pennsylvania river." In 2023 that classic volume was updated with a new edition, followed in 2025 by this book of photos and narrative focusing on the Youghiogheny's natural environment.

Tim has received the National Conservation Achievement Award as Communicator of the Year from the National Wildlife Federation, the Ansel Adams Award for excellence in photography from the Sierra Club, the Lifetime Achievement Award from American Rivers, the Wild and Scenic Rivers Award from the River Management Society, and the Distinguished Alumni Award from the Penn State University College of Arts and Architecture. His books of text and photos have garnered the National Outdoor Book Award, Independent Publishers Award, Adventure Book of the Year, Benjamin Franklin Book Award, and others.

Tim has written extensively about Pennsylvania, American rivers, conservation history, and nature, including guest editorials in major newspapers nationwide. He speaks throughout the Youghiogheny region and nationwide to recreation, academic, conservation, and professional audiences about rivers, forests, and adventure travel. See www.timpalmer.org.

Index